RESEARCH IN MATHEMATICS EDUCATION IN AUSTRALASIA 2004-2007

NEW DIRECTIONS IN MATHEMATICS AND SCIENCE EDUCATION
Volume 12

Scope

Mathematics and science education are in a state of change. Received models of teaching, curriculum, and researching in the two fields are adopting and developing new ways of thinking about how people of all ages know, learn, and develop. The recent literature in both fields includes contributions focusing on issues and using theoretical frames that were unthinkable a decade ago. For example, we see an increase in the use of conceptual and methodological tools from anthropology and semiotics to understand how different forms of knowledge are interconnected, how students learn, how textbooks are written, etcetera. Science and mathematics educators also have turned to issues such as identity and emotion as salient to the way in which people of all ages display and develop knowledge and skills. And they use dialectical or phenomenological approaches to answer ever arising questions about learning and development in science and mathematics.

The purpose of this series is to encourage the publication of books that are close to the cutting edge of both fields. The series aims at becoming a leader in providing refreshing and bold new work—rather than out-of-date reproductions of past states of the art—shaping both fields more than reproducing them, thereby closing the traditional gap that exists between journal articles and books in terms of their salience about what is new. The series is intended not only to foster books concerned with knowing, learning, and teaching in school but also with doing and learning mathematics and science across the whole lifespan (e.g., science in kindergarten; mathematics at work); and it is to be a vehicle for publishing books that fall between the two domains—such as when scientists learn about graphs and graphing as part of their work.

Research in Mathematics Education in Australasia 2004-2007

Edited by

Helen Forgasz
Anastasios Barkatsas
Alan Bishop
Barbara Clarke
Stephen Keast
Wee Tiong Seah
Peter Sullivan
Monash University, Australia

SENSE PUBLISHERS
ROTTERDAM / TAIPEI

A C.I.P. record for this book is available from the Library of Congress.

ISBN 978-90-8790-499-9 (paperback)
ISBN 978-90-8790-500-2 (hardback)
ISBN 978-90-8790-501-9 (ebook)

Published by: Sense Publishers,
P.O. Box 21858, 3001 AW Rotterdam, The Netherlands
http://www.sensepublishers.com

Printed on acid-free paper

CONTENTS

HELEN FORGASZ, TASOS BARKATSAS, ALAN BISHOP, BARBARA CLARKE, STEPHEN KEAST, WEE TIONG SEAH, & PETER SULLIVAN

INTRODUCTION: REVIEW OF MATHEMATICS EDUCATION RESEARCH IN AUSTRALASIA 2004-2007

HISTORICAL CONTEXT

The series of four-yearly reviews of research in mathematics education in Australasia produced by the Mathematics Education Research Group of Australasia (MERGA) had its beginning in 1984, when the then MERGA Executive commissioned the publication of *Summary of research in mathematics education in Australia* (Briggs, 1984) to coincide with Australia's hosting of the fifth quadrennial International Congress on Mathematical Education (ICME).

Over the years, the visionary idea which culminated in the publication of the first review took on a life of its own, renewing itself every four years. In 1992, the scope of the research reviewed extended beyond Australia to encompass all of Australasia; this reflected the expanding membership of MERGA. In 2008, the present review of Australasian research is, for the first time, published by an international academic publisher – SENSE – enabling the recognised, outstanding mathematics education research from Australasia to reach a broader international audience.

The series of MERGA four-yearly reviews has been unique in the sense that it has allowed Australasian mathematics education researchers to take stock of Australasian research in the field over the preceding four years. This provides an intellectual and academic breather from the hustle and bustle of the regime of regularly-published journal articles and papers for annual conferences, including MERGA. This is by no means inward-looking. In fact, in celebrating what has been researched in Australasia and by Australasians, work conducted by colleagues in an international context, either through collaborative research efforts or through their physical presence out of Australasia, is also documented in the review series.

The dates of publication and the editors of the MERGA series of four-yearly reviews to date are summarised in Table 1. With the publication of this, the seventh four-yearly review, the commitment of the series to promote and publicise Australasian mathematics education research is affirmed. To clarify to authors what should be included in their chapters, the following definition of what comprises Australasian mathematics education research was provided:

> The editors have defined "Australasian research" as research conducted in Australasia, about the Australasian context, or by Australasians. Australasia

H. Forgasz et al. (eds.), Research in Mathematics Education in Australasia 2004-2007, 1-7.

comprises: Australia, New Zealand, PNG, and the Pacific Islands closely allied to Australia and/or New Zealand. Australasian research published in Australasia and in international publications should be included in the review.

Table 1. MERGA 4-yearly reviews: dates of publication and editors

Year of publication	Editors		
1984	Jack Briggs		
1988	Dudley Blane	Gilah Leder	
1992	Bill Atweh	Jane Watson	
1996	Bill Atweh	Kay Owens	Peter Sullivan
2000	Kay Owens	Judy Mousley	
2004	Bob Perry	Glenda Anthony	Carmel Diezmann

Some of the themes included in the MERGA four-yearly reviews have been repeated over successive volumes (e.g., gender and mathematics has appeared in every review since 1984), while others have been introduced or disappeared over the years. The themes found in the various reviews reflect the ways in which mathematics education research has developed and matured in Australasia, and the foci of research interest during particular four-yearly time periods.

CHAPTER IDENTIFICATION PROCESS AND STRUCTURE OF THE BOOK

An open request was sent to members of MERGA inviting expressions of interest to contribute chapters in which a theme or area of the research literature was reviewed. Broad sections were identified including: a focus on teachers as learners; teachers' contemporary practices; what is learnt and how (K-12); non-cognitive dimensions of learning; special learners; and history, policy, and curriculum. Team contributions were encouraged. The purpose of the review is to showcase the research efforts of MERGA members and there is a sense in which the process of soliciting chapters itself reflects the research emphases and priorities of the community. One of the challenges of reviewing the complex body of work that comprises 'mathematics education research' is the necessary categorisation or division that must take place, that is, how do you *slice the pie*? While there was an effort to be comprehensive, this was counter-balanced by the preferences set by the authors through the themes put forward, and the subsequent chapters submitted.

Prospective authors were asked to submit an abstract describing the area of interest and samples of the literature to be reviewed. Once the submissions were received, the editors determined potential overlaps and gaps. In some cases a revised chapter with joint authorship was suggested, or a slight refocus to provide a more coherent and comprehensive presentation across all chapters.

While the present review of Australasian mathematics education research includes a number of chapters with similar themes to those found in the most recent MERGA four-yearly review (Perry, Anthony, & Diezmann, 2004 – see

Diezmann, Anthony, & Perry in this volume for more detail on the content of that review), prospective authors did not put forward suggestions for separate chapters with traditional mathematics content. This, in part, reflects a shift in both research funding opportunities and emphases on levels of education, such as early childhood, and subsequently on the learning of specific mathematics content within these contexts. Thus, in this review, the research on content is predominantly found within the chapters focussing on specific grade levels (e.g., early years, middle years).

Surprisingly, an invitation had to be issued for an additional chapter with a focus on research in technology use in mathematics to be written. The reference list from that chapter (see Thomas & Chinnappan in this volume) indicates the extent of the research in this area and the importance of its inclusion. It was intended that a chapter be included that focused on exceptional students, including gifted students and those with special needs. The prospective authors of the chapter subsequently withdrew, citing limited published research in the area during the period, an issue identified by Perry et al. (2004) in the most recent MERGA 4-yearly review of the Australasian mathematics education literature. The area of exceptional children and mathematics learning would appear to represent a gap in the emphasis on this dimension of research in Australasia; it was not an oversight in the process for identifying research to be included in this review. A proposed chapter on history, policy, and curriculum suffered a similar fate.

The authors of the previous MERGA 4-yearly review (Perry et al., 2004) were invited to write a chapter reflecting on the four year period, 2000-2003 in which the research reported in their review took place. They were asked to do this to set the scene for the reader with respect for the work reported in the present volume. Ken Clements, a founding member of MERGA and long-time observer of Australasian mathematics education, was invited to write a final chapter to highlight aspects from specific chapters included in the current review, to provide an overview and synthesis of the research presented, and to identify future research directions.

Further to the discussion above of the specific research themes *privileged* through the chapter foci identified and included in this volume, it is important for readers to recognise that there are some idiosyncrasies associated with the conduct of research programs in Australasia. Unlike in some international locations where there are substantial programs addressing specific aspects of content such as fractions or algebra, for example, in Australasia there are often fewer studies addressing specific content areas. This may create an impression that Australasian research in mathematics education is somewhat eclectic. Also contributing to this view is an interest in the social dimensions of education, such as teacher learning or student disadvantage, with some content focussed studies undertaken under such umbrellas. These emphases do not, however, detract from the quality of the research within specific, traditional, mathematics content areas. Readers are reminded that they may need to search across the various chapters for discussions of research within particular mathematical education research fields and content domains of interest.

POLICY AND RESEARCH CONTEXT

One of the characteristics of mathematics education research in the Australasian context has been that many of the studies are unfunded or, if funded, the funding is relatively small. The shortage of funding for sophisticated equipment and additional research assistance can limit the scale of many research studies, and this is reflected in a number of the reports of the research summarised in the following chapters. Also reported here is outstanding research conducted by graduate students. Such studies are generally small scale, both in sample size and in duration. In whichever category the research might be found, neither the scale nor the available financial or resource support limits the rigor with which the research was conducted. This is clearly reflected in all of the chapters included in this review of the Australasian mathematics education literature for the years 2004-2007.

Within the policy context in which the researchers have undertaken their studies, there have been some interesting shifts over the past four years. It is not easy to document all of these. However, considering the field generally, the editors share certain perceptions.

In the period 2004-2007, the national governments in New Zealand and Australia spearheaded the development of accountability measures for research funding and research activity. It should be noted that the New Zealand government has implemented many of the measures described at a faster pace than has the Australian government, but similar directions and outcomes are anticipated. With these developments necessarily comes an increase in bureaucratisation – more forms to complete, more adoption of standardised terminology, for example, more use of sanctioned definitions of "research" etc. The related shifts in the policy context of research in terms of the dichotomies and complementarities can be characterised as described below.

1. *A decrease in creative and idiosyncratic research v. an increase in programmatic research.*

In a healthy national research-supportive environment one would expect that both kinds of research would receive adequate funding and similar emphasis – programmatic research may produce useful data, but creative research is also needed, particularly in a field like mathematics education where the context and conditions are changing with increased rapidity. Developing creative research is always much more difficult but is equally necessary.

2. *A decrease in individual research v. an increase in group or team research.*

Both for purposes of funding, and also because of governmental pressures to define research groupings within universities, the pressure is on to reduce the significance of individual research. There are of course good reasons for encouraging group or team research, particularly in a mixed discipline field such as mathematics education. However PhD research, one of the main sources of research ideas, is emphasised as a largely individual matter, and whilst research training is a strong part of a PhD program, it is still possible, and we would think highly desirable, to be developing a next generation of individual, and creative, researchers.

3. *A decrease in funding for basic research v. an increase in funding for practice-oriented projects.*

With a growing concern about the teacher shortages in mathematics, together with a (dubious) perception that standards of mathematics learning in schools are lower than governments would wish, has come an increase in funding for in-service and practice-oriented projects, such as the Australian School Innovation in Science, Technology and Mathematics (ASISTM) project. A related development can be seen in New Zealand where in-service initiatives are more closely linked to research and where there is an increased expectation that any findings will be published in research contexts. One example comes from the annual publication on developments with the national Numeracy Development Project, which are now called *Findings from the New Zealand Numeracy Development Projects*. Whilst this overall shift is no bad thing, and many researchers are assisting with these programs, it is still the case that imaginative, basic research needs continuing support. In-service, practice-based work is not research, and whilst in-service is aimed at the immediate teaching context, research has inevitably a longer term perspective.

4. *A decreasing concern with the quantity of research v. an increasing concern with the quality of research.*

This has been formalised nationally in New Zealand where the Performance-Based Research Fund (PBRF) regime was trialled in 2003, and the first full 6-year cycle was completed in 2006. In Australia also there has been much preparation for the introduction of a Research Quality Framework system, which has now been put on hold by the newly elected national government. It is however likely that it will return in some form in the future. Whilst these moves have brought many concerns about ways to judge research quality, on a more positive note, there may well be useful lessons to be learnt about weaknesses of peer evaluation of research and about worthwhile criteria for judging quality. Within universities, academics have been used to accumulating their research references and credentials, almost with the 'publish or perish' maxim implying that more is better. Now we see a greater concern for judging one's own 'best outputs', the 'impact' of one's research and the development of similar qualitative criteria. No bad thing, one may say.

We believe that growth, not a decline, is likely to be seen in these shifts over the next four years, and it will be interesting indeed to speculate on how this will affect the research outcomes over the next four years.

FINAL COMMENTS

We hope that in reading this review, insights into the range of research undertaken in Australasia in the period 2004-2007 will be gained. The process of summarising, reviewing, and critiquing research over an extended period provides the potential for building informed, cohesive, and focused future research agendas. This is particularly important in a context where individual researchers are trying to work across institutions or with international colleagues, and when students are often isolated from a substantive research community. Such a process can highlight the gaps that exist in the foci and extent of the research undertaken, and thus what has been highlighted in the various chapters in this volume. We encourage readers to examine critically the research presented, recognise the constraints under which the researchers have worked, and ensure that these issues are taken into consideration in planning future research projects in which needs, not tradition or external demands, are guiding principles.

Finally we commend the research efforts of our colleagues, the high quality of their published research, and their willingness to author chapters in this volume, continuing a long and important tradition established by MERGA of sharing and critiquing Australasian mathematics education research.

ACKNOWLEDGMENTS

The editors wish to acknowledge the financial support of MERGA in the production of this volume. We thank Anne Bult for her assistance in working with the Sense template and her careful proofreading of each of the chapters. We are also most grateful to reviewers for their constructive comments that contributed to the final quality of each chapter included in the book.

REFERENCES

Atweh, B., Owens, K., & Sullivan, P. (Eds.). (1996). *Research in mathematics education in Australasia 1992-1995*. Sydney: MERGA.

Atweh, B., & Watson, J. (Eds.). (1992). *Research in mathematics education in Australasia 1988-1991*. Brisbane: MERGA.

Blane, D., & Leder, G. C. (Eds.). (1988). *Research in mathematics education in Australia: A selection of recent research*. Melbourne: MERGA.

Briggs, J. (Ed). (1984). *Summary of research in mathematics education in Australia*. Brisbane: MERGA.

Owens, K. & Mousley, J. (Eds.). (2000). *Research in mathematics education in Australasia 1996-1999*. Sydney: MERGA.

Perry, B., Anthony, G., & Diezmann, C. (Eds.). (2004). *Research in mathematics education in Australasia 2000-2003*. Flaxton, Qld: Post Pressed.

AFFILIATIONS

Helen Forgasz
Tasos Barkatsas
Alan Bishop
Barbara Clarke
Stephen Keast
Wee Tiong Seah
Peter Sullivan

Monash University, Australia

CARMEL DIEZMANN, GLENDA ANTHONY, & BOB PERRY

REFLECTIONS ON THE MERGA RESEARCH REVIEW 2000-2003

INTRODUCTION

The tradition of 4-yearly reviews of Australasian research in mathematics education commenced in 1984 when MERGA commissioned its first review (Briggs, 1984) to coincide with the 4th International Congress for Mathematical Education [ICME-4] held in Adelaide, South Australia and has continued every four years from that time.

The review previous to the present volume was published in 2004 and was edited by the three authors of this chapter (Perry, Anthony, & Diezmann, 2004). Authors were brought together to synthesise and critique Australasian mathematics education research in the period 2000-2003. As part of the critique, chapter authors were asked to reflect not only on the current highlights in research in their areas but also to look forward and bravely predict what they saw as the most significant areas of future research. In this chapter we reflect on the 2000-2003 review and the directions it foreshadowed in order to provide a context for the present review.

To a large extent, the research themes identified in the original MERGA review – curriculum, attitudes to mathematics, learning and teaching of mathematics, problem solving, learning and cognitive development, calculators and computers, tertiary and teacher education in mathematics, and gender in mathematics education – have continued throughout the series of six reviews up until 2004. As might be expected, themes have been added from time to time and the emphases have changed.

In the final chapter of the 2004 MERGA review, Jones (2004) reprised his original contribution to the first review (Jones, 1984) and asked in what areas of mathematics education research has Australasian work had significant impact. Through a careful critique of the Australasian mathematics education research in six reviews up to the one published in 2004, Jones selected six themes in which Australasian mathematics education research has made "an impact in the international mathematics education research community during the last 20 years" (Jones, p. 343): cognitive aspects of mathematical learning, gender and mathematics, language and mathematics, socio-cultural aspects of mathematics teaching and learning, early childhood education, and teacher education and the teaching of mathematics.

While Australasian mathematics education researchers publish and present their findings in the same forums as their colleagues from other regions of the world, there is no doubt that the synthesis and critique of their research in the 4-yearly MERGA reviews have greatly increased the impact of their work, not only in the region but also internationally.

H. Forgasz et al. (eds.), Research in Mathematics Education in Australasia 2004-2007, 9-16.

THE 2000-2003 MERGA 4-YEARLY REVIEW

The chapter headings in the 2000-2003 review were as follows:

- Social justice and sociocultural perspectives in mathematics education
- Gender and mathematics
- Exceptional students in mathematics
- Problem solving in learning and teaching mathematics
- Learning to teach mathematics
- University mathematics learning and teaching
- Number learning in the primary years
- Statistics education
- Learning and teaching algebra
- Calculators and computer algebra systems
- Computers, multi-media and the internet in mathematics education
- Research trends in mathematics education
- Affective issues in mathematics education
- Mathematics in early childhood education
- The impact of 20 years of research

While Atweh, Meaney, McMurchy-Pilkington, Neyland, and Trinick (2004) noted that there was "a healthy growth in issues as well as theoretical development in sociocultural research in mathematics education" (p. 45), in his final chapter of the MERGA 2000-2003 review, Jones (2004, p. 371) noted his belief "that research dealing with sociocultural aspects of mathematics education will assume far greater importance in the next two decades". He continued by suggesting:

> that mathematics education research in the next 20 years will take a more multifaceted yet integrated approach to learning and teaching; one that uses multiple lenses to look at teachers and learners from cultural, social, political and psychological perspectives. In essence, the research problematique will unite objects of research like equity, access, and context with extant objects like learning and teaching. (p. 372)

Effectively, Jones was predicting an increasing emphasis on socio-cultural dimensions within the framework of the teaching and learning of mathematics curriculum.

What was being researched?

A feature of the 2000-2003 review was a chapter on research trends (Walshaw & Anthony, 2004) involving an analysis of papers published in MERGA conference proceedings. In order to profile the research interests, methodologies and contexts of research that MERGA members participated in over the review period, questions were asked about which content areas and issues were of interest to researchers in Australasia, who was being researched and within what context, and how the research was being conducted. In addressing the content of research the authors noted the trend for research studies to "incorporate a range of wider interests [than mathematics content] such as the social and cultural milieu of learning, professional development, and program evaluation" (p. 6). Only 50% of papers featured a distinctly mathematical focus, and of these, research involving studies of rational number and numeracy were by far the most common. During the review period, much of the research on number appeared to be both informing (e.g., Jacob & Willis, 2003) and informed (e.g., Mitchelmore & White, 2003) by the development and implementation of large-scale systemic numeracy projects. When looking at educational issues that were of interest for the first part of the 21st century, a focus on cognition – including learner effects, visualization, problem solving and thinking practices associated with mathematical content – was prominent.

Chapter authors of the 2000-2003 review all identified areas that remained under-researched and under-theorised. Wood (2004) noted that research in the tertiary environment tended to focus on the problem student, rarely interrogating the lecturers' own practice and beliefs; Pfannkuch and Watson (2004) highlighted the need for studies to address adult statistical literacy; and Diezmann, Faragher, Lowrie, Bicknell, and Putt (2004) were concerned with the scarcity of research with exceptional students. An over-riding concern expressed by Jones (2004) was that despite advances in our research capability and increased awareness of socio-cultural issues, there remained an interminable challenge "to provide equitable mathematical access to all children irrespective of culture, ethnicity, gender, economic and social position" (p. 371). Atweh et al. (2004) called for the need to move from the identification of factors of disadvantage to consideration of solutions to the problems. Adopting a similar position, Schuck and Grootenboer (2004) in their chapter on affective issues argued that there was a need "to engage in research studies which offer substantive interventions which enhance attitudes, and interrupt those beliefs and values that do not promote effective mathematical teaching and learning" (p. 69).

In the 2000-2003 review period approximately 16% of papers focused on effective pedagogy and professional development. Walshaw and Anthony (2004) noted that interest in professional development rose from the low level of 1% in 2000 to 7% in 2003. Considering the current climate of research accountability – with calls for research to inform practice, and for professional development to measurably impact on learning (Timperley, Wilson, Barrar, & Fung, 2007) – there are signs of an increased research focus on professional development. Whilst some research reflects an ongoing interest in sustaining large scale primary school based numeracy initiatives that began in the previous MERGA review period (e.g., Knight, 2005), significantly, the focus appears to be widening to include reforms within the secondary sector (e.g., Goos, Dole, & Makar, 2007). It is expected that these trends will be reflected in the four-yearly review period covered by the current MERGA review.

While Australasian researchers have made significant international contributions to assessment practices – "exhorting educators to exploit assessment as a powerful means of developing mathematical competence rather than as an end in itself" (Jones, 2004, p. 367) – research focused on assessment, according to Walshaw and

Anthony's (2004) analysis, declined over the previous review period. Given the increased calls for teachers to use student achievement data to inform their instruction, significant changes in national assessment practices within Australasia, and increased recognition of the role of formative assessment in learning, will the trend for a paucity of research on assessment persist?

What tools and designs were being used?

The field of mathematics education research continues to change over time. In the 2000-2003 review period Walshaw and Anthony (2004) noted that a small number of research papers provided exemplars of innovative tools for data collection and frameworks for data analysis. New forms of research are important; they offer ways to "draw attention to previously unseen aspects of practice" (Klein, 2002, p. 395). The movement of theoretical perspectives from the traditional theories and constructs of psychology to a greater application of theories from sociological perspectives, as noted by Anthony (2004), continues to be evident. While papers that proposed a framework based on the adaptation of differing theoretical perspectives were the exception in the previous review, the exploration of multiple theoretical perspectives as a way of understanding research would be invaluable, as theoretical "choice necessarily influences what is looked for and noticed" (Beswick, Watson, & De Geest, 2007, p. 121).

A range of data generation strategies were noted in the 2000-2003 review. It is expected that research – such as Wilson's (2007) use of bibliotherapy to provide a new approach to eliciting and understanding the affective response of pre-service teachers, Chick, Pham, and Baker's (2006) novel framework for investigating teachers' pedagogical content knowledge, and Muir's (2006) piloted observation schedule for documenting aspects of effective numeracy teaching – within the current review will continue to showcase innovative data tools.

Walshaw and Anthony (2004) noted the prevalence of studies based on test items or task assessment performance of students or teachers, with some research studies utilising pre- and post-test items. When combined with ethnographic and case study enquiry these research studies accounted for approximately 60% of the research papers in the 2000-2003 review period. However, a notable feature of the 2000-2003 period was the increase in papers designated as design studies – a category described as including work that engaged teachers in collaboration with researchers, classroom action research, and teaching experiments. Walshaw and Anthony suggested that the increase reflected a greater involvement of "school professionals [in] research from the inside, navigating between their practice and their research through experiences, such as teacher study groups, teacher narratives, and classroom-based action researcher studies" (p. 17). There is evidence of this continuing trend, for example, Geiger and Goos' (2006) description of their successful collaborative experiences, and Pritchard and Bonne's (2007) details of some of the challenges faced by both the teacher researchers and university researchers within their collaborative study of teacher questioning practices.

Who was involved in the research?

In Walshaw and Anthony's (2004) analysis of MERGA papers from the previous review period it was noted that students were the main research participants. Most often student involvement was in the form of completing test/task items or interviews; in relatively few studies were students' perceptions of their learning explored. Including student voice, as did Doig, Groves, Campbell, Mousley, and Williams (2007) with examples from primary, secondary, and tertiary students would highlight how "information about teaching and learning that would otherwise be inaccessible to mathematics educators" (p. 884) could be made available.

Learners have been a central focus of research studies, and the school sector the main setting in which they are studied. The 2000-2003 review revealed a scarcity of research reported at MERGA conferences that were contextualized within the home, the community, pre-school settings, or the tertiary sector. While the number of researchers working in these contexts still appears to be small, it was significant that the 2007 MERGA conference featured a symposium focused on early years mathematics research (e.g., Fox, 2007).

In association with the increased focus on design research studies noted in the 2000-2003 review, and with the expectation that this will persist, it is anticipated that teachers will become an increasing focus of research attention, particularly in studies into effective professional development models. In the current review period, there are signs of a growing interest in research on teacher learning and teacher change, with more teachers actively involved as research collaborators/participants, and more self studies by teacher educators (e.g., Cheeseman, 2007).

CONCLUSION

The current review of Australasian mathematics education research is the seventh in the short history of MERGA. That such a series exists at all is a credit to the organisation and its members. There have been many similarities in the contents of these reviews, much the direct result of dedicated advocates for particular areas of mathematics education research who have not only undertaken significant research in their own right but have also stimulated quality research in their areas by colleagues and younger researchers. This trend appears to have continued into the present review. Of particular note is the area of gender and mathematics that has featured in the review series since the first review in 1984. Also of note – and following the predicted trend highlighted by Jones (2004) – is the presence of chapters dealing with sociocultural contexts. As Australia and New Zealand strive to provide their indigenous groups and some other culturally, socially and linguistically diverse groups with outcomes in mathematics equivalent to those enjoyed by their white middle-class populations, mathematics education researchers will more and more need to be seeking their evidence through sociocultural lenses.

However, the contexts of the 2000-2003 MERGA review (Perry, Anthony, & Diezmann, 2004) and the 2004-2007 MERGA review (Forgasz et al., 2008) are more dissimilar than similar. In both periods, there was a concern with the quality of mathematics teaching and the shortage of mathematics teachers and mathematics teacher educators. However, in the latter period there has been considerable change in the Australasian research context, with the foci on the assessment of research quality, and the substantial funding change in favour of practitioner projects. The nature of many MERGA members' work will have changed as they have responded to and been successful in these new contexts. According to Sternberg (2003), successful intelligence involves the ability to work effectively within the context by "adapting to, shaping and selecting environments" (p. 152). In the chapters in the present 4-yearly review of mathematics education research, the research reported will build on the past, and the ways that researchers, or their participants, have adapted to, shaped, and selected mathematics education environments will be reported. Within Australasia, such adaptations will result in the enhanced learning and teaching of mathematics as well as a continued stream of significant quality mathematics education research.

REFERENCES

Anthony, G. (2004). MERGA: A community of practice. In I. Putt, R. Faragher, & M. McLean (Eds.), *Mathematics education for the third millennium: Towards 2010* (Proceedings of the 27th annual conference of the Mathematics Education Research Group of Australasia, Townsville, Vol. 1, pp. 2-15). Sydney: MERGA.

Atweh, B., Meaney, T., McMurchy-Pilkington, C., Neyland, J., & Trinick, T. (2004). Social justice and sociocultural perspectives in mathematics education. In B. Perry, G. Anthony, & C. Diezmann (Eds.), *Research in mathematics education in Australasia 2000-2003* (pp. 29-52). Flaxton, Qld: Post Pressed.

Beswick, K., Watson, A., & De Geest, E. (2007). Describing mathematics departments: The strengths and limitations of complexity theory and activity theory. In J. Watson, & K. Beswick (Eds.), *Mathematics: Essential research, essential practice* (Proceedings of the 30th annual conference of the Mathematics Education Research Group of Australasia, Hobart, Vol. 1, pp. 113-122). Adelaide: MERGA.

Briggs, J. (Ed.). (1984). *Summary of research in mathematics education in Australia.* Brisbane: MERGA.

Cheeseman, L. (2007). Mathematical reform: What does the journey entail for teachers? In J. Watson & K. Beswick (Eds.), *Mathematics: Essential research, essential practice* (Proceedings of the 30th annual conference of the Mathematics Education Research Group of Australasia, Hobart, Vol. 1, pp. 201-206). Sydney: MERGA.

Chick, H. L., Pham, T., & Baker, M. K. (2006). Probing teachers' pedagogical content knowledge: Lessons from the case of the subtraction algorithm. In P. Grootenboer, R. Zevenbergen, & M. Chinnappan (Eds.), *Identities, cultures and learning spaces* (Proceedings of the 29th annual conference of the Mathematics Education Research Group of Australasia, Canberra, Vol. 1, pp. 139-146). Adelaide: MERGA.

Diezmann, C., Faragher, R., Lowrie, T., Bicknell, B., & Putt, I. (2004). Exceptional students in mathematics. In B. Perry, G. Anthony, & C. Diezmann (Eds.), *Research in mathematics education in Australasia 2000-2003* (pp. 175-195). Flaxton, Qld: Post Pressed.

Doig, B., Groves, S., Campbell, C., Mousley, J., & Williams, G. (2007). Listening to students' voices in mathematics education. In J. Watson, & K. Beswick (Eds.), *Mathematics: Essential research, essential practice* (Proceedings of the 30th annual conference of the Mathematics Education Research Group of Australasia, Hobart, Vol. 2, p. 884). Adelaide: MERGA.

Forgasz, H. J., Barkatsas, T., Bishop, A. J., Clarke, B., Keast, S., Seah, W. T., & Sullivan, P. (Eds.). (2008). *Research in mathematics education in Australasia: 2004-2007.* Rotterdam: SENSE Publishers.

Fox, J. (2007). International perspectives on early years mathematics. In J. Watson, & K. Beswick (Eds.), *Mathematics: Essential research, essential practice* (Proceedings of the 30th annual conference of the Mathematics Education Research Group of Australasia, Hobart, Vol. 2, pp. 865-869). Adelaide: MERGA.

Geiger, V., & Goos, M. (2006). Living in the gap: A tale of two different types of researchers. In P. Grootenboer, R. Zevenbergen, & M. Chinnappan (Eds.), *Identities, cultures and learning spaces* (Proceedings of the 29th annual conference of the Mathematics Education Research Group of Australasia, Canberra, Vol. 1, pp. 254-261). Adelaide: MERGA.

Goos, M., Dole, S., & Makar, K. (2007). Supporting an investigative approach to teaching secondary school mathematics: A professional development model. In J. Watson, & K. Beswick (Eds.), *Mathematics: Essential research, essential practice* (Proceedings of the 30th annual conference of the Mathematics Education Research Group of Australasia, Hobart, Vol. 1, pp. 325-334). Adelaide: MERGA.

Jacob, L., & Willis, S. (2003). The development of multiplicative thinking in young children. In L. Bragg, C. Campbell, G. Herbert, & J. Mousley (Eds.), *Mathematics education research: Innovation, networking, opportunity* (Proceedings of the 26th annual conference of the Mathematics Education Research Group of Australasia, Geelong, Vol. 2, pp. 460-467). Sydney: MERGA.

Jones, G. A. (1984). A review of MERGA research in mathematics education in Australia. In J. Briggs (Ed.), *Summary of research in mathematics education in Australia* (pp. 6-35). Brisbane: MERGA.

Jones, G. A. (2004). The impact of 20 years of research. In B. Perry, G. Anthony, & C. Diezmann (Eds.), *Research in mathematics education in Australasia 2000-2003* (pp. 339-386). Flaxton, QLD: Post Pressed.

Klein, M. (2002). A postructuralist perspective on the productive power of *process* in mathematics education: Practical implications for pedagogy and research. In B. Barton, K. Irwin, M. Pfannkuch, & M. Thomas (Eds.), *Mathematics Education in the South Pacific* (Proceedings of the 25th annual conference of the Mathematics Education Research Group of Australasia, Rotorua, Vol. 1, pp. 390-397). Sydney: MERGA.

Knight, N. (2005). The contested notion of sustainability: Possibility or pipe dream for numeracy reforms in New Zealand. In P. Clarkson, A. Downton, D. Gronn, M. Horne, A. McDonough, R. Pierce, & A. Roche (Eds.), *Building connections: Research, theory and practice* (Proceedings of the 28th annual conference of Mathematics Education Research Group of Australasia, Melbourne, Vol. 2, pp. 467-474). Sydney: MERGA.

Mitchelmore, M., & White, P. (2003). Count Me In Too and the Basic Skills Test in New South Wales. In L. Bragg, C. Campbell, G. Herbert, & J. Mousley (Eds.), *Mathematics education research: Innovation, networking, opportunity* (Proceedings of the 26th annual conference of the Mathematics Education Research Group of Australasia, Geelong, Vol. 2, pp. 460-467). Sydney: MERGA.

Muir, T. (2006). What does effective teaching for numeracy look like? The design of an observation schedule. In P. Grootenboer, R. Zevenbergen, & M. Chinnappan (Eds.), *Identities, cultures and learning spaces* (Proceedings of the 29th annual conference of the Mathematics Education Research Group of Australasia, Vol. 2, pp. 368-375). Adelaide: MERGA.

Perry, B., Anthony, G., & Diezmann, C. (Eds.). (2004). *Research in mathematics education in Australasia 2000-2003.* Flaxton, Qld.: Post Pressed.

Pfannkuch, M., & Watson, J. (2004). Statistics education. In B. Perry, G. Anthony, & C. Diezmann (Eds.), *Research in mathematics education in Australasia 2000-2003* (pp. 265-289). Flaxton, Qld: Post Pressed.

Pritchard, R., & Bonne, L. (2007). Teachers research their practice: Developing methodologies that reflect teachers' perspectives. In J. Watson, & K. Beswick (Eds.), *Mathematics: Essential research, essential practice* (Proceedings of the 30th annual conference of the Mathematics Education Research Group of Australasia, Hobart, Vol. 2, pp. 621-630). Adelaide: MERGA.

Schuck, S., & Grootenboer, P. (2004). Affective issues in mathematics education. In B. Perry, G. Anthony, & C. Diezmann (Eds.), *Research in mathematics education in Australasia 2000-2003* (pp. 53-74). Flaxton, Qld: Post Pressed.

Sternberg, R. J. (2003). A broad view of intelligence: A theory of successful intelligence. *Consulting Psychology Journal: Practice and Research, 55*, 139-154.

Timperley, H., Wilson, A., Barrar, H., & Fung, I. (2007). *Teacher professional learning and development (schooling): Best evidence synthesis iteration.* Wellington: Ministry of Education.

Walshaw, M., & Anthony, G. (2004). Research trends in mathematics education. In B. Perry, G. Anthony, & C. Diezmann (Eds.), *Research in mathematics education in Australasia 2000-2003* (pp. 3-28). Flaxton, Qld: Post Pressed.

Wilson, S. (2007). My struggle with maths may not have been a lonely one: Bibliotherapy in a teacher education number theory unit. In J. Watson, & K. Beswick (Eds.), *Mathematics: Essential research, essential practice* (Proceedings of the 30th annual conference of the Mathematics Education Research Group of Australasia, Hobart, Vol. 2, pp. 815-823). Sydney: MERGA.

Wood, L. (2004). University mathematics learning and teaching. In B. Perry, G. Anthony, & C. Diezmann (Eds.), *Research in mathematics education in Australasia 2000-2003* (pp. 219-242). Flaxton, Qld: Post Pressed.

AFFILIATIONS

Carmel Diezmann
Queensland University of Technology, Australia

Glenda Anthony
Massey University, New Zealand

Bob Perry
Charles Sturt University, Australia

BOB PERRY, JENNY YOUNG-LOVERIDGE, SUE DOCKETT, & BRIAN DOIG

THE DEVELOPMENT OF YOUNG CHILDREN'S MATHEMATICAL UNDERSTANDING

Early childhood mathematics education research is burgeoning in Australasia. This chapter highlights and critiques key research in the area that has been published between 2004 and 2007. In particular, it considers specific mathematical topics such as number and numeracy, space and measurement, and structure and patterning; contextual matters such as links among home, school, prior-to-school settings and community, indigenous learners and mathematics learning as children start school; assessment of mathematics learning in early childhood settings and the professional development of early childhood teachers of mathematics. It concludes with some suggestions for fruitful areas of future research.

Key words: assessment, contexts, early childhood, indigenous, number, numeracy, professional development, transition to school

INTRODUCTION

The internationally accepted definition of *early childhood* refers to the period of a child's life between birth and age 8 years (Bredekamp & Copple, 1997; OECD, 2001). In this chapter, we adhere to this definition by including reference to studies in mathematics education research which refer, exclusively or at least substantially, to children in this age range, their educators or the contexts in which the children learn mathematics.

Since the publication of the previous *MERGA Review of Research* (Perry, Anthony, & Diezmann, 2004), the impetus and focus of Australasian early childhood mathematics education research has strengthened and broadened. One of the impacts of this has been the involvement of Australasian early childhood mathematics education researchers in national and international reviews of research (for example, Anthony & Walshaw, 2007; Fox, 2007; Mulligan & Vergnaud, 2006; Perry & Dockett, in press; Young-Loveridge & Peters, 2005) and special issues of international journals such as *Early Childhood Research Quarterly* and *Mathematics Education Research Journal*. Another perspective on change can be seen in the efforts Australasian mathematics education researchers have made to have their work published not only in mathematics education research journals but also in more generic early childhood research journals. This has broadened the impact of their work.

Another aspect of the broadening of impetus and focus has been the increased diversity of early childhood mathematics education fields in which Australasian researchers are working. While the influence of systemic numeracy programs in

H. Forgasz et al. (eds.), Research in Mathematics Education in Australasia 2004-2007, 17-40.

the early years of school can still be seen in the directions of this research, changes in emphasis towards, for example, early algebra development, assessment and mathematics learning in the years prior-to-school have been added to the collective repertoire of Australasian early childhood mathematics education researchers. However, not only has the breadth of early childhood mathematics education research increased. Recognition of key Australasian researchers by international journals and publishers suggests that the quality of their work is also on the march.

The aim of this chapter is to highlight and critique key Australasian early childhood mathematics education research published between 2004 and 2007. The chapter seeks to incorporate what the authors see as the key research directions in the field over the stated period of time. It is divided into subsections covering specific mathematical topics; contextual matters such as links among home, school, prior-to-school setting and community, indigenous learners and mathematics learning as children start school; assessment of mathematics learning in early childhood settings and the professional development of early childhood teachers of mathematics.

NUMBER AND NUMERACY

Systemic numeracy programs (particularly *Count Me In Too* (*CMIT*) and *Early Numeracy Research Program* (*ENRP*) in Australia and the *Early Numeracy Project* (*ENP*) in New Zealand) continue to have major impact both on the mathematics learning of young children and early childhood mathematics education research in New Zealand and Australia (Bobis et al., 2005; Wright, Stanger, Stafford, & Martland, 2005; Young-Loveridge & Peters, 2005).

Analysis of the extensive database collected by *ENRP* (for example, Clarke, Clarke, & Horne, 2006; Gervasoni, 2004a, 2004b, 2005, 2006; Horne & Levy, 2006) has continued to provide insights into young children's numeracy development. Clarke, Clarke and Horne (2006) use this database as the foundation of a longitudinal study of children's mental computation strategies as they move through primary school. They suggest that in spite of the gains younger children make through adopting efficient mental strategies for computation in the first years of school, a significant proportion of them still rely on inefficient counting strategies to solve arithmetical problems mentally in the upper years of primary school.

Horne and Levy (2006) consider the development of place value knowledge in Year 1 children. By analysing data from 200 children, they report that 75% of the children could read and order at least two-digit numbers, 60% could write them on a calculator and 37% could interpret them in terms of a tens structure. The authors were concerned about whether the children who did display knowledge of all four aspects of place value might "have to mark time" in later years as they were already ahead of the grade-based curriculum. They were also concerned about the likely learning trajectories of children who did not display the four aspects of place value knowledge.

The considerable body of work that Gervasoni has built from her involvement in *ENRP* is concerned with children who are vulnerable in their number learning. Using the notion of *on the way growth points* (Gervasoni, 2004a, 2004b), vulnerable children from Years 1 (n=576) and 2 (n=679) were identified from the *ENRP* database. Their vulnerability in the domains of counting, place value, addition and subtraction and multiplication and division was determined and it was found that there was great diversity among the identified vulnerable children in terms of their learning needs:

> ... children who are vulnerable in aspects of learning school mathematics have diverse learning needs, and this calls for particular customized instructional responses from teachers. It is likely that teachers will need to make individual decisions about the instructional approach for each child, and that there is no 'formula' that will meet all children's instructional needs. Further, the diversity of children's mathematical knowledge in the four domains suggests that knowledge in any one domain is not necessarily prerequisite for knowledge construction in another domain. (Gervasoni, 2004b, p. 253)

After identifying vulnerable children from the entire *ENRP* cohort, Gervasoni (2005, 2006) chose 42 Year 1 and 60 Year 2 children to undertake a mathematics intervention program. Many of these vulnerable children were unable to use efficient reasoning-based strategies and often relied on count-all to solve addition tasks. Of particular concern was that many of the Year 2 children seemed to have already entrenched these inefficient strategies, making successful intervention quite difficult, or, at least, suggesting that initial intervention at Year 1 might be more profitable for these children than waiting for intervention in Year 2.

Gervasoni's work has been extended to incorporate using the numeracy development of primary school children to inform the professional development of teachers in the Ballarat Diocese of the Catholic education system (Gervasoni, Hadden, & Turkenburg, 2007). Data have confirmed earlier findings highlighting the diverse mathematical learning needs of young children.

New Zealand's *Numeracy Development Project* (*NDP*) is a collective term encompassing the four individual projects catering for students at different points in their schooling (*Early Numeracy Project* (*ENP*) for students in Years 0-3, *Advanced Numeracy Project* (*ANP*) for students in Years 4-6, *Intermediate Numeracy Project* (*INP*) for students in Years 7-8, and *Secondary Numeracy Project* (*SNP*) for students in Years 9-10). Initially each of the individual projects comprising the *NDP* was evaluated annually (for example, Higgins, 2004; Irwin, 2004; Thomas & Tagg, 2004). However, since 2005, the quantitative data sent by primary school teachers to a secure website (i.e., *ENP*, *ANP*, & *INP* data) have been analysed by one researcher (Young-Loveridge, 2005).

Thomas and Tagg (2004) reported on data from more than 31,000 students in Years 0-3 (5 to 8-year-olds) in English-medium schools. While 20% of students at the end of Years 0-1 were able to use counting on or back to solve addition and subtraction problems, by the same time in Year 2 this had increased to 58%.

Fifteen percent of Year 2 students were able to use part-whole strategies to solve simple addition and subtraction problems and this increased to 41% by the end of Year 3. Students' proficiency with strategies varied as a function of ethnicity and socio-economic status (as reflected in the decile ranking of the school: decile 1 – schools serving the lowest 10% of the population in terms of income and education, decile 10 serving communities with the highest levels of education and income). The Asian and European cohorts had the greatest proportion of students using higher-level strategies (e.g., 26% and 22% using simple part-whole strategies, respectively), while Māori and Pasifika cohorts had the least students at this stage (13% and 8%, respectively). Similarly, 24% of students in high-decile schools were using part-whole strategies compared with only 13% of those in low-decile schools. Age was a factor as older students tended to make greater progress on the number framework central to NDP than younger students. Gender was an issue at higher stages on the framework, where boys made slightly greater progress than girls. As well, in the higher stages of the framework, European and Māori students made greater progress than Pasifika students.

Young-Loveridge (2004b) reported on an analysis of data collected from more than 200,000 students between 2001 and 2003. Her analysis showed that while all students benefited from participating in NDP, regardless of ethnicity, gender and socio-economic status, the relative differences between subgroups were virtually identical at the end of the project, and, if anything, there was a slight widening of the various 'gaps'. European and Asian students tended to make greater progress on the framework than did Māori or Pasifika students. At lower stages on the framework, girls made slightly better progress than boys. Students at higher decile schools started the project at higher framework stages than those at low- and medium-decile schools, and made larger gains over the course of the projects. Ethnicity, gender, and socio-economic status had a combined effect on students' performance and progress, with being a female Māori or Pasifika student attending a low-decile school more disadvantageous than any one of those factors on its own.

The New Zealand Ministry of Education has introduced many Schooling Improvement Projects – "planned interventions designed to raise overall academic achievement of targeted students. … A priority group of students in New Zealand targeted for involvement in schooling improvement are Māori (New Zealand's indigenous people) and Pasifika in origin" (Annan, 2007, p. 116), many of whom are students at low-decile schools. NDP clearly has a Schooling Improvement Project perspective. For example, Young-Loveridge (2005) noticed that students at low-decile schools in 2004 who began NDP at lower stages on the framework made greater progress than comparable students from medium- and high-decile schools. Effect sizes (calculated by comparing younger students after the project with slightly older students before they began the project) were almost half a standard deviation for multiplication/division and proportion/ratio, and about a quarter of a standard deviation for addition/subtraction.

Heirdsfield and her colleagues (Heirdsfield, 2005; Heirdsfield & Cooper, 2004a, 2004b; Heirdsfield, Dole, & Beswick, 2006; Heirdsfield & Lamb, 2005, 2006a, 2006b) have used extensive small-scale case studies to explore the development of mental addition and subtraction strategies in young children as an alternative to what is felt to be the too early introduction of written algorithms. Through the introduction of mental computational practices and teacher development around these practices, strong advances have been made not only in the accuracy of young children's calculations but also in the sophistication and flexibility of their methods (Heirdsfield & Lamb, 2005).

The diversity of mental computation skills, approaches and needs among young learners has also been explored by Heirdsfield and her colleagues (Heirdsfield & Cooper, 2004a; Heirdsfield et al., 2006; Heirdsfield & Lamb, 2006a). As in Gervasoni's (Gervasoni, 2005; Gervasoni et al., 2007) work, a wide diversity in skills and knowledge was described as was the reliance on inefficient counting strategies by many 'at risk' mathematics learners.

The nature of numeracy in the early childhood years has been explored by a number of researchers (Howell & Kemp, 2005; Killdeery & Yelland, 2005; Wood & Frid, 2005; Young-Loveridge, 2004a). Using a modified Delphi procedure with a small group of Australasian academics published in early mathematics or in the study and development of number sense, Howell and Kemp sought consensus on the skills comprising number sense as children start school. In the first round of the procedure, 24 of the 25 skills listed were rated by the 'experts' in the range 'strongly agree' to 'strongly disagree', suggesting that there was little agreement. In Round 2 of the procedure, there was still a lot of disagreement. The authors conclude that:

> … it is essential to establish exactly which skills reflect number sense and whether or not these skills can be effectively taught. The lack of genuine consensus reached in the present study raises questions about the emphasis currently being placed on number sense within curriculum documents. Without a clear definition of number sense it may be that current assessment and teaching practices are premature in their claim to reflect number sense. (Howell & Kemp, 2005, p. 568)

Wood and Frid (2005) examined numeracy teaching and learning practices in a multiage (pre-primary to Year 2) classroom. Through extensive observation, including video and audio records, over almost 5 months, Wood and Frid (2005, p. 96) noted that "the key numeracy teaching and learning practices that emerged as themes in this case study of a multiage early childhood classroom were: teacher planning, teacher 'assisted performance', peer sharing and tutoring, and peer regulation". Further, they noted that the multiage grouping of the students did not, of itself, support learning. "Rather, effective learning was dependent upon the teacher's capacities to develop productive discussion among children, as well as implement developmentally appropriate curricula that addressed the needs of the different children" (Wood & Frid, 2005, p. 80).

In New Zealand, before the introduction of a system-wide initiative in numeracy, Young-Loveridge (2004a) explored the effectiveness of a program designed to improve the number skills of 5-year-olds from low-decile schools by using number books and games with pairs of children guided by an expert teacher.

The numeracy levels of children who participated in the program increased significantly and showed an effect size of almost two standard deviations. Fifteen months after the program ceased the effect size had reduced to half a standard deviation but the advantage for the children who had participated in the program remained statistically significant.

By filming 10 Prep grade children from a Melbourne school for a day in their school lives, Killdeery and Yelland (2005) have been able to analyse the children's school activities in terms of the mathematics skills and knowledge that they used and the ways in which they used these. The "Prep children were observed applying their numeracy skills in a range of activities throughout the school day" (Killdeery & Yelland, 2005, p. 117). Beyond the diversity of the nature of these numeracy experiences is the realization that, inter alia, "ICT provides meaningful contexts in which young children's numeracy explorations are possible and they can support children to 'form strong, valid mathematical ideas' (Clements, 2002, p. 165) as well as improve the children's ICT literacy and skills" (Killdeery & Yelland, 2005, p. 120). Not only do ICT enhance the possibilities for children learning expected numeracy ideas, they also provide opportunities to learn different mathematical ideas and to apply these in their lives.

SPACE AND MEASUREMENT

Clarke (2004) outlined the importance of the *ENRP* framework of growth points within the space or geometry area through linking it to the work of van Hiele (1986) and the database of *ENRP* outcomes from both children and teachers. She showed exploration, investigation and instruction based upon experiences informed by the framework could result in a significant growth in children's understanding of geometric ideas. Clarke (2004) also emphasized the need to ensure that children explore the properties of a range of shapes rather than spending most of their geometry learning time "naming large numbers of formal geometric shapes" (p. 124). The importance of children investigating shapes in their environment and discussing why certain shapes are used for certain functions was also emphasized. Just as has been the case in working with number, the *ENRP* framework of growth points for geometric learning has facilitated the analysis of children's learning and provided guidance for the teaching of geometric concepts.

Curry and Outhred (2005) considered pedagogical links among the measurement of length, area and volume. As part of the development of the *Count Me Into Measurement* program, they designed and implemented a 45-minute clinical interview to 96 Year 1-4 students in Sydney schools. The interview was designed to assess students' knowledge of the measurement of length, area and two aspects of volume – filling and packing. Important findings included the following:

> … students were able to measure volume by filling as well as they did length, using a similar unit iteration procedure; measurement of length was not a prerequisite for measurement of area, partly because both seem to be affected by a general tendency towards precision in recording the unit iteration; and understanding of the unit structure for area provides the foundation for

understanding measurement of volume by packing. (Curry & Outhred, 2005, p. 271)

The consequences of these findings for the teaching of length, area and measurement are significant. The differences in understandings of the two aspects of volume measurement suggest that packing be introduced to students much later than filling which, on the basis of this study, could be introduced at the same time as length. These findings indicate that there may need to be consideration of curriculum changes in the order of introduction of certain measurement concepts.

A cross-cultural study by Irwin, Vistro-Yu, and Ell (2004) also considered young children's measurement of length. Through interviews with 48 children aged 8 and 9 years from the Philippines and 43 children of the same age from New Zealand, the authors have been able to make cross-cultural comparisons as well as investigate the relationships between the children's informal and formal measurement experiences. Using a specifically constructed interview schedule consisting of five tasks, Irwin et al. illustrated the importance of prior experiences on children's measurement of length, with a striking example from the Filipino children of the impact of their rounding of monetary transactions to include only whole numbers on their tendency to round their measurement of a length to the whole number of centimetres. Other differences in the student's performance on the measurement tasks may be related to differences in curricula. For example, students from both countries achieved equally on tasks requiring the use of a standard measuring tool but Filipino children were not as successful as the New Zealand children on tasks requiring visualisation. The authors emphasise the importance of informal measurement experiences and lament the lack of these experiences in both countries. While agreeing with Irwin et al. concerning the importance of informal experiences as children learn about various measurement concepts, Curry and Outhred (2005) suggest that the choice of particular informal units that reflect the property being measured is critical as is the order in which experiences are encountered.

STRUCTURE AND PATTERNING

A group of researchers based at Macquarie University has implemented the *Australian Pattern and Structure Mathematics Awareness Program* (*PASMAP*) with the broad aim of describing "general characteristics of structural development and how pattern and structure are integral to [mathematical] concept development" (Mulligan, Mitchelmore, & Prescott, 2006, p. 4-209). Through interventions with both prior-to-school and school learners, *PASMAP* has developed a framework of broad stages of structural development (Mulligan, Mitchelmore, & Prescott, 2005) and an individual interview assessment instrument – *Pattern and Structure Assessment* (*PASA*) – which is very highly correlated with these stages (Mulligan, Mitchelmore, & Prescott, 2006). Through both intervention and longitudinal studies (Mulligan, 2007; Mulligan, Mitchelmore, & Prescott, 2006; Mulligan, Prescott, Papic, & Mitchelmore, 2006), the relevance of pattern and structure to young children's mathematical development has been indicated:

> It is not conclusive from our data whether the awareness and appropriate use of pattern and structure is a good predictor, or a consequence of, successful acquisition of basic mathematical concepts and skills. ... Our findings support our initial hypothesis that the more a child's internal representational system has developed structurally, the more coherent, well-organised, and stable in its structural aspects will be their external representations, and the more mathematically competent the child will be. (Mulligan, Mitchelmore, & Prescott, 2006, pp. 4-213-4-214)

Two researchers have investigated patterning in prior-to-school settings both for its own sake and as a possible pre-cursor to the development of algebraic reasoning (Fox, 2005, 2006; Papic & Mulligan, 2005, 2007; Waters, 2004). Based in the research of *PASMAP*, Papic and Mulligan (2005, 2007) have investigated the development of patterning skills in two groups of children (total n=53) in two 'matched' preschools. One of the groups of children participated in a six month intervention promoting these patterning skills while the other group acted as a non-intervention control. Individual interviews were used to assess the children's patterning skills on eight separate tasks both before and after the instructional intervention. On the pre-intervention assessment, the children in the non-intervention group were more successful than the intervention group on all but one of the tasks. However, after the intervention, the intervention group was more successful than the other group on all the tasks. Further assessment after a year of schooling has shown that this trend has been maintained (Papic & Mulligan, 2007).

In her initial investigation of patterning in early childhood settings, Waters (2004) observed patterning experiences in two settings in Brisbane. While she saw patterning experiences – both teacher- and child-initiated – in both settings, she also observed a number of instances where the possibility of discussion about patterns was missed by the teachers. She concludes that there is a need to enhance prior-to-school teachers' knowledge of patterning and skills in both planning for patterning experiences and eliciting discussion about patterning from young children. In follow-up analyses of her data, Fox (previously publishing as Waters) considers both child-initiated (Fox, 2005) and teacher-initiated (Fox, 2006) patterning experiences in the two settings. She concludes that "child-initiated experiences can be powerful learning opportunities with the potential to develop children's knowledge of mathematical patterning in meaningful contexts" but "teachers need to have knowledge of mathematical patterning and be capable of capitalizing on children's interests" (Fox, 2005, p. 2-319). The need for teachers to develop strong understandings about mathematical patterning and its importance in later mathematical learning, especially in algebra, is emphasized by both Fox (2006) and Papic and Mulligan (2007).

This work on patterning, structure and early algebra is a significant 'new' field for Australasian early childhood mathematics education researchers and one in which both output and conceptual framing are significant. It fits well with the work of English (2004, 2005, 2006, 2007; English & Watters, 2005) and Warren (2005a, 2005b, 2006) with slightly older children.

CONTEXTS FOR EARLY CHILDHOOD MATHEMATICS

The *Catch the future* project was designed to help raise literacy and numeracy outcomes for children living in low socio-economic families in an area south-east of Melbourne (Clarke & Robbins, 2004; Fleer & Robbins, 2005; Kennedy & Ridgeway, 2005; Kennedy, Ridgeway, & Sunman, 2006). Families in the *Catch the future* project were invited to use disposable cameras to photograph their preschool child undertaking literacy and numeracy activities outside of their early childhood setting. These photographs were discussed with the families and the different responses of people to the literacy and numeracy experiences illustrated were noted through three lenses that were originally formulated by Rogoff (1998):

- what everyone could see in the photographs about literacy and numeracy (personal lens);
- what only the family could see in the photographs as literacy and numeracy (interpersonal lens); and
- what literacy and numeracy practices were so much part of everyone's everyday life that you could no longer see them (institutional/cultural lens) (Clarke & Robbins, 2004).

The analysis of the photographs from 52 families revealed "the extensive numeracy enactments occurring within many families in lower socio-economic circumstances" (Clarke & Robbins, 2004, p. 180) and the high value that families put on these enactments. "The challenge for preschool and early years teachers is to connect and build upon this rich base of mathematical experiences in ways that acknowledge and support the family's role" (Clarke & Robbins, 2004, p. 181).

Fourteen early childhood educators from the centres attended by the families in the project had the opportunity to be involved in professional development activities designed to build pathways for learning literacy and numeracy between homes and early childhood settings. Initially, these teachers seemed to hold a deficit view about the families' knowledge and understanding of literacy and numeracy but, after viewing the photographs taken by the families and listening to the accompanying comments (Fleer & Robbins, 2005), the educators' views seemed to change although there still seemed to be a reluctance to work with families to build pathways in literacy and numeracy. This theme is taken up by Kennedy et al. (2006) who suggest pathways to literacy and numeracy which link home and early childhood centre experiences through play, using computers, family routines and reading and writing texts.

Many of the findings arising from *Catch the future* have also been noted by Goos and her colleagues in a national project designed to explore the links between home, school and community that support young children's numeracy development (Goos et al., 2004; Goos & Jolly, 2004). Through a mixed-mode methodology including seven detailed case studies, critical changes required to build long-term and meaningful relationships among educators, families and communities that support children's numeracy development were identified. These included:

- forging parental and community involvement in mathematics education development and change;
- recognizing social, cultural and gender differences and relations of power in building partnerships;
- enhancing communication between teachers and parents;
- improving teachers', parents', and communities' understanding of the nature of numeracy and numeracy learning; and
- understanding the nature of partnerships and participant roles (derived from Goos, 2004, p. 20).

One of these case studies is exemplified more fully in Lowrie (2005) through a detailed description of the establishment of a numeracy culture in an environment serviced by distance education. Echoing the findings from *Catch the future*, Goos and Jolly (2004, p. 285) note that "even though some schools endeavour to help parents recognize numeracy learning opportunities in the home environment, we suspect that the rich variety of numeracy events embedded in home and family contexts remains invisible to most parents and teachers". There is a clear need for continuing work in this area.

EARLY CHILDHOOD MATHEMATICS AND INDIGENOUS CHILDREN

In Papua New Guinea, education reform has encouraged the use of local vernaculars in teaching mathematics in elementary schools (Matang, 2005; Vagi & Green, 2004). Many parents are concerned about this move as they believe that the way ahead for their children is through English or Tok-Pisin medium schooling. In a study of 52 children in three schools, two of which use the local vernacular as the medium of instruction and one which uses mainly Tok-Pisin and some English, Matang (2005) implemented a modified version of the *Schedule of Early Number Assessment* from *Count Me In Too.* He found that students from the vernacular-medium schools showed no significant difference in their responses to the English-medium numeracy assessment tasks from those shown by students from the Tok-Pisin/English-medium school. Hence, "those children learning early number knowledge in Kâte language using its counting system are not in any way disadvantaged in learning the formal English arithmetic strategies normally taught in schools when compared to [the] children not taught in Kâte" (Matang, 2005, pp. 509-510).

The issue of contextualization of mathematical experiences for indigenous children has received a lot of attention in recent years (Cronin & Yelland, 2004; Macmillan, 2004; Watego, 2005). McMillan's paper describes a project that offered access to numeracy knowledge for preschool Koori children in an urban regional area of New South Wales. The children were given numeracy kits to play with at their early childhood centre, and to take home to play with their families. While the materials and suggested activities in the kits were not specifically related to the Koori culture of the children and their families, the children responded well to the kits. Further,

> Being able to take the kit home meant that the families of the children could support them in their play, they could develop a better understanding of their children's numeracy potential, and they could perceive a tangible connection between centre and home. (Macmillan, 2004, p. 36)

Watego (2005) also discusses the issue of contextualization of numeracy experiences in preschools and what it means for Aboriginal people. Her initial analysis of the contextualisation observed in one inner-city preschool raises many issues around what is meant by contextualization and how it might be enacted in such a setting. She argues that while this setting implements a play-based curriculum for its children, the subtle differences resulting from an awareness and understanding of Aboriginal ways of knowing, and the incorporation of "specific language, meanings and behaviours" result in distinctly Aboriginal activities that help empower the children "by building the base of confidence and identity that is necessary" (Watego, 2005, p. 773).

Cronin and Yelland (2004) investigated teaching practices designed to enhance numeracy learning outcomes for young Aboriginal students through case studies of four teachers working in the preschool to Year 3 classes of a school with 100% Aboriginal enrolment in a rural community in Queensland. It was found that "a 'special approach' was not required [for the Aboriginal students] but rather effective pedagogy as for all students. Teachers of Aboriginal students need to be culturally sensitive just as all teachers should" (Cronin & Yelland, 2004, p. 106). However, because the potential for teacher and students to connect may be lessened in the context of an Aboriginal community:

> the professional teacher, who is more interested in student learning outcomes, will develop an interest in the child's cultural and social context – and consider it in their planning of learning experiences. In fact, it is argued here that if a teacher is not prepared to engage with the student's social and cultural context, then there is little relevance engaging with the syllabus either. It is further argued that those who are not prepared to engage with the student's social and cultural context will severely disadvantage outcomes for Indigenous students. (Cronin & Yelland, 2004, p. 108)

Taken together, the papers of Cronin and Yelland (2004), Macmillan (2004), Matang (2005), and Watego (2005) clearly establish the pre-eminence of context and culture when considering the mathematics education of young indigenous children.

In New Zealand, the emphasis on cultural and linguistic appropriateness is reflected in the provision of *Te Poutama Tau* the Māori-medium equivalent of the *NDP*. The challenge for *Te Poutama Tau* has been to focus on Māori pedagogy and patterns of discourse that are authentic to the culture and the syntax of the language, at the same time as retaining the integrity of key aspects of the *NDP* such as the number framework and the diagnostic interview. These patterns of discourse include such aspects as the ways that language is used to talk about mathematics (including the ways teachers give instructions), how various concepts are explained, and the scaffolding of interaction in ways that lead to understanding.

The goal of *Te Poutama Tau* is to advance the teaching and learning of mathematics while taking cognisance of linguistic concerns, such as ensuring that the authenticity of the language is maintained and items of vocabulary are simple, concise and enhance understanding. It is important that throughout the pāngarau/mathematics curriculum, Māori pedagogy and tikanga (customs), Māori language, and Māori contexts are used and a student's Māori identity is reinforced (Christensen, 2004). The guiding principles for *Te Poutama Tau* evaluation research include a unique Māori worldview, culturally safe research practices, challenges to existing power relationships, accountability to and mediation by a Māori supervisory group/whanau (*Te Ropu Whaiti*), and concern by the researcher to make a contribution to Māori development and advancement. It is important to recognise that *Te Poutama Tau* is not just a version of the *NDP* in the language of Māori; it is part of a much greater goal, which is the maintenance and revitalisation of the Māori language (te reo Māori) (Trinick & Stephenson, 2006). As such it emphasises the importance of cultural understanding and language in ways that are reinforced by the studies discussed earlier from Papua New Guinea and Australia.

MATHEMATICS LEARNING AS CHILDREN BEGIN SCHOOL

Issues around continuity and change as children move from prior-to-school contexts to school contexts have been well-researched over the last 10 years (see, for example, Dockett & Perry, 2007). On both sides of the Tasman, this general interest has been exemplified by studies in relationships between mathematics education in prior-to-school and the first years of school. The work of Papic and Mulligan (2005, 2007) and the developing project in indigenous communities reported by Warren, Young, and de Vries (2007) are examples.

In New Zealand, two studies have particular relevance to this section of the review. Peters' (2004) doctoral study focused on children's general transition to school, with mathematics learning one of the many aspects explored. Her findings underline the complexity of children's learning, including an appreciation of the different 'cultures' of home, early childhood centre, and school, negotiated by children as they learn to take on the role of school pupils. Peters argues that instead of eliminating the differences between settings, teachers in both early childhood centres and schools should support children to ensure that they can capitalise on the developmental possibilities of coping with change. The lack of flexibility and the focus on performance in schools seemed to exert strong pressure on children to conform to expectations and in the process gain status in the classroom. The Peters' study has important implications for early childhood teachers who want to foster communities of inquiry within their classrooms, and want mathematics learning to occur as part of a collaborative, shared process among members of such a community.

Belcher (2006) looked at the way that five new entrant (5-year-old) children and their parents experienced and interpreted the transition from an early childhood education centre to primary school in terms of numeracy learning. Belcher used participant observation in the classroom and unstructured interviews to gather data on the perspectives and experiences of the children and their parents. Parents were unaware of the differences in philosophy among the teachers in their child's early childhood centre and those in the primary school, and the impact of those differences on their child's learning experiences in the two settings. A greater need for communication between the two sectors, as well as between teachers and parents, was identified (Belcher, 2006).

In Australia, *ENRP* has considered various aspects of children's prior-to-school numeracy abilities and their consequences for teachers in the first year of school (Clarke, Clarke, & Cheeseman, 2006) and beyond (Horne, 2005). Through the application of their *First Years of School Mathematics Interview* and, where appropriate, the full *ENRP Interview*, with over 1400 students, Clarke, Clarke, and Cheeseman have provided a very detailed picture of the mathematics knowledge and skills of Victorian children as they commence school, and at the end of their first year of school. Traditional expectations about the mathematics children bring to the first year of school are questioned.

> Clearly most Prep children arrive at school with considerable skills and understandings in areas that have been traditional mathematics content for that age. As acknowledged by many ENRP Prep teachers, this means that expectations could be raised considerably in terms of what can be achieved in that first year. (Clarke, Clarke, & Cheeseman, 2006, p. 97)

In spite of the strength of their findings in terms of the overall strengths of young children as they start school, Clarke, Clarke, and Cheeseman (2006) make the point that there is a danger of losing the individual in such considerations of the typical and that the "mathematical thinking of the child should be the starting point" (Clarke, Clarke, & Cheeseman, 2006, p. 99) in the development of learning activities.

Project Good Start (Thomson, Rowe, Underwood, & Peck, 2005) investigated practices in prior-to-school settings that seemed to benefit children most in their first year of school. More than 1300 children in 80 early childhood settings across Australia were tested twice in their year before school using *Who am I* (de Lemos & Doig, 1999) and *I can do maths* (Doig & de Lemos, 2000). In their first year of school, 1620 children in 44 schools, including 282 children from the prior-to-school cohort, were assessed with *I can do maths*. Apart from these quantitative measures, surveys of pre-school and school teachers' attitudes and practices in numeracy were conducted. Parents were involved in focus groups, and site visits were made to the pre-schools involved in every state and territory.

The main findings of the prior-to-school phase of the study were that effective numeracy programs involved teachers who:

- were interested in numeracy, and had high expectations of their children;
- have formal and explicit plans for numeracy learning;
- provide numeracy-focussed materials;
- use and develop mathematical language;
- have the ability to engage children in mathematical activity; and
- conduct regular assessment and feedback on children's abilities and progress.

The findings for the children in the prior-to-school settings were that:
- there were no sex differences in achievement;
- indigenous children performed less well than non-indigenous children; and
- children from higher SES achieved better than children from lower SES levels.

Overall findings from *Project Good Start* were that:
- an effective practice for numeracy development was regular formative and summative assessment, combined with feedback to children and parents;
- high expectations by parents were a strong influence on achievement;
- nearly half of the first year of school teachers had not had the opportunity to engage in quality professional development in numeracy in the previous two years;
- the mean achievement of indigenous students was lower than that of non-indigenous students;
- about a third of the indigenous group achieved above the median of non-indigenous students;
- many children had highly developed numeracy skills in the year before school;
- lower SES level children gained from a pre-school experience, reducing the 'gap' in numeracy between them and higher level SES children;
- many pre-school and school teachers were unaware of the achievements of children, mainly due to the lack of assessment of children's numeracy skills and understandings; and
- much of the transition to school information sent on to schools was neither consistent nor valued by FYS teachers.

Perry and Dockett (2004a, 2005) have taken the approach of small case studies to illustrate similar points to those of Clarke, Clarke, and Cheeseman (2006). Their story of Harry and the lamentable approach taken to his school mathematics education reinforces the need for all children to be treated as individuals as they move into school. In spite of meeting, in his first year of school, little mathematics that he was not already capable of completing with ease, Harry did maintain

> the outward semblance of being interested in the work and, at least, being willing to complete it. However, it seems that the strongest lesson he learned in his Kindergarten mathematics experience is that 'you do not have to work hard at it'. (Perry & Dockett, 2005, p. 4-71)

Taken together, the studies highlighted in this section have shown that many young children have access to powerful mathematical ideas as they commence school and that, often, this access is not recognised by their teachers. On the other hand, not all children have developed their mathematical ideas to the level expected of beginning school children and these children need to be nurtured from their own individual starting points. In fact, a valuable starting point for teachers in the first year of school would be the recognition of the individual mathematical strengths that each child brings to school and a willingness to build on these strengths.

ASSESSMENT OF EARLY CHILDHOOD MATHEMATICS LEARNING

The key assessment approaches of the systemic numeracy projects mentioned earlier continue to dominate the accepted approaches to assessment of early childhood mathematics learning, particularly in the first years of school (Bobis et al., 2005; Clarke, Clarke, & Cheeseman, 2006; Mulligan, Prescott, Papic, & Mitchelmore, 2006). However, two other emerging approaches deserve mention.

The work of *Project Good Start* (Thomson et al., 2005) is based on two assessment instruments developed through the Australian Council for Educational Research – *Who am I?* and *I can do maths.* The psychometric credentials of these instruments and a summary of their usefulness in many different contexts are presented in Doig (2005) who concludes that:

> The use of these assessment tools by researchers internationally suggest that there exists a need for standardized, early years mathematics assessments. While not all early years professionals need or want such tools for their particular contexts, there are others whose interests lie in mapping children's mathematical abilities. Quality assessment tools provide a means of achieving a mapping over time, place, or culture. (p. 116)

A different approach to the planning and assessment of mathematics learning in the prior-to-school year has been developed through the *Southern Numeracy Initiative* in South Australia. A group of preschool teachers worked with researchers to develop an extensive numeracy matrix which they now use in the assessment of their children's mathematics learning (Harley, Perry, & Hentschke, 2006; Perry, Dockett, Harley, & Hentschke, 2006). The approach of narrative assessment pioneered in New Zealand by Carr (2001, 2006) is used in conjunction with the numeracy matrix so that the assessment of young children's mathematics knowledge and skills is accompanied by the assessment of their learning dispositions such as courage, curiosity, trust, perseverance, confidence and responsibility, all of which have clear applicability to the learning of mathematics. Mathematical thinking can include all of these actions or behaviours indicative of learning dispositions. Hence, even children whose mathematical skills per se might be considered limited from a more traditional skills-based assessment approach, may exhibit many of these behaviours, indicating that they are competent and confident learners. The variety of assessment tools available to early childhood teachers means that choices can be made based on the needs of the children and teachers and the purposes of the assessment.

PROFESSIONAL DEVELOPMENT OF EARLY CHILDHOOD TEACHERS

The relative lack of confidence of many early childhood teachers in their knowledge of mathematics and its pedagogy is a challenge for their professional development in this area. One of the common ramifications of this is that teachers often revert to the 'safety' of teaching as they were taught. According to Vagi and Green (2004), such an approach is quite common among the trainers of elementary school teachers in Papua New Guinea. Elementary schools have been established in each village and community-selected trainee teachers need to undergo teacher preparation. Faced with an innovative community-based curriculum in such schools, including the use of the local vernacular, the training of the elementary teachers is conducted by district-based experienced primary trained teachers. However, many of these experienced teachers remain uncertain of appropriate approaches in the elementary years and tend to revert to traditional teacher-centred approaches.

Professional development was a central component of *ENRP* and other systemic numeracy projects (Bobis et al., 2005). Through a combination of extensive statewide, regional and local input from *ENRP* team members, cluster coordinators and participant teachers, well-informed 'professional learning teams' were established at the local level. These teams were key to the success of the project. As well as previously recorded professional learning in terms of the research-based framework of growth points in children's mathematics learning, there is evidence that *ENRP* teachers grew in their practice and changed their beliefs about mathematics and its learning and teaching (McDonough & Clarke, 2005).

Of particular interest in the professional development model enacted by *ENRP* is the role of the school early numeracy coordinators. Cheeseman and Clarke (2005) have analysed two forms of open-ended written data from the 36 coordinators to reveal the complexity of the task and the variety of ways in which the coordinators carried out their roles, given variations in expertise, school factors and available time. The early numeracy coordinators were clearly the focal points for much of the project work carried out in each school and "exhibited considerable personal professional growth, provided a substantial amount of mutual support, and contributed to the professional development of their team members, thus increasing the chance of improved student learning outcomes" (Cheeseman & Clarke, 2005, p. 232).

To determine which approach should be taken to the professional development of teachers in a particular context can be a challenge. From her perspective as a District Mathematics Consultant, McPhail (2004) tried three variations on the amount of support afforded 17 teachers who were implementing a sequence of specifically designed lessons on area measurement. The teachers chose their mode of support during the 6-week implementation of the lessons:

- no consultant support;
- fortnightly team meetings with the consultant; or
- release for 30 minutes after each lesson for student interviews, fortnightly team meetings with the consultant, and a 1-day team meeting.

With student learning outcomes as the dependent variable, it was found that some support from the consultant was better than none but that the groups who only met with the consultant on a fortnightly basis obtained better student learning outcomes than the groups who also had opportunities for after lesson interviews and extended team meetings. No claim is made about the generalisability of this finding but it does point to an interesting question about whether there might be optimal levels of support and even diminishing returns from requiring too much involvement in professional development.

It is well established that early childhood educators' mathematical knowledge and dispositions are key to effective mathematics learning in early childhood settings (Anthony & Walshaw, 2007; Haynes, Cardno, & Craw, 2007; Perry & Dockett, in press). It is also clear that "low levels of content knowledge and the resulting lack of confidence about mathematics limit teachers' ability to maximise opportunities for engaging children in the mathematical learning embedded within existing activities" (Anthony & Walshaw, 2007, p. 47). This further reinforced in the key finding of the Haynes et al. study:

> By far the outstanding finding of the project was the realisation of the importance of teachers' own mathematical knowledge and their personal disposition towards mathematics, leading to improved strategies for providing mathematically stimulating learning opportunities for children. (Haynes et al., 2007, p. 75)

While these findings are not new, they do provide support for continued efforts to improve the professional development of early childhood educators such as those outlined in this section.

CONCLUSION AND SUGGESTIONS FOR FUTURE RESEARCH

After reviewing the Australasian early childhood mathematics education research in the period 2000-2003, Perry and Dockett (2004b) listed what they saw as fruitful areas for future research:

1. approaches to assessment and teaching / learning in numeracy and possible mismatches between these;
2. successful approaches to the mathematics education of young indigenous students;
3. successful approaches to the mathematics education of young children from culturally and linguistically diverse backgrounds;
4. technology in the mathematics education of young children;
5. play in the mathematics education of young children;
6. development of mathematics concepts among children before they start school;
7. continuities and discontinuities of learning in children as they move from prior-to-school to school settings; and
8. recognition of young children as capable learners of mathematics and the results of such recognition in their mathematical outcomes in the first years of school. (Perry & Dockett, 2004b, pp. 119-120)

It is pleasing to see that many of these areas have been featured in the current review. There has been continuing work emanating from the systemic numeracy projects in both Australia and New Zealand and other projects dealing with assessment, the development of mathematics concepts and issues around numeracy at the time children start school. Young indigenous students are being catered for in more meaningful ways in our schools but there is still a long way to go before they perform, on standardised mathematics assessment instruments, as well as non-indigenous students.

Little research seems to have been undertaken in the areas of technology and early childhood mathematics, even though there continue to be numerous computer-based learning programs being made available. While a lot of research has been undertaken on the needs of indigenous children in both New Zealand and Australia, little seems to have happened with other culturally and linguistically diverse groups.

The emergence of pattern, structure and early algebra as a key field of research, along with innovative approaches to the assessment of young children's mathematical learning provides fertile ground for further work, as does the ongoing need for early childhood educators' professional development in mathematics education.

Clearly, the early childhood mathematics education researchers of Australasia have achieved a great deal in the period 2003-2007 both in the consolidation of the work done is the systemic numeracy projects and in the development of new areas of endeavour. However, much remains to be done in order to further enhance the mathematics education of our young children within a range of contexts. We exhort our colleagues to continue the work.

REFERENCES

Annan, B. (2007). The Numeracy Development Projects: A successful policy-research-practice collaboration. In B. Annan, F. Ell, J. Fisher, N. Hawera, J. Higgins, K. C. Irwin, A. Tagg, G. Thomas, T. Trinick, J. Ward, & J. Young-Loveridge (Eds.), *Findings from the New Zealand numeracy development projects 2006* (pp.116-127). Wellington: Ministry of Education.

Anthony, G., & Walshaw, M. (2007). *Effective pedagogy in mathematics/pāngarau.* Wellington, NZ: Ministry of Education.

Belcher, V. (2006). *"And my heart is thinking": Perceptions of new entrant children and their parents on transition to primary school numeracy.* Unpublished master's dissertation, Christchurch College of Education, Christchurch.

Bobis, J., Clarke, B., Clarke, D., Thomas. G., Wright, R., Young-Loveridge, J., & Gould, P. (2005). Supporting teachers in the development of young children's mathematical thinking: Three large scale cases. *Mathematics Education Research Journal, 16*(3), 27-57.Bredekamp, S., & Copple, C., (Eds.), (1997). *Developmentally appropriate practice in early childhood programs (rev. ed.).* Washington, DC: National Association for the Education of Young Children.

Bredekamp, S., & Copple,C. (Eds.). (1997). *Developmentally appropriate practice – Revised.* Washington, DC: National Association for the Education of Young Children.

Carr, M. (2001). *Assessment in early childhood settings: Learning stories.* London: Paul Chapman.

Carr, M. (2006). Learning dispositions and key competencies: A new curriculum continuity across the sectors. *Set: Research Information for Teachers, 2*, 23-27.

Cheeseman, J., & Clarke, D. (2005). Early numeracy coordinators in Victorian primary schools: Components of the role, highlights and challenges. In P. Clarkson, A. Downton, D. Gronn, M. Horne, A. McDonough, R. Pierce, & A. Roche (Eds.), *Building connections: Research, theory and practice* (Proceedings of the 28th annual conference of the Mathematics Education Research Group of Australasia, Melbourne, pp. 225-232). Sydney: MERGA.

Christensen, I. (2004). *An evaluation of Te Poutama Tau 2003.* Wellington: Ministry of Education.

Clarke, B. (2004). A shape is not defined by its shape: Developing young children's geometric understanding. *Journal of Australian Research in Early Childhood Education, 11*(2), 110-127.

Clarke, B., Clarke, D. M., & Cheeseman, J. (2006). The mathematical knowledge and understanding young children bring to school. *Mathematics Education Research Journal, 18*(1), 78-103.

Clarke, B., Clarke, D. M., & Horne, M. (2006). A longitudinal study of children's mental computation strategies. In J. Novotná, H. Moraová, M. Krátká, & N. Stehlíková (Eds.), *Proceedings of the 30th conference of the International Group for the Psychology of Mathematics Education* (Vol. 2, pp. 329-336). Prague: Charles University.

Clarke, B., & Robbins, J. (2004). Numeracy enacted: Preschool families' conceptions of their children's engagements with numeracy. In I. Putt, R. Faragher, & M. McLean (Eds.), *Mathematics education for the third millennium: Towards 2010* (Proceedings of the 27th annual conference of the Mathematics Education Research Group of Australasia, Townsville, pp. 175-182). Sydney: MERGA.

Cronin, R., & Yelland, N. (2004). Achieving positive outcomes in numeracy for young indigenous students. *Journal of Australian Research in Early Childhood Education, 11*(2), 99-109.

Curry, M., & Outhred, L. (2005). Conceptual understanding of spatial measurement. In P. Clarkson, A. Downton, D. Gronn, M. Horne, A. McDonough, R. Pierce, & A. Roche (Eds.), *Building connections: Research, theory and practice* (Proceedings of the 28th annual conference of the Mathematics Education Research Group of Australasia, Melbourne, pp. 265-272). Sydney: MERGA.

de Lemos, M., & Doig, B. (1999). *Who am I? Developmental assessment.* Melbourne: Australian Council for Educational Research.

Dockett, S., & Perry, B. (2007). *Starting school: Perceptions, experiences and expectations.* Sydney: University of New South Wales Press.

Doig, B. (2005). Developing formal mathematical assessment for 4- to 8-year-olds. *Mathematics Education Research Journal, 16*(3), 100-119.

Doig, B., & de Lemos, M. (2000). *I can do maths.* Melbourne: Australian Council for Educational Research.

English, L. D. (Ed.), (2004). *Mathematical and analogical reasoning of young learners.* Mahwah, NJ : Lawrence Erlbaum Associates.

English, L. D. (2005). Combinatorics and the development of children's combinatorial reasoning. In G. A. Jones (Ed.), *Exploring probability in schools: Challenges for teaching and learning* (pp.121-141). Dordrecht: Kluwer.

English, L. D. (2006). Mathematical modelling in the primary school: Children's construction of a consumer guide. *Educational Studies in Mathematics, 63*, 303-323.

English, L. (2007). Interdisciplinary modelling in the primary mathematics curriculum. In J. Watson & K. Beswick (Eds.), *Mathematics: Essential research, essential practice* (Proceedings of the 30th annual conference of the Mathematics Education Research Group of Australasia, Hobart,,pp. 275-284). Adelaide: MERGA.

English, L. D., & Watters, J. J. (2005). Mathematical modelling in the early school years. *Mathematics Education Research Journal, 16*(3), 58-79.

Fleer, M., & Robbins, J. (2005). 'There is much more to this literacy and numeracy than you realise …': Family enactments of literacy and numeracy versus educators' constructions of learning in home contexts. *Journal of Australian Research in Early Childhood, 12*(1), 23-41.

Fox, J. (2005). Child-initiated mathematical patterning in the pre-compulsory years. In H. L. Chick & J. L. Vincent (Eds.), *Proceedings of the 29th conference of the International Group for the Psychology of Mathematics Education* (Vol. 2, pp. 313-320). Melbourne: University of Melbourne.

Fox, J. (2006). Connecting algebraic development to mathematical patterning in early childhood. In J. Novotná, H. Moraová, M. Krátká, & N. Stehlíková (Eds.), *Proceedings of the 30th conference of the International Group for the Psychology of Mathematics Education* (Vol. 3, pp. 89-96). Prague: Charles University.

Fox, J. (2007). International perspectives on early years mathematics. In J. Watson & K. Beswick (Eds.), *Mathematics: Essential research, essential practice* (Proceedings of the 30th annual conference of the Mathematics Education Research Group of Australasia, Tasmania, pp. 865-869). Adelaide: MERGA.

Gervasoni, A. (2004a). Exploring the diversity of grade 1 and grade 2 children who are vulnerable in learning mathematics. In M. Hoines & M. Fugelstad (Eds.), *Proceedings of the 28th conference of the International Group for the Psychology of Mathematics Education* (Vol. 1, pp. 376-383). Bergen, Norway: Bergen University College.

Gervasoni, A. (2004b). The challenge of meeting the instructional needs of grade 1 and grade 2 children who are at risk in mathematics. In I. Putt, R. Faragher, & M. McLean (Eds.), *Mathematics education for the third millennium: Towards 2010* (Proceedings of the 27th annual conference of the Mathematics Education Research Group of Australasia, Townsville, pp. 247-254). Sydney: MERGA.

Gervasoni, A. (2005). The diverse learning needs of young children who were selected for an intervention program. In H. L. Chick & J. L. Vincent (Eds.), *Proceedings of the 29th conference of the International Group for the Psychology of Mathematics Education* (Vol. 3, pp. 33-40). Melbourne: University of Melbourne.

Gervasoni, A. (2006). Insights about the addition strategies used by grade 1 and grade 2 children who are vulnerable in number learning. In J. Novotná, H. Moraová, M. Krátká, & N. Stehlíková (Eds.), *Proceedings of the 30th conference of the International Group for the Psychology of Mathematics Education* (Vol. 3, pp. 177-184). Prague: Charles University.

Gervasoni, A., Hadden, T., & Turkenburg, K. (2007). Exploring the number knowledge of children to inform the development of a professional learning plan for teachers in the Ballarat Diocese as a means of building community capacity. In J. Watson & K. Beswick (Eds.), *Mathematics: Essential research, essential practice* (Proceedings of the 30th annual conference of the Mathematics Education Research Group of Australasia, Tasmania, pp. 305-314). Adelaide: MERGA.

Goos, M. (2004). Home, school and community partnerships to support children's numeracy. *Australian Primary Mathematics Classroom, 9*(4), 18-20.

Goos, M., & Jolly, L. (2004). Building partnerships with families and communities to support children's numeracy learning. In I. Putt, R. Faragher, & M. McLean (Eds.), *Mathematics education for the third millennium: Towards 2010* (Proceedings of the 27th annual conference of the Mathematics Education Research Group of Australasia, Townsville, pp. 279-286). Sydney: MERGA.

Goos, M. E., Lincoln, D., Coco, A., Frid, S., Galbraith, P., Horne, M., Jolly, L., Kostogriz, A., Lowrie, T., Short, P. M., & Gholam, M. (2004). *Home, school and community partnerships to support children's numeracy.* Canberra: Department of Education, Science and Training.

Harley, E., Perry, B., & Hentschke, N. (2006). Using children's learning stories to assess their mathematics learning in preschools and schools. *Proceedings of the annual conference of the Australian Association for Research in Education, Adelaide.* Retrieved April 2, 2007 from http://www.aare.edu.au/06pap/HAR06339.htm

Haynes, M., Cardno, C., & Craw, J. (2007). *Enhancing mathematics teaching and learning in early childhood settings.* Wellington, NZ: Teaching and Learning Research Initiative.

Heirdsfield, A. (2005). One teacher's role in promoting understanding in mental computation. In H. L. Chick & J. L. Vincent (Eds.), *Proceedings of the 29th conference of the International Group for the Psychology of Mathematics Education* (Vol. 3, pp. 113-120). Melbourne: University of Melbourne.

Heirdsfield, A., & Cooper, T. (2004a). Factors affecting the process of proficient mental addition and subtraction: Case studies of flexible and inflexible computers. *Journal of Mathematical Behavior, 23*(4), 443-463.

Heirdsfield, A., & Cooper, T. (2004b). Inaccurate mental addition and subtraction: Causes and compensation. *Focus on Learning Problems in Mathematics, 26*(3), 43-65.

Heirdsfield, A., Dole, S., & Beswick, K. (2006). Instruction to support mental computation development in young children of diverse ability. *Proceedings of the annual conference of the Australian Association for Research in Education, Adelaide.* Retrieved April 2, 2007 from http://www.aare.edu.au/06pap/HEI06407.htm

Heirdsfield, A., & Lamb, J. (2005). Mental computation: The benefit of informed teacher instruction. In P. Clarkson, A. Downton, D. Gronn, M. Horne, A. McDonough, R. Pierce, & A. Roche (Eds.), *Building connections: Research, theory and practice* (Proceedings of the 28th annual conference of the Mathematics Education Research Group of Australasia, Melbourne, pp. 419-426). Sydney: MERGA.

Heirdsfield, A., & Lamb, J. (2006a). Year 2 inaccurate but flexible mental computers: Teacher actions supporting growth. *Proceedings of the annual conference of the Australian Association for Research in Education, Adelaide.* Retrieved April 2, 2007 from http://www.aare.edu.au/06pap/HEI06408.htm

Heirdsfield, A., & Lamb, J. (2006b). Teacher actions: Enhancing the learning of mental computation in Year 2. In J. Novotná, H. Moraová, M. Krátká, & N. Stehlíková (Eds.), *Proceedings of the 30th conference of the International Group for the Psychology of Mathematics Education* (Vol. 3, pp. 281-288). Prague: Charles University.

Higgins, J. (2004). *An evaluation of the Advanced Numeracy Project 2003.* Wellington: Ministry of Education.

Horne, M. (2005). The effects of number knowledge at school entry on subsequent number development: A five-year longitudinal study. In P. Clarkson, A. Downton, D. Gronn, M. Horne, A. McDonough, R. Pierce, & A. Roche (Eds.), *Building connections: Research, theory and practice* (Proceedings of the 28th annual conference of the Mathematics Education Research Group of Australasia, Melbourne, pp. 443-450). Sydney: MERGA.

Horne, M., & Levy, S. (2006). Young children developing place value understandings. In J. Novotná, H. Moraová, M. Krátká, & N. Stehlíková (Eds.), *Proceedings of the 30th conference of the International Group for the Psychology of Mathematics Education* (Vol. 3, pp. 313-320). Prague: Charles University.

Howell, S., & Kemp, C. (2005). Defining early number sense: A participatory Australian study. *Educational Psychology, 25*(5), 555-571.

Irwin, K. C. (2004). *An evaluation of the Numeracy Projects for Years 7-9 2003.* Wellington: Ministry of Education.

Irwin, K. C., Vistro-Yu, C. P., & Ell, F. R. (2004). Understanding linear measurement: A comparison of Filipino and New Zealand children. *Mathematics Education Research Journal, 16*(2), 3-24.

Kennedy, A., & Ridgeway, A. (2005). 'Catch the future' literacy and numeracy pathways for young children: Voices from the field. *Journal of Australian Research in Early Childhood, 12*(1), 43-57.

Kennedy, A., Ridgeway, A., & Sunman, L. (2006). 'Boundary crossing': Negotiating understandings of early literacy and numeracy. *Australian Journal of Early Childhood, 31*(4), 15-22.

Killdeery, A., & Yelland, N. (2005). A day in the life: Rethinking numeracy in the information age. *Journal of Australian Research in Early Childhood Education, 12*(1), 113-121.

Lowrie, T. (2005). Establishing a numeracy culture in a distance education learning environment: A case study. In P. Clarkson, A. Downton, D. Gronn, M. Horne, A. McDonough, R. Pierce, & A. Roche (Eds.), *Building connections: Research, theory and practice* (Proceedings of the 28th annual conference of the Mathematics Education Research Group of Australasia, Melbourne, pp. 498-504). Sydney: MERGA.

McDonough, A., & Clarke, B. (2005). Professional development as a catalyst for changes in beliefs and practice: Perspectives from the Early Numeracy Research Project. In P. Clarkson, A. Downton, D. Gronn, M. Horne, A. McDonough, R. Pierce, & A. Roche (Eds.), *Building connections: Research, theory and practice* (Proceedings of the 28th annual conference of the Mathematics Education Research Group of Australasia, Melbourne, pp. 521-528). Sydney: MERGA.

Macmillan, A. (2004). Take-home numeracy kits for pre-school children. *Australian Journal of Early Childhood, 29*(2), 29-40.

McPhail, D. (2004). Professional learning in the teaching of area. In I. Putt, R. Faragher, & M. McLean (Eds.), *Mathematics education for the third millennium: Towards 2010* (Proceedings of the 27th annual conference of the Mathematics Education Research Group of Australasia, Townsville, pp. 359-366). Sydney: MERGA.

Matang, R. A. (2005). Formalising the role of indigenous counting systems in teaching the formal English arithmetic strategies through local vernaculars: An example from Papua New Guinea. In P. Clarkson, A. Downton, D. Gronn, M. Horne, A. McDonough, R. Pierce, & A. Roche (Eds.), *Building connections: Research, theory and practice* (Proceedings of the 28th annual conference of the Mathematics Education Research Group of Australasia, Melbourne, pp. 505-512). Sydney: MERGA.

Mulligan, J. (2007). Pattern and structure in the K-2 classroom. *Reflections, 32*(1), 18-20.

Mulligan, J., Mitchelmore, M., & Prescott, A. (2005). Case studies of children's development of structure in early mathematics: A two-year longitudinal study. In H. L. Chick & J. L. Vincent (Eds.), *Proceedings of the 29th conference of the International Group for the Psychology of Mathematics Education* (Vol. 4, pp. 1-8). Melbourne: University of Melbourne.

Mulligan, J., Mitchelmore, M., & Prescott, A. (2006). Integrating concepts and processes in early mathematics: The Australian Pattern and Structure Mathematics Awareness Program (PASMAP). In J. Novotná, H. Moraová, M. Krátká, & N. Stehlíková (Eds.), *Proceedings of the 30th conference of the International Group for the Psychology of Mathematics Education* (Vol. 4, pp. 209-216). Prague: Charles University.

Mulligan, J., Prescott, A., Papic, M., & Mitchelmore, M. (2006). Improving early numeracy through the Pattern and Structure Mathematics Awareness Program (PASMAP). In P. Grootenboer, R. Zevenbergen & M. Chinnappan (Eds.), *Identities, cultures and learning spaces* (Proceedings of the 29th annual conference of the Mathematics Education Research Group of Australasia, Canberra, pp. 376-383). Sydney: MERGA.

Mulligan, J., & Vergnaud, G. (2006). Research on children's early mathematical development. In A. Gutiérrez & P. Boero (Eds.), *Handbook of research on the psychology of mathematics education* (p. 117-146). Rotterdam: Sense Publishers.

OECD (2001). *Starting strong: Early childhood education and care.* Paris: OECD.

Papic, M., & Mulligan, J. (2005). Preschoolers' mathematical patterning. In P. Clarkson, A. Downton, D. Gronn, M. Horne, A. McDonough, R. Pierce, & A. Roche (Eds.), *Building connections: Research, theory and practice* (Proceedings of the 28th annual conference of the Mathematics Education Research Group of Australasia, Melbourne, pp. 609-616). Sydney: MERGA.

Papic, M., & Mulligan, J. (2007). The growth of early mathematical patterning: An intervention study. In J. Watson & K. Beswick (Eds.), *Mathematics: Essential research, essential practice* (Proceedings of the 30th annual conference of the Mathematics Education Research Group of Australasia, Tasmania, pp. 591-600). Adelaide: MERGA.

Perry, B., Anthony, G., & Diezmann, C. (Eds.). (2004). *Research in mathematics education in Australasia: 2000-2003.* Flaxton, QLD: PostPressed.

Perry, B., & Dockett, S. (2004a). Early childhood numeracy. *Journal of Australian Research in Early Childhood Education, 9*(1), 62-73.

Perry, B., & Dockett, S. (2004b). Mathematics in early childhood education. In B. Perry, G. Anthony, & C. Diezmann (Eds.), *Research in mathematics education in Australasia: 2000-2003* (pp. 103-125). Flaxton, QLD: PostPressed.

Perry, B., & Dockett, S. (2005). "I know that you don't have to work hard": Mathematics learning in the first year of primary school. In H. L. Chick & J. L. Vincent (Eds.), *Proceedings of the 29th conference of the International Group for the Psychology of Mathematics Education* (Vol. 4, pp. 65-72). Melbourne: University of Melbourne.

Perry, B., & Dockett, S. (in press). Young children's access to powerful mathematical ideas. In L. D. English (Ed.), *Handbook of international research in mathematics education: Directions for the 21st century* (2nd ed.). Mahwah, NJ: Lawrence Erlbaum.

Perry, B., Dockett, S., Harley, E., & Hentschke, N. (2006). Linking powerful mathematical ideas and developmental learning outcomes in early childhood mathematics. In P. Grootenboer, R. Zevenbergen & M. Chinnappan (Eds.), *Identities, cultures and learning spaces* (Proceedings of the 29th annual conference of the Mathematics Education Research Group of Australasia, Canberra, pp. 408-415). Sydney: MERGA.

Peters, S. A. (2004). *"Crossing the border": An interpretive study of children making the transition to school.* Unpublished doctoral thesis, University of Waikato, Hamilton.

Rogoff, B. (1998). Cognition as a collaborative process. In W. Damon, D. Kuhn, & R. S. Siegler (Eds.), *Cognition, perceptions and language* (5th ed.) (pp. 679-744). New York: John Wiley & Sons.

Thomas, G., & Tagg, A. (2004). *An evaluation of the Early Numeracy Project 2003.* Wellington: Ministry of Education.

Thomson, S., Rowe, K., Underwood, C, & Peck, R. (2005). *Numeracy in the early years: Project Good Start final report.* Melbourne: Australian Council for Educational Research.

Trinick, T., & Stephenson, B. (2006). An evaluation of Te Poutama Tau 2005. In F. Ell, J. Higgins, G. Thomas, T. Trinick & J. Young-Loveridge (Eds.), *Findings from the New Zealand Numeracy Development Projects 2005* (pp. 34-45). Wellington: Ministry of Education.

Vagi, O., & Green, R. (2004). The challenges in developing a mathematics curriculum for training elementary teachers in Papua New Guinea. *Early Child Development and Care, 174*(4), 313-319.

van Hiele, P. M. (1986). *Structure and insight.* Orlando, Fl: Academic Press.

Warren, E. (2005a). Patterns supporting the development of early algebraic thinking. In P. Clarkson, A. Downton, D. Gronn, M. Horne, A. McDonough, R. Pierce, & A. Roche (Eds.), *Building connections: Research, theory and practice* (Proceedings of the 28th annual conference of the Mathematics Education Research Group of Australasia, Melbourne, pp. 759-766). Sydney: MERGA.

Warren, E. (2005b). Young children's ability to generalize the pattern rule for growing patterns. In H. L. Chick & J. L. Vincent (Eds.), *Proceedings of the 29th conference of the International Group for the Psychology of Mathematics Education* (Vol. 4, pp. 305-312). Melbourne: University of Melbourne.

Warren, E. (2006). Teacher actions that assist young students write generalizations in words and in symbols. In J. Novotná, H. Moraová, M. Krátká, & N. Stehlíková (Eds.), *Proceedings of the 30th conference of the International Group for the Psychology of Mathematics Education* (Vol. 5, pp. 377-384). Prague: Charles University.

Warren, E., Young, J., & de Vries, E. (2007). Australian indigenous students: The role of language and representations in the negotiation of mathematical understanding. In J. Watson & K. Beswick (Eds.), *Mathematics: Essential research, essential practice* (Proceedings of the 30th annual conference of the Mathematics Education Research Group of Australasia, Tasmania, pp. 775-784). Adelaide: MERGA.

Watego, L. (2005). An indigenous perspective on mathematics contextualisation in a pre-school: From safety to empowerment. In P. Clarkson, A. Downton, D. Gronn, M. Horne, A. McDonough, R. Pierce, & A. Roche (Eds.), *Building connections: Research, theory and practice* (Proceedings of the 28th annual conference of the Mathematics Education Research Group of Australasia, Melbourne, pp. 767-774). Sydney: MERGA.

Waters, J. (2004). Mathematical patterning in early childhood settings. In I. Putt, R. Faragher, & M. McLean (Eds.), *Mathematics education for the third millennium: Towards 2010* (Proceedings of the 27th annual conference of the Mathematics Education Research Group of Australasia, Townsville, pp. 565-572). Sydney: MERGA.

Wood, K., & Frid, S. (2005). Early childhood numeracy in a multiage setting. *Mathematics Education Research Journal, 16*(3), 80-99.

Wright, R. J., Stanger, G., Stafford, A. K., & Martland, J. (2005). *Teaching number in the classroom with 4–8 year olds.* London: Sage.

Young-Loveridge, J. (2004a). Effects on early numeracy of a program using number books and games. *Early Childhood Research Quarterly, 19*, 82-98.

Young-Loveridge, J. (2004b). *Patterns of performance and progress on the Numeracy Projects 2001-2003: Further analysis of the Numeracy Project data.* Wellington: Ministry of Education.

Young-Loveridge, J. (2005). Patterns of performance and progress. In J. Higgins, G. Thomas, T. Trinick & J. Young-Loveridge (Eds.), *Findings from the New Zealand Numeracy Development Projects 2004* (pp. 5-20, 115-127). Wellington: Ministry of Education.

Young-Loveridge, J., & Peters, S. (2005). Mathematics teaching and learning in the early years in Aotearoa / New Zealand. *Australian Journal of Early Childhood, 30*(4), 19-24.

AFFILIATIONS

Bob Perry
Murray School of Education, Charles Sturt University, Australia

Jenny Young-Loveridge
School of Education, University of Waikato, New Zealand

Sue Dockett
Murray School of Education, Charles Sturt University, Australia

Brian Doig
School of Education, Deakin University, Australia

ALLAN L. WHITE

LEARNING MATHEMATICS IN THE MIDDLE YEARS

This chapter is dedicated to Beth Southwell, my colleague and friend, who passed away as we began working on it. She was an inspiration and will be greatly missed, in particular by the NSW Education community.

This chapter focuses on Australasian research into the mathematical understandings of students in the middle years of school (years 4-9), a key stage in students' mathematical development in terms of engagement with mathematics and future study choices. Research focusing on student thinking across a range of content areas including rational number, space, algebra, and chance and data is reviewed. Studies on related issues including primary to secondary transition and teacher professional learning are also discussed. A range of areas for further research is presented.

Key words: algebra, chance and data, mental computation, numeracy, problem-solving, professional learning, rational number, space, transition

INTRODUCTION

The middle years of schooling constitute the last two years of primary school and the first 2-3 years of secondary school (typically 9-14 year olds). As well as the natural growth of young people entering adolescence, the transition from primary to secondary school can be problematic for several reasons. These include the differences in curricula, the different organization of the school site, and the different pedagogical practices. The Victorian Middle Years and Development Project (MYRAD) (Centre for Applied Educational Research, 2002) states that a review of research into mathematics in the middle years reveals considerable disengagement with, and underachievement in, mathematics on the part of lower secondary students. As well, there is a reported increase in early adolescence of truancy, disruptive behaviour, alienation, and isolation (Sullivan, McDonough, & Turner Harrison, 2004). Other Australian states have produced documents setting out some of the difficulties and possible effects of this disengagement.

Anthony (2004), in her analysis of research trends between 1994 and 2003, does not mention middle school as a context for research, with primary and secondary schooling listed as separate entities, and a further category for early childhood. Yet there is evidence of a general growing interest in middle schooling and research has developed in the area; for example, substantial research on the middle years has been conducted by the State of Victoria's Department of Education and Training (2004), and the Australian Journal of Middle Schooling was first published in 2001.

H. Forgasz et al. (eds.), Research in Mathematics Education in Australasia 2004-2007, 41-72.

This general interest in the middle years of schooling is very evident in mathematics education. In Australia, the Australian Association of Mathematics Teachers [AAMT] sponsored a conference in 2005 as part of the Quality Mathematics in the Middle Years project. The proceedings of the conference contain many references to school based initiatives such as the Tasmanian Kingston Primary School's *Scaffolding Numeracy in the Middle Years* project (Morony & Stocks, 2005). AAMT also supported the publication "Research Pointers to Practice: A Review of Research to Inform Middle Years Mathematics Teaching and Learning" (Doig, 2005). As well, a growing number of schools have introduced a middle school organization, and some educational systems have built separate middle schools, such as in South Australia's Seaford 6-12 school.

In this chapter research on the learning of mathematics in the middle years is reported, first from a general viewpoint – transition, numeracy, and problem solving – and then in terms of specific topics that have been investigated – mental computation, rational number, space, algebra, and chance and data, with a final section on professional teacher learning.

Many educational systems and schools are developing policies to cope with the transition from primary to secondary school. Some of this is reflected in the review that follows. While selecting reports for this review, it became obvious that researchers do not always categorise their participants as belonging to the middle years nor do they cover the full range from grades 5-9. In general, studies that fall within the total range of the middle years of schooling have been included.

GENERAL ISSUES

Transition

The middle years have been described as a transitional stage in the lives of school students that cause them to have "special needs that are products of, or exacerbated by, changes in a student's social and emotional life" (Doig, 2005, p. 10). The transition includes moving from being children to becoming adults which means:

> they are becoming more able to deal with abstraction; they have a growing tendency to be reflective; they have ideas and the urge to express them; they want to develop relationships that include a range of growing expectations including that they and their ideas are valued; they will be able to make choices and take control in their lives and learning. Unlike many adults (including many of their teachers) many current middle years students are relatively 'fearless' users of information and communication technology (ICT). (Morony & Stocks, 2005, p. 8)

In the primary school, the students have been in the care of mainly one teacher who is their guide, mentor, and general care-giver for sometimes more than one year. They also have a home room from which they seldom move except for lessons in personal development, health or physical education, for example. In the secondary school, by contrast, students usually meet a different teacher for each subject and move from room to room as they change classes. When, too, physiological changes of adolescence are taking place as well as those associated with social groupings, there develops a disinclination to be serious about study, and particularly about subjects such as mathematics that require some rigorous thinking. While there is some emerging evidence that students' brains undergo changes during puberty which may partly contribute to these changes in behaviour towards mathematics learning, yet it seems an inadequate explanation for the overall trend. There are other reasons for what has become known as a plateau or dip in mathematics achievement, such as the nature of the content itself (Centre for Applied Educational Research, 2002). Students who have been working confidently with whole numbers stumble when required to think relationally and algebraically. The cumulative effect of such a complex mix of influences results in many students finding the transition from primary to secondary school too traumatic and consequently they disengage themselves from the learning process.

Sullivan and McDonough (2007) developed an instrument focussing on students' orientation to, and motivation for, learning mathematics that was administered to 205 Victorian grade 8 students, 101 male and 104 female, from government and Catholic schools; selected students were subsequently interviewed. The results indicated that grade 8 students showed more confidence in their ability to learn mathematics and in their persistence than was evident in observations of their classes. The students reported a negative influence of peers on some classmates but less of an influence for themselves, and had modest career aspirations. The researchers stated the implications are that teachers can assist students by becoming aware of their orientations to learning, their perceptions of the value of schooling, and their further vocational aspirations, and by finding ways to overcome factors inhibiting their engagement in school.

Just how different is mathematical practice in primary and secondary schools? This is the question Doig, Groves, Tytler, and Gough (2005) studied. They used the "Improving Middle Years Mathematics and Science (IMYMS) Components of Effective Teaching and Learning" mapping procedures with 99 teachers (55 primary and 44 secondary) to audit teacher practices. The application of a Rasch procedure to the data revealed there were differences on sub-components between the teachers' practices for mathematics teaching in primary and secondary schools. For example, the most difficult sub-component for secondary teachers to implement was contained in the statement: "The learning program provides opportunities to connect with local and broader communities". This contrasted with the primary school teachers who reported it as less difficult. Other sub-components that secondary teachers reported having greater difficulty with compared to primary teachers were:

- "The learning environment provides a safe place for students to take risks with their learning";
- "Learners engage with a rich, contemporary view of mathematics and science knowledge and practice";
- "Persistence and effort are valued and lead to a sense of accomplishment"; and

- "The teacher builds positive relationships through knowing and valuing each student".

The last of these showed a considerable difference as teachers felt this was much more difficult to achieve in secondary schools. Sub-components on which secondary teachers, compared to primary teachers, found it easier to score had to do with students developing 'meaningful understandings'. Primary and secondary teachers shared common difficulty scores for: "Students are challenged to extend their understandings" and "Students are encouraged to see themselves as mathematical and scientific thinkers".

A reason behind the differences in primary and secondary schools may be manifested in the role of play (Caswell, 2005). Caswell's study involved 27 students (9-12 years old) who identified and described elements of play in mathematical learning. The students believed that play activities engaged them in using their mathematical knowledge to solve open ended problems. The activities were engaging, challenging, diverse and non repetitive. The researcher asserted that play was forgotten in the transition from primary to secondary where there was evidence of greater procedural instruction. In a ten week study, data were collected from transcripts of weekly class discussions, student journals and conceptual models created by the class as well as classroom anecdotes. Caswell found that the students' concept of play changed over the ten weeks from focussing on the materials of play to play being of value as a means of conceptual understanding and growth. There was an increase in engagement with mathematical thinking, a growth in confidence, and a non-judgmental environment had evolved. The value of pursuing the use of play beyond the primary years was recognised.

Textbooks assume greater importance in the middle years of schooling. Shield (2005) reported on the development of a methodology for comparing and evaluating mathematics textbooks for the middle years. Based on the principles underlying the syllabus in mathematics, further development and extension of this methodology should provide a useful adjunct for teachers in selecting textbooks for their students, particularly when new curricula are being introduced. The procedural complexity and mathematical solving processes required by problems on two topics in seven grade 8 textbooks from four Australian states were studied by Vincent and Stacey (2007) using definitions from the 1999 TIMSS Video Study. Although variation existed between textbooks, the majority of problems were of low procedural complexity, requiring only the practice of procedures. The general picture was consistent with that painted by the Video Study, with a somewhat stronger emphasis on procedural work.

Numeracy

Definitions of numeracy abound, including the one used in the *Numeracy Education Strategy Development Conference*:

> In school education, numeracy is a fundamental component of learning, discourse and critique across all areas of the curriculum. It involves the disposition to use, in context, a combination of:

– underpinning mathematical concepts and skills from across the discipline (numerical, spatial, graphical, statistical and algebraic);
– mathematical thinking and strategies;
– general thinking skills; and
– grounded appreciation of context. (AAMT, 1997, p. 15)

Other definitions have demanded a different focus. Zevenbergen (2004), for example, argued for the reconceptualising of numeracy, with school mathematics given a central role in the development of new forms involving predicting, confirming, problem solving, and mathematical modelling. Morony and Stocks (2005), however, sought a greater emphasis on the cross-curriculum aspects, with numeracy being "what 'students need as the next generation', for purposes now and in their lives beyond schooling" (p. 12).

There is some confusion over numeracy and mathematical literacy. Generally mathematical literacy has gained wider international acceptance. It has been defined as:

> An individual's capacity to identify and understand the role mathematics plays in the world, to make well-founded judgements, and to use and engage with mathematics in ways that meet the needs of that individual's life as a constructive, concerned and reflective citizen (Organisation for Economic Co-operation and Development (OECD), 2003, p. 15).

The importance of numeracy is evident in the existence of a number of frameworks and programs that have already been established in Australian states and in New Zealand. Generally students who perform at the lowest levels are usually targeted for carefully focused numeracy programs. For example, in Tasmania there is the *Essential Learnings Framework (ELF)* and the follow-up program, *Being Numerate in the Middle Years.* The *Numeracy Project* in New South Wales (NSW) was part of a Federal government funded initiative to look at what makes a difference in numeracy learning. Mulligan and Busatto (2004) evaluated the project and several factors were identified as contributing to numeracy growth, and the importance of the school 'owning' its numeracy plan. The limitation of not being able to implement a longitudinal study was also raised. In 2006, Gervasoni, Hadden, and Turkenburg (2007) conducted a large study of number learning of over 7000 Victorian children in Ballarat for the purpose of identifying issues that could inform the development of a professional learning plan. A notable number of students (31%) who were beginning grade 6 was found not yet able to read, write, order, and interpret four-digit numbers, nor use reasoning-based strategies for calculations in addition and subtraction, and multiplication and division.

The understanding of the concept of a positional, or place value system, is central to developing number sense and is also the basis for the four fundamental operations. Nataraj and Thomas (2007) conducted a case study of a class of 27 New Zealand grade 9 (age 13 years) secondary school students that was focused on deepening students' understanding of positional notation by looking at place value in different number bases as a way of enhancing students' understanding of the decimal number system. Students were given a range of tasks, based on the historical development of number systems, to work on using appropriate concrete materials. Data were obtained via pre- and post-standardised tests. The results suggest that the combination of a historical and a concrete approach helped the students to understand the place value system to the extent that they could generalise it to other bases.

Pegg, Graham, and Bellert (2005) used a *Quicksmart* intervention program, replacing the usual program of 25 weeks, focusing on memorisation of basic facts to determine whether this could free up short term memory to be used for more complex tasks. The participants were 12 low-achieving mathematics students, six from grade 5 and six from grade 7. The intervention program was preceded and followed by achievement tests in order to gauge any improvement that had taken place. The intervention made a marked difference for the grade 5 students in achievement. The grade 7 students also benefited as shown on standardised test scores as well as improvements on State-wide testing measures that were not the focus of instruction. Interestingly, in a follow-up study with the same pupils it was found that they had not regressed one year after the intervention program. The pupils were able to maintain the gains made, and performance was sustained across all the mathematics tasks tested.

Numeracy demands practical understandings of the ways in which information is gathered. The information is often presented *visually.* Visual reasoning differs substantively from sequential reasoning. Diezmann, Lowrie, and Kozak (2007) conducted a study in which a set of graphically-oriented numeracy items was used to compare the thinking of a mix of ten high and low performing participants aged between 10 and 11 years. The data were collected via interviews and video-taping of students working on a set of tasks, and generally showed that the students' thinking differed in four ways:

- high performers drew on existing knowledge and skills which low performers appeared to lack;
- high performers used multiple cues to complete tasks whereas low performers worked from a single cue or overlooked cues;
- high performers used simple solution procedures correctly while low performers used more mentally demanding procedures with limited success; and
- high performers were more knowledgeable than low performers about everyday graphics.

An important implication, listed by the researchers, is that it is essential that low performers are encouraged to develop and use mathematical and graphical knowledge, to generate information from graphics, to build repertoires of strategies, and to select and use these strategies judiciously.

Due to the link between teachers' beliefs and behaviours, research has concentrated on uncovering existing beliefs concerned with aspects of numeracy and with developing ways of changing or influencing the development of new beliefs. For example, Skalicky (2007) presented findings from part of a larger qualitative study designed to investigate the positioning of numeracy by teachers of middle years classrooms (grades 5-8) in Tasmania. Teacher interviews designed to gain insight into teachers' beliefs and practices were semi-structured. The results indicated that the teachers positioned numeracy as a cross-curricular construct in the curriculum. Nisbet and Grimbeek (2004) used factor analysis to explore the beliefs of teachers about the compulsory numeracy tests given in grades 3, 5, and 7. Although the response rate to their survey was small (24.1%), they found that most teachers did not believe the tests were valid measures of students' mathematical ability, and that only 47% agreed with the tests in principle. Teachers' lack of faith in tests was reinforced in a study by Ellerton and Olson (2005). Findings from 83 grade 7 and 8 American students who completed a test comprising items from the Illinois Standards Achievement Tests confirmed that students' scores on tests do not necessarily reflect their levels of understanding of mathematical concepts and relationships. The results indicated a 35% mismatch with students who gave correct answers with little or no understanding and others who gave incorrect answers but possessed some understanding. The authors cast doubt on the use of large scale testing programs as a means of making comparisons or being used as basis for the allocation of resources.

Beswick (2005) studied a program designed to improve teachers' beliefs towards numeracy teaching of children with mathematics learning difficulties. While the number of participants was small (22 teachers) and the program was brief, there was a change in beliefs on the part of the participants towards being more aligned to the values underpinning the program. The researcher stated that attention to the relevant teacher beliefs is needed if inclusive policy is to have a real impact on students with mathematics learning difficulties.

A common aspect in most definitions of numeracy is the ability to apply mathematical knowledge across various contexts. It is not only necessary to teach the mathematical content but also to provide students with strategies for recognising and applying mathematics in a range of contexts. Hurst (2007) conducted a qualitative multiple case study involving a group of eight grade 6 and 7 Western Australian female primary school students aged 11 or 12 years in which the students' capacity to recognise, apply, and question the use of mathematical ideas embedded in a range of contexts was investigated. Also of interest was the extent to which students' capacity to connect mathematical knowledge to other contexts could motivate them to learn mathematics. In the interviews, students were shown articles, maps, and advertisements, about which they were asked questions to probe the development of their thinking. Findings indicated that student thinking about mathematics and their attitudes towards it could be enhanced by targeting mathematical connections.

The need to apply mathematical knowledge across various contexts is often motivated by the need to solve a problem. In the next section research into problem solving and the middle years of schooling is discussed.

Problem solving

Anthony's (2004) analysis is again useful to indicate the trends in the topics of research. She indicated that between 1994 and 2003 there had been a sharp decline from 9% to 1% in the number of studies dealing with problem solving, and an increase from 5% to 12% in studies dealing with number and computation. There was no indication as to what proportion of studies were predominantly in the middle schooling range. This trend may be continuing. However, there seems to be a re-awakening of interest in research into problem solving, urged by Lesh (2006), particularly related to non-routine problems, collaborative problem solving, discourse communities, and mathematical modelling. Thus several of the studies reported here focus on more than one aspect of the study of problem solving.

Muir and Beswick (2005) gave 20 grade 6 students from five different schools six non-routine problems to solve during a semi-structured interview. The theoretical framework used was consolidated to three stages: *understanding the problem, choosing and implementing a solution strategy,* and *getting the answer*. The students' responses were analysed and marked according to rating scales that took into consideration the degree to which the student had used metacognition, as perceived by the researcher, at each of the stages of problem solution. This led to the conclusion that the students needed most help at the execution and verification stages. Using the same data source, Beswick and Muir (2004) investigated the problem solving performance of grade 6 students in relation to written or verbal responses. They found that above average, average, and below average students could all express their problem solving processes better verbally than in writing. The implication drawn was that all students might benefit from instruction that helped them use meaningful methods of recording their thinking.

Problem solving in the mathematics classroom requires teachers to develop discourse communities in which all students have equitable opportunities to engage in productive discourse. Brown and Renshaw (2004) explored replacement and interweaving patterns of discourse used by lower secondary students in doing tasks and justifying their results. They found that these patterns could quite easily be part of the students' preparation in mathematics and replace current approaches. In another study, R. Hunter (2007) reported on a one teacher case study involving a one-year collaborative teaching experiment focused on the interactional strategies used by the teacher to scaffold the participation of a diverse range of students in small group problem solving interactions. The study was conducted at a New Zealand urban primary school. The teacher aimed to shift the patterns of discourse from a disputational form to one in which the students collectively constructed group explanations and justifications. The results showed that effecting change in the small group interactions was a lengthy process requiring ongoing attention to the discourse used in the groups, and that the teacher should model interaction patterns in the group and highlight student behaviour to demonstrate valued interaction patterns. Of key importance in the development of productive group processes and discourse in the classroom was the emphasis placed on group members' responsibilities to each other. Further research to examine other factors important in enacting and maintaining diverse learners' use of productive discourse was suggested.

The relationship between the nature of the task and the impact on students in terms of mathematics anxiety and attitudes towards mathematics was studied by Anderson (2007) using six mathematics problems and an online survey to collect the responses of 43 students from grades 4, 5, and 6. Although the sample was small, a weak relationship was found between mathematics anxiety and attitude to the six mathematics problems presented to the students. That is, as anxiety increased a student's attitude towards mathematics declined. However a small number of responses showing positive attitude and high anxiety response indicated that anxiety was not an effective motivator for some students' performance. Some differences were observed between boys and girls in their responses to the rich tasks, with girls exhibiting a stronger association between high anxiety and poor attitude.

Mathematical modelling has long been regarded as a specific manifestation of problem solving and was usually confined to the upper secondary and tertiary mathematics curricula. This is no longer the case, English and Fox (2005) reported a study based on a grade 7 class in the third year of a three year program of mathematical modelling to solve problems. The results indicated that the students had developed a range of understandings and knowledge as they worked through modelling problems and had learnt to argue about the relative value of different aspects of the problem. This study was a development on an earlier one (English, 2004). English and Doerr (2004) used mathematical modelling as the context to investigate the effect of teacher interaction with students. Based on one teacher, they concluded that modelling activities that foster different reasoning patterns provide opportunities for teachers to learn about their students' development of mathematical ideas. English (2007) used a multilevel collaborative design to study students using modelling within an interdisciplinary context in the primary curriculum. Three classes of grade 5 Queensland children worked on a modelling problem, Australia's settlement, situated within the curriculum domains of science and studies of society and environment (SOSE). The results showed that children developed mathematisation processes that extended beyond their regular curriculum. These findings resonate with White (2006) who mapped Carpenter and Lehrer's (1999) model of five forms of mental activity upon the modelling process. Thus the mathematical modelling process provided a structure and direction for students to follow in solving problems while deepening their understanding and learning of mathematics.

Stillman, Galbraith, Brown, and Edwards (2007) developed a framework to support successful implementation of mathematical modelling in the grade 9 secondary classroom based on the transitions between stages in the modelling process and the associative cognitive activities. The implications are that teachers and curricula designers wishing to implement a series of modelling tasks over the course of a year can use the framework to ensure that, although not all elements will be addressed in every task, every element is included in as many tasks as necessary to develop students' modelling competencies. A more detailed review of the research on mathematical modelling across all levels of schooling is found in chapter 8.

With a greater focus on discourse, Doyle (2007) reported part of a larger study in which the changing nature of mathematics teaching and tasks was examined. A total of 57 students from two grade 4 classes in Queensland were compared after mathematical modelling tasks were undertaken with, and without, top-level structuring (TLS), a literacy strategy that aids comprehension of oral, textual, or graphic information. Results indicate that mathematical-modelling and TLS tasks can advance mathematical literacy. Recommendations for teaching were provided, particularly on the vital role of the teacher in creating a positive learning environment through facilitating discourse and literacy development in mathematics students.

SPECIFIC TOPIC AREAS

Mental computation

Mental computation has been a focus of change of emphasis in mathematics education, with syllabus documents advising teachers to take an approach that focuses more on mental computation as part of a range of computational strategies and less on traditional written algorithms. In the studies reported here, teachers' responses to this change and the effect of context upon mental computation have been examined.

Hartnett (2007) was concerned that teachers have been slow to adopt mental computation strategies in their classroom planning, postulating that it was due to their lack of knowledge about possible computation strategies and a lack of clear organisation of school programs. Harnett (2007) discussed a framework for the categorisation of mental computation strategies which provides a common language for teachers and students to discuss strategies used, and for the development of school and classroom programs. Caney (2004) studied thirty-four middle school Australian teachers and 172 middle school students. Using a Likert scale, the teachers were asked what emphasis they placed on mental computation in their teaching, and the students were asked to say how often they used mental computation when working mathematically. The results were then compared. Overall, there was little difference between the responses of the primary and secondary teachers, but secondary teachers tended to emphasise mental computation less than primary teachers. In general, the students indicated that they used mental computation less frequently than reported by the teachers.

Students tend to have their own ways of doing mental computation that are not necessarily the same as those that have been taught. Callingham (2005) advocated a whole school approach to systematically develop mental computation strategies, while White and Singh (2005) presented the computation triangle (with vertices – mental computation, written computation and calculator) as a *plan of attack* for teachers and students that could be used in a variety of different contexts demanding mathematical computation. At the top of the computation triangle, in a place of pre-eminence, is mental computation. Before any computation is attempted, whether written or using a calculator, students are required to use mental strategies to either compute a correct answer or produce an estimation of the answer. Thus mental computation strategies are always meant to precede either written methods or calculator use.

Paterson and Bana (2005) investigated the effect of context in mental computation. The 64 participants were from a Perth secondary school and two of its feeder primary schools. Four male and four female students were selected from two classes from each of grades 3, 5, 7, and 9. The participants were given the same question in two forms: one in context and one not. Money was used as the context because it was assumed that most students would have some experience of that context. The context was found to make no difference to the performance of the students except in grade 3, perhaps because these younger students would have had less experience with money. Another interesting result was that grade 7 students performed better than grade 9 students, perhaps because in grade 9 less time than in the lower grade is spent on emphasising mental computation. In a small scale exploratory study using a different context in game playing, Asplin, Frid, and Sparrow (2006) investigated the game's role in the development of mental computation. They observed a grade 6 class and, using observation and interviews, were able to document the strategies the students used. The key aspects of the game playing observed were the structure of the use of the game, participation and communication, and mental computation methods. Thus there was a social effect as well as a cognitive one. The researchers concluded that game playing has many benefits, some of which are those of playing with different people and exposing the students to new ideas and verbalisations. These findings provide further support for Caswell's (2005) contention about game playing, albeit the games involved were of different types.

Rational number

Rational number forms an important part of the middle years mathematics curriculum, underpinning the development of proportional reasoning, and is important for later topics in mathematics, such as algebra and probability. McGee, Kervin, and Chinnappan (2006) highlighted the strong bond between numeracy and understanding of fractions, and recommended that fractions be studied in authentic contexts. However, it is a topic which many teachers find difficult to understand and teach and which is reflected in student learning outcomes. It remains a fertile area for research. The research studies reported here involved investigating ways to improve student understanding using various models and to overcome the dominance of whole number thinking.

Part-whole numbers are a source of considerable confusion among middle years students. A commonly reported misconception of students dealing with part-whole numbers is to confuse fractions with ratios (instead of part to whole they report part to part). An area divided into four equal parts with one part shaded can be reported as representing one quarter or mistakenly one third. Chinnappan (2005) used Javabars to present visual models to grade 5 students to test their understanding of part-whole representations embodied in fractions. He reported that students had difficulty mapping the symbolic representation to the area analogs.

In a large cross-sectional study of students' qualitative concepts of fractions similar findings were reported (Gould, Outhred, & Mitchelmore, 2006). Using a total of 1676 students in grades 4 to 8 the researchers focussed on students' concept images of various fractions. Students' area diagrams revealed confusion; the area was ignored in preference to the number of pieces approach. This confusion was also evident in the pie task reported in Clarke, Sukenik, Roche and Mitchell (2006). To overcome this, Gould et al. (2006) recommend that teachers promote wider experience of regional models of fractions and greater use of compare and contrast questions to reduce misconceptions.

Callingham and Watson (2004) used a Rasch modelling technique on mental computation tests of fractions, decimals, and percents to determine a six-point developmental scale. The participants in their study came from grades 3 to 10 and had varying experience in the three topics. The scale highlighted the progression of part-whole mental computation proficiency across grades 3 to 10 and highlighted apparent difficulties. This proved to be an innovative and productive link between mental computation and part-whole numbers, and the researchers were able to give a clear developmental scale with distinct characteristics at each level.

In a study of 323 Victorian grade 6 students, Clarke, Roche, and Mitchell (2007) used one-to-one interviews and a range of rational number assessment tasks. Their data indicated clearly that Victorian students, and possibly their teachers, needed greater exposure to the sub-constructs of rational number (sometimes described as part-whole, measure, quotient, operator, ratio, and rate) and related models. The researchers encouraged teachers to use some of the tasks in one-to-one interviews to develop deeper insights into their students' understanding and common misconceptions. Wong and Evans (2007) investigated students' conceptual understanding of equivalent fractions by examining their responses to questions using symbolic and pictorial representations. Two hundred and thirteen students in grades 3 to 5 from three Sydney primary schools were studied. Differences emerged between the responses of students with limited general mathematics achievement and those of their more competent peers. The differences between the students in the limited and the high general mathematics achievement groups seem to lie not in the errors they made, as similar types of errors were observed, but in the depth of their procedural and declarative knowledge and the strength of the connections they made between procedures and concepts. The results corroborated the findings of Siemon, Izard, Breed, and Virgona (2006) who demonstrated that students with developing fraction knowledge were able to perform simple fraction tasks, but were unable to explain or justify their thinking in writing. A further aspect studied by Gould (2005b) was how NSW grade 6 students compared the size of fractions. He was able to uncover a diverse range of student strategies which were also revealing of the students' understanding of the concept of fractions.

Another common problem involves whole number learning dominating over any new experiences, therefore increasing the chances that students generalise their learning from whole numbers to rational numbers, often inappropriately. In an attempt to facilitate student conceptual learning of fractions and the operations with fractions that moved away from whole number dominance, Gould (2005a) used several NSW grade 6 classes to investigate different uses of diagrams to record and share methods of representing and calculating with fractions. The researcher made a number of recommendations for teachers such as: children sharing diagrams to support thinking; and, using diagrams in realistic contexts as a way of enabling students to make sense of fractions.

The diagrams used in realistic settings can become a tool for supporting thinking and enabling students to make sense of fractions. Pearn and Stephens (2004) found the persistence of whole number thinking in fractional situations among Victorian grade 8 students. For example, a number of students compared 2/3 and 3/5 by considering the difference between the numerator and denominator in each fraction, rather than using "strategies which preserve the fundamental ratio between numerator and denominator" (p. 431). They did note however that uncovering such thinking was difficult and required careful probing and recommended using number lines to assist in correcting student thinking. Pearn and Stephens (2007) followed their own recommendation and conducted a teaching experiment with Victorian students in grade 5 and 6 classes at one school using number lines to highlight and develop students' understanding of whole numbers then fractions. Data were collected from screening tests and interviews and showed that successful students had easily accessible whole number knowledge and recognised the relationship between the whole and the parts, whereas the weakest students had poor number knowledge and could not see the connections.

Diezmann and Lowrie (2006) studied 67 NSW grade 5 students' knowledge of number lines. The results indicated that students were not proficient users even for simple tasks, and the need for teachers to emphasise the linearity rather than the cardinality of the line. The rational number domain is more like the measurement domain where continuous properties are measured, than like the whole number domain where discrete items are counted. Mitchell (2005) interviewed 26 students across grades 5 to 8 on their fraction knowledge and found that the use of measurement and spatial language was evident.

Decimal fractions are open to a number of misconceptions and the relationship between them and whole numbers may be an important teaching strategy. Task assessment was used by Irwin and Britt (2004) in a 15 item test given to 1356 students to explore the compensation and decomposition strategies used in working with whole numbers and with decimal fractions. They found that students who were able to use the compensation strategy with whole numbers could also use the decomposition strategy with problems involving decimal fractions. They proposed that this might provide an alternative direction for teaching decimal fractions in the primary school.

Over many years Steinle and Stacey have studied students' decision making in comparing decimals. In the period of interest covered by this review they reported a continuing very low rate of expertise, and a disturbing persistence of misconceptions, across grades 4 to 10 (Steinle & Stacey, 2004). While targeted teaching and teacher knowledge of misconceptions have been shown to improve student understanding, it appears that normal teaching is inadequate to remove misconceptions. In fact, much teaching seems to cover over misconceptions rather than overcome them.

Informed by the work of Steinle and Stacey (2004), Roche and Clarke (2004) interviewed 48 grade 3 to 6 students about their understanding of decimals, focussing on their explanations of how to compare them. Explanations reflecting greater place value knowledge, rather than applying procedures such as adding zeros to equalise the length of decimals, demonstrated greater conceptual understanding and were stronger predictors of success on more difficult and novel tasks.

A group of NSW researchers has investigated teaching programs that have produced varying levels of success in the areas of percentages, rates, and ratios. White and Mitchelmore (2005) constructed and trialled a unit for teaching percentages to grade 6 students that focused on percentage as a multiplicative relationship. While it was effective for some students, the results were not as high as expected. Later Mitchelmore, White, and McMaster (2007) studied a group of practising NSW teachers who implemented the *Teaching for Abstraction* (TfA) method for the grade 8 topic of *ratio and rates*. TfA assumes that most elementary mathematical ideas are abstractions from experience, and consists of four steps in which the teacher helps students to: familiarise themselves with the structure of a variety of relevant contexts; recognise the similarities between these different contexts; reify the similarities to form a general concept; and then apply the abstract concept to solve problems in related contexts. The students and teachers from six grade 8 classes in four schools participated. The results indicated that teachers experienced considerable difficulties adapting the approach to the abilities and interests of their particular classes, but that all students showed evidence of learning. White, Wilson, Faragher, and Mitchelmore (2007) investigated five practising teachers in regional NSW implementing TfA for the grade 6 topic *percentages* in three primary schools. The students explored familiar percentage contexts, searched for similarities in their mathematical structures, and then applied their learning to more abstract situations. Particular emphasis was given to additive versus multiplicative approaches in different percentage situations. The results show that this approach has the potential to benefit student engagement, learning, and attitudes of students and teachers.

The ability to read scales is an important application of rational number thinking, as scales appear in a wide variety of contexts including measurement, algebra, and statistics. Drake (2007) reported one aspect of a larger study into student understanding of scale. The sample comprised 13 New Zealand students from grades 7 and 8 from an urban intermediate school. A range of student strategies were identified, most of which were classed as informal knowledge, and were sorted into a progression that related to the level of number thinking involved. In many cases there was consistency, where a number line was used for mathematically similar items, and a familiar context evoking the same solution strategy. Yet, fraction questions caused students to change their strategies. Also, the context for developing an understanding of scale sometimes results in *contextual pollution*, for which context-situated knowledge interferes with the intended learning about another topic.

Using the context of technology, Norton (2006) investigated 56 grade 7 students' understanding of ratio while constructing simple Lego machines involving the use of cogs and pulleys. The context provided a rich environment for student learning.

Space

There is a distinct paucity of research in the area of space and geometry. Hansen and Pratt (2005) investigated the inclusivity and exclusivity of geometrical concepts in a practical activity in which students were asked to place shapes on an empty Venn diagram. The nine and ten year old students were video taped and the teacher gave immediate feedback as to the correctness of their efforts. Four pairs of students were selected to participate in this activity. The researchers found that the students had great difficulty in distinguishing between class and instance, that is, the students labelled each instance rather than the class to which several instances belong, and the inclusive nature of geometrical definitions proved difficult for them.

The need to help students develop ways of expressing their geometrical reasoning was emphasised by Brown, Jones, Taylor, and Hirst (2004). In reporting on the work of a group of teachers developing teaching ideas, they noted that students from lower secondary schools were able to verbally express their arguments as they investigated several geometrical situations involving properties and definitions of shapes, but had more difficulty in communicating them in writing. The researchers also emphasised the structure of the tasks and the role of the teacher.

Lowrie (2004b) argued for school mathematics to make greater use of the social and cultural contexts of learning, and explored the use of authentic artefacts used by 28 grade 5 students in solving problems (Lowrie, 2004a). The students were given the situation of a family visit to a theme park that involved number, measurement, and space ideas. Students responded by modifying the situation to make it more authentic, and used a variety of mathematical processes and skills, though not necessarily recognising the task as a mathematical activity since they had not had anything like it in their normal classroom.

At a small inner city school in Tasmania, Scott (2007) examined 23 grade 5 and 6 students' changes in their perceptions of themselves as learners, their knowledge and skills in an aspect of Space, and their ability to use specific tools and strategies over a 2-week period. The focus of the study was on whether the use of the following three specific questions at the commencement of reflective writing sessions gave evidence of development in children's thinking and mathematical understandings:

- What have you learnt which is somewhat different to what you already knew about **mathematics**? Give examples.
- What have you learnt which is somewhat different to what you already knew about the **program, tools and games** used?

- What have you learnt which is somewhat different to what you already knew about **yourself or the way you learn**?

Preliminary findings suggested that students with limited skills in English were able to communicate their thought processes, and some students deepened their mathematical understandings about aspects of Space in four lessons (each of 60-80 minutes duration) over the two weeks. However the researchers stated that reflective writing is a text-type and a generic skill that needs to be explicitly taught.

Algebra

If algebra is generalised arithmetic consisting of operational and structural phases then many students come to the study of early algebra with poor understandings of the structures of arithmetic. There is a greater awareness of this area in the early years of school with Perso (2005) making a plea for greater consideration of the pre-algebra understanding that pushes beyond work in number and basic patterning. In this section of the chapter, research in which attempts were made to bridge number and algebraic thinking using relational thinking are reported, the importance of language, symbol, and discourse is highlighted, and student understanding of variable and function is examined.

Stephens (2005, 2006) studied a way to bridge number and algebraic reasoning with number sentences by using relational thinking. The participants were Japanese students from grades 3 to 8 who demonstrated a capacity to generate equivalent numerical expressions that do not rely on computation. This appears to be a pre-cursor to formal algebraic thinking. Some of the findings were concerned with language, as success at relational thinking was linked to the students' ability to read passages involving literal expressions correctly Stephens (2007) expanded the study to investigate how well students in grades 5, 6, and 7 in three countries were able to use relational thinking to solve different types of number sentences. The study involved 227 students from Japan, 301 students from Australia and 194 from Thailand who were a given a pencil-and-paper questionnaire. The results revealed that students' use of relational thinking to solve number sentences is evident in all three countries by the end of primary school. The extent of its acquisition varied between the countries and between the schools and, in all three countries, there were students who appeared to rely solely on computational methods to solve the same number sentences. In a further study, J. Hunter (2007) used a questionnaire and a scoring rubric devised by Stephens (2006) to conduct a study of 361 New Zealand primary and intermediate school students' use of relational thinking to solve equivalence problems. The results revealed that most students used computational, rather than relational, strategies to solve the equivalence problems. Many of the common errors students made indicated a lack of understanding of the equals sign, and reflected their view of the equals sign as an indication to carry out an operation. The implications of the findings described by the researchers suggest that an emphasis on increasing numerical reasoning is not adequate to develop deep powerful understandings of essential algebraic concepts. Explicit attention needs to be given to developing relational forms of thinking and to developing primary and secondary students' notions of the equals sign as representing relational equivalence.

The ability to work mathematically using algebra is closely linked to the acquisition and use of the appropriate language of mathematical discussion. Falle (2005a) used audio-taped interviews with 33 students from grades 8, 9 and 10 in NSW secondary students to capture their algebraic conceptualisations of questions involving cancelling. She found that the source of many misconceptions could be traced to the use of natural language in contrast to formal mathematical use. In a further study Falle (2005b) investigated students' responses to algebraic items and proposed that their explanations revealed not only the degree of their mathematical thinking, but also that the linguistic features used may serve as indicators of their levels of understanding. Falle (2005b) noted that less successful students resorted to "parroting" mathematical rules, even though they may not have been able to use them. In contrast, students who were more mathematically capable tended to experiment with logic and have greater control over the language needed to express themselves.

Many students find the demands of shifting their thinking from arithmetic to algebra difficult and challenging. Falle (2007) studied grade 8 and 9 students' responses to a test of introductory algebra items. The study consisted of two phases, one being a test consisting of forty items based on beginning algebra. Falle applied Rasch modelling to the data and three distinct "ability" clusters emerged. The data revealed that the tendency to conjoin terms inappropriately (e.g., $5x + 3 = 8x$) occurred with different frequencies in each of the three groups. Students of lower ability tended to conjoin terms more often than other students. Interview data and error analyses provided further insight into the students' thinking and suggested that the use of informal language in the classroom may serve to obscure mathematical ideas. Implications for classroom practice also included the need for students to experience, and use, a variety of mathematical language and symbols in order to explore and express mathematical meanings.

Student algebraic thinking is often confused or guided by serious misconceptions. In a study of written responses to six linear inequation tasks, Vaiyatvutjamai and Clements (2004) analysed the errors made by 231 Form 3 (grade 9) Thai students in two Chiang Mai government secondary schools. Students completed tasks before and immediately after a series of 13 lessons on linear equations and inequations. Several misconceptions were revealed and although some were clarified as a result of the lessons, there were others that remained and seemed to be "fossilised". A fossilised misconception was used to denote the situation where a student maintains a faulty conception despite having been specifically taught the "official" defining characteristics of the relevant concept. Associated with this then is the absence of cognitive change over time or even resistance to change over time, so that cognitive inertia persists despite the individual having been taught the "proper" view of the concept. While the study involved students across the range of abilities, Vaiyatvutjamai and Clements (2004) challenged the use of the term misconception for many of the errors made by low performing students: "A misconception can be regarded as a fairly stable, but inappropriate, way of thinking ... analysing the errors made by low performers in this study, was that the word 'stable' was not one that could sensibly be used" (p. 181).

Student difficulties with the study of algebra include lack of understanding of variables and formal symbolic manipulation, and this acts as a barrier to success in mathematics study. Callingham, Falle, and Clark (2004) asked two grade 5/6 classes in a selective school to prove equality in the expression g + b – b = g. They reported that the majority of students used number substitution in the task, though there was an indication that some students were moving towards more abstract thinking. Norton and Irvin (2007a) conducted a case study of a classroom teacher who used a concrete approach to teaching algebra to 18 students in a Queensland grade 9 class over a 6 week algebra course. Results indicated that the students struggled due to a lack of understanding of arithmetic concepts including those associated with equivalence, operations with negative integers, the distributive law, and fraction concepts. Once these difficulties were addressed through the explicit teaching of the links between materials and symbols, materials and language, and language and symbols, the students made considerable progress in writing, simplifying expressions, and solving equations with variables on both sides. Norton and Irvin (2007b) also collected data on student perceptions about their learning experiences and found that the students valued the classroom discourse much more than they did the normal mathematics learning experiences. They concluded that the use of concrete materials, games, and explicit language should underpin middle school mathematics teaching and learning in order to foster positive perceptions of algebra.

Brown (2007) studied the notions of function held by grade 9 students as they begin to study functions in a technology rich teaching and learning environment (TRTLE). Students were given extended tasks that provided opportunities to develop and consolidate their concept images and where they were required to make choices about solution paths. The results indicated that the tasks provided opportunities for, and evidence of, thinking about and making connections between functions in each of the numerical, graphical, and algebraic representations. The fragility of these notions was evident at times through the choices of representations selected in solving tasks. However, evidence of the strength of student understanding was more likely to be detected through attending to classroom discourse at a private level between two or three members of the TRTLE, rather than through written work, as little trace of this was recorded by students on their task scripts.

Chance and data

An important part of numeracy, made more important by the information explosion via the internet, is the ability to work with "Chance and Data" and to develop critical thinking skills that allow individuals to interrogate data. This ability is sometimes called statistical literacy which is closely tied to the use of language and reasoning. Allied to this is the explosion of statistical software programs available to schools. The research reported here explores statistical literacy and the thinking of middle years students from both wide and close lenses.

Watson (2005b) stressed to teachers the importance of statistical literacy and its three main areas of student learning: understanding language, understanding language in context, and being able to criticise. She has provided a number of research-based assessment items and their rubrics for teachers to use to assess their students' learning. Earlier, Watson (2005a) examined the evolution of statistics in Australian curriculum from 1991 and focused on the place of expectation and variation. She argued that variation is a pre-eminent concept and students should develop an appreciation of variation before expectation, a reverse of current curriculum design. Watson and Kelly (2005) investigated an instructional program on sampling bias with 630 Tasmanian students in grades 3, 5, 7, and 9 divided into two groups. After two years there was little difference between the instructional group and the control group. Chick and Baker (2005a, b) examined the complex interplay of content knowledge and pedagogical content knowledge in the teaching of chance to grade 5 students, and the importance of teachers having appropriate knowledge in order to avoid creating misconceptions among students. The researchers highlighted the important role of teacher content knowledge and teacher pedagogical content knowledge in fostering student understanding.

Bill and Watson (2007) studied student responses to three selected tasks on the statistical concept of distribution in a grade 9 class of 29 students. The tasks involved the exclusion of outliers, analysis of data using a semi-formal framework (GICS) developed for this study, and comparing two distributions; the software, Fathom™, a product of Key Curriculum Press was used. The results indicated that all three tasks and the use of Fathom™ contributed to "sense-making" and the development of statistical intuitions. Fitzallen (2006) developed a theoretical framework for students' statistical thinking and reasoning. The framework was embedded in her work in an ICT context using the data analysis program TinkerPlots™. In the resultant design and assessment instrument particular attention was paid to the variety of forms of data representation available through the software.

Watson, Caney, and Kelly (2004) conducted a longitudinal study of students' beliefs about chance (probability). It involved 148 grade 3 and 117 grade 6 students. They described the changes in the two cohorts over the four years, and highlight a small change in chance beliefs, a smaller change in the beliefs about the measurement of chance, and the weak association between the two belief scales. In a further report of the study, Watson, Kelly, and Izzard (2006) collected data from 5514 Tasmanian students across grades 3 to 11 to measure cohort change and performance trends against the curriculum. The use of Rasch analysis allowed the placing of students from different grades and different times on the same scale in order to make comparisons. The longitudinal data suggested that the greatest improvement in understanding occurred between grade 3 and grade 5, whereas after a period in grades 6 and 7, medium degrees of improvement were again found.

PROFESSIONAL TEACHER LEARNING ABOUT MIDDLE YEARS

Researchers have focused on professional learning programs aimed at supporting middle years teachers of mathematics. The research reported here ranges across programs for implementing problem solving, numeracy, teachers' content knowledge, technology, and student engagement and perceptions.

A practical approach to the problem of implementing problem solving in the mathematics classroom was taken by Judy Anderson (2005). She asked groups of NSW teachers from grades 4 and 8 what professional development programs they felt they needed. From their responses she was able to identify and categorise their perceived needs into four categories: i) those related to the teachers themselves, such as knowledge and motivation, ii) those related to time, such as accountability and curriculum, iii) those related to classroom practices, and iv) those related to the students themselves. She suggested that professional development programs focussing on these themes would assist teachers to implement more effective problem solving approaches in their classrooms.

Davies and Walker (2005) report on the first year of a two year project on New Zealand teachers' knowledge in numeracy. The aim was to develop a small scale supportive community of learners including teachers, facilitators, and researchers. The focus was on the content knowledge teachers bring to their teaching, and data were collected from a wide range of sources. Because the teachers were more aware of their own content knowledge they were able to provide students with more appropriate challenges, and they also listened more closely to the ideas students raised. In their follow-up study, Davies and Walker (2007) investigated how four teachers from two New Zealand primary schools listened to, and made sense of, their students' ideas and the influence that teachers' content knowledge had on their capacity to listen. The study revealed that the depth of teachers' subject matter knowledge and pedagogical content knowledge mediated their enactment of effective listening practices. The teachers' content knowledge became a central organiser for the lessons and a defining feature of effective teaching. Another study by Julie Anderson (2005) was associated with involvement in the New Zealand *Intermediate Numeracy Project*. The case study reported was of a single teacher who gained considerable insight into her own content knowledge, her pedagogical approaches, her beliefs, and her ability to implement change. She recognised a gap between her espoused beliefs and her ability to use developmental thinking in her teaching. She also recognised that her lack of content knowledge limited her ability to help students make sense of their mathematics.

Children's perceptions of teachers' roles in helping them learn mathematics was the subject of a study by Taylor, Hawera, and Young-Loveridge (2005). They interviewed 183 grade 5 and grade 6 students at six schools. The children were Maori (over half), European (about a third), Pacific Islanders (about a tenth), and the rest were Asian, and covered a range of ability. Four roles were identified: the teacher as transmitter of information, the teacher as mentor, the teacher as classroom manager, and the teacher as arbiter of decisions. A large percentage of students (46%) opted for the teacher as a transmitter of information. Another surprising result was that 21% did not know. The researchers suggested that effort is needed to encourage students to question, verify, and justify their thinking. In a study by Woodward and Irwin (2005), the discourse of one teacher in the New Zealand *Numeracy Project* was examined and compared to that of another teacher. The study teacher was found to exercise considerable wait time, respond to students' statements, and challenge them to talk among themselves to justify their mathematical thinking. In comparison, the other class was dominated by everyday language with limited use of mathematical vocabulary, students were not encouraged to make mathematical statements, and they received minimal help in learning the mathematics vocabulary. The researchers recommended that the way the teacher challenged the class was a means of promoting mathematical thinking and argued that it assisted students to determine more clearly the teacher's role in helping them learn mathematics.

A technology that allows students to communicate their mathematical ideas incorporating the spoken word and still and moving images in everyday mathematics classroom settings, may have the ability to influence positively the teaching and learning process of mathematics. Thus Hilton (2007) studied whether the context of filmmaking could improve learner outcomes in mathematics. The participants were 13 girls and 14 boys from a grade 7 class at a Queensland state primary school. It was concluded that mini-documentary makers retained information at least as effectively as, if not better than, students who used pen-and-paper strategies. This suggests that mathematics educators should be confident of positive effects on students' knowledge retention through student filmmaking in mathematics; this resonates with findings about the relationship between student engagement and improved performance mentioned earlier in this chapter.

In a small scale study with a class of 30 students, Ainley and Luntley (2005) developed a methodology for measuring attention in the classroom. The knowledge that enables teachers to respond effectively to what happens during the lesson was called *attention-dependent* knowledge. They conjectured that understanding the performance of experienced teachers required an account of the interplay between subject and pedagogic knowledge evident in learning objectives and lesson plans, as well as attention-dependent knowledge that can only be revealed in the classroom. In the study, two researchers observed a teacher several times. From videos of the class, and the use of several stimulated recall techniques, they identified teaching episodes that indicated the role of attention on the part of the teacher. Using a grounded theory model, they coded the episodes at four levels (cognitive or affective, conceptual or non-conceptual, reacting or responding, and, noting or interrogating) and argued that attention-dependent knowledge is what causes experienced teachers to make the kind of decisions they make in the classroom.

Teachers' meta-awareness is seen by Caswell and Nisbet (2005) as a means of enhancing students' mathematical understanding. Based on a small scale study with one grade 6 class, it was reported that greater understanding can be achieved though an emphasis on self-assessment and self-regulation. Besides greater understanding of the mathematical concepts being embedded in class activities, students also displayed greater confidence in working mathematically.

In recent years there has been an explosion of websites created to promote the teaching and learning of mathematics and to provide data and related information that has potential for use in middle school classrooms. The challenge for teachers is to select and integrate what is helpful and reject what is not. Handal and Herrington (2005) articulated a theoretical framework that may be helpful in endeavouring to harness the potential benefits and avoid chaos. A guide on how the framework might be used was provided by Handal, Handal, and Herrington (2005), who discussed interface design, navigation, and user control of an online resource.

CONCLUSION

The middle years of schooling offer unique challenges, particularly with respect to the disengagement of students from mathematics. Under-participation in the learning of mathematics is a multifaceted issue and may be a product of some students either lacking confidence, giving up readily, not connecting current learning opportunities with their future goals, or experiencing discontinuities between the curriculum, pedagogy, assessment regimes, and their own culture and family expectations. While middle years education had received considerable attention before the time-frame of this review, there is evidence of on-going interest and research by mathematics education researchers in research on this level of schooling. Clearly researchers are aware of the issues and the variety and number of studies are testament to them taking up the on-going challenges offered by teaching and learning in this area.

However there are areas that require further research. Participants at the 2005 Quality Mathematics in the Middle Years conference produced a comprehensive list of issues (Morony & Stocks, 2005) and argued for a productive and focused research agenda. A summary of that list follows, with some annotations that reflect the progress to date. It serves as a challenge to researchers in this area:

- Indigenous education – Research ways to incorporate an indigenous ‘experience’ and address the factors that cause differential achievement in Indigenous education for the middle years. *The research on mathematics learning for Indigenous students is reviewed by Meaney et al. in another chapter of this book. While some progress has been made, particularly in New Zealand, this is an area of continued disadvantage that requires further study.*
- Engagement – Research factors that lead to disengagement in mathematics and identify effective approaches to the problem. *The dip in mathematics achievement in the middle years may have its roots in this. A number of researchers have been working in this area but it continues to be a major factor in these key years of schooling.*
- Teacher Education – Research factors behind appropriate middle years teacher preparation and recruitment programs. *While there are specially designated Middle Schools, there are few designated Middle School pre-service teacher education programs and even less research into this area.*

- Professional development – Research effectiveness and economics. *The working context of a teacher is 'time poor', yet teacher professional learning is at the core – an on-going challenge to research.*
- ICTs – Research factors of relevance, curriculum integration, and catering for diversity in students' experience, interests and ability. *ICTs have already influenced all facets of the teaching and learning process, and have been the focus of considerable research. Some examples have appeared within this chapter and others involving different age cohorts will appear in other chapters of this book. ICTs will continue to be a major area of research.*
- Remote and rural – Research the discontinuity experienced by remote and rural students.
- Boys' education – Research issues of achievement, attitudes, and self perceptions.
- Curriculum integration – Research existing and alternative approaches that include mathematics that students need to encounter.
- Transition – Research the nature and impact of the move from primary to secondary school with a focus on teacher and parental expectations. *The dip in mathematics achievement in the middle years may also have its roots in this. A number of researchers have been working in this area and our knowledge is increasing but more work is needed.*
- Class organisation – Research the effectiveness of streaming, setting, and mixed ability class organisation for mathematics in the middle years. *There has been considerable early research involving students across the range of ages and abilities. Research specifically targeted towards the Middle Years would be a welcome addition to this body. Research on gifted and talented students could also inform this area.*
- Specialist teachers – Research the effectiveness of specialist mathematics teachers in primary and secondary settings.
- Curriculum – Research the factors and differentiated pedagogy enabling clarity and significant learning of the 'big ideas' of mathematics in the middle years.
- Teacher knowledge – Research the impact of teachers' practices on student learning, with a focus on assessment feedback and student attitudes.
- Communication – Research strategies for effective dissemination of research to teachers.
- Impact – Research the impact of middle years mathematics on the 'final examination' at the end of schooling.

There is fertile ground for research into the areas listed above which have not appeared in this chapter. It is hoped that the enormous energy and resources focused upon the mathematics of the early years will now be focused upon the mathematics of the middle years.

REFERENCES

Ainley, J., & Luntley, M. (2005). The role of attention in classroom practice: Developing a methodology. In P. Clarkson, A. Downton, A. Gronn, M. Horne, A. McDonough, R. Pierce, & A. Roche (Eds.), *Building connections: Research, theory and practice* (Proceedings of the 28th annual conference of the Mathematics Education Research Group of Australasia, Melbourne, pp. 73-80). Sydney: MERGA.

Anderson, Judy. (2005). Implementing problem solving in the mathematics classroom: What help do teachers want? In P. Clarkson, A. Downton, D. Gronn, M. Horne, A. McDonough, R. Pierce, & A. Roche (Eds.), *Building connections: Research, theory and practice* (Proceedings of the 28th annual conference of the Mathematics Education Research Group of Australasia, Melbourne, pp. 89-96). Sydney: MERGA.

Anderson, Julie. (2005). I didn't know what I didn't know: A case study of growth in teacher knowledge within the Intermediate Numeracy Project. In P. Clarkson, A. Downton, D. Gronn, M. Horne, A. McDonough, R. Pierce, & A. Roche (Eds.), *Building connections: Research, theory and practice* (Proceedings of the 28th annual conference of the Mathematics Education Research Group of Australasia, Melbourne, pp. 97-104). Sydney: MERGA.

Anderson, V. (2007). An online survey to assess student anxiety and attitude response to six different mathematical problems. In J. Watson & K. Beswick (Eds.), *Mathematics: Essential research, essential practice* (Proceedings of the 30th annual conference of the Mathematics Education Research Group of Australasia, Hobart, Vol. 1, pp. 93-102). Adelaide: MERGA.

Anthony, G. (2004). MERGA: A community of practice. In I. Putt, R. Faragher, & M. McLean (Eds.), *Mathematics education for the third millennium: Towards 2010* (Proceedings of the 27th annual conference of the Mathematics Education Group of Australasia, Townsville, pp. 2-15). Sydney: MERGA.

Asplin, P., Frid, S., & Sparrow, L. (2006). Game playing to develop mental computation: A case study. In P. Grootenboer, R. Zevenbergen, & M. Chinnappan (Eds.), *Identities, cultures and learning spaces* (Proceedings of the 29th annual conference of the Mathematics Education Research Group of Australasia, Canberra, pp. 46-53). Adelaide: MERGA.

Australian Association of Mathematics Teachers (AAMT). (1997). *Numeracy = everybody's business.* (Report of the Numeracy Education Strategy Development Conference). Adelaide: Author.

Beswick, K. (2005). It depends on the students: Influencing teachers' beliefs about ends and means of numeracy teaching. In P. Clarkson, A. Downton, D. Gronn, M. Horne, A. McDonough, R. Pierce, & A. Roche (Eds.), *Building connections: Research, theory and practice* (Proceedings of the 28th annual conference of the Mathematics Education Research Group of Australasia, Melbourne, pp. 137-145). Sydney: MERGA.

Beswick, K., & Muir, T. (2004). Talking and writing about the problem solving process. In I. Putt, R. Faragher, & M. McLean (Eds.), *Mathematics education for the third millennium: Towards 2010* (Proceedings of the 27th annual conference of the Mathematics Education Group of Australasia, Townsville, pp. 95-102). Sydney: MERGA.

Bill, A., & Watson, J. (2007). Three student tasks in a study of distribution in a "best practice" statistics classroom. In J. Watson & K. Beswick (Eds.), *Mathematics: Essential research, essential practice* (Proceedings of the 30th annual conference of the Mathematics Education Research Group of Australasia, Hobart, Vol. 1, pp. 123-132). Adelaide: MERGA Inc.

Brown, J. (2007). Early notions of functions in a technology-rich teaching and learning environment (TRTLE). In J. Watson & K. Beswick (Eds.), *Mathematics: Essential research, essential practice* (Proceedings of the 30th annual conference of the Mathematics Education Research Group of Australasia, Hobart, Vol. 1, pp. 153-162). Adelaide: MERGA.

Brown, M., Jones, K., Taylor, R., & Hirst, A. (2004). Developing geometrical thinking. In I. Putt, R. Faragher, & M. McLean (Eds.), *Mathematics education for the third millennium: Towards 2010* (Proceedings of the 27th annual conference of the Mathematics Education Group of Australasia, Townsville, pp. 127-134). Sydney: MERGA.

Brown, R., & Renshaw, P. (2004). Integrating everyday and scientific ways of knowing mathematics through forms of participation in classroom talk. In I. Putt, R. Faragher, & M. McLean (Eds.), *Mathematics education for the third millennium: Towards 2010* (Proceedings of the 27th annual conference of the Mathematics Education Group of Australasia, Townsville, pp. 135-142). Sydney: MERGA.

Callingham, R. (2005). A whole school approach to developing mental computation strategies. In H. L. Chick, & J. L. Vincent, (Eds.), *Proceedings of the 29th annual conference of the International Group for the Psychology of Mathematics Education* (Vol. 2, pp. 201-207). Melbourne: University of Melbourne.

Callingham, R., Falle, J., & Clark, G. (2004). Primary students' understanding of proof. In I. Putt, R. Faragher, & M. McLean (Eds.), *Mathematics education for the third millennium: Towards 2010* (Proceedings of the 27th annual conference of the Mathematics Education Group of Australasia, Townsville, pp. 151-158). Sydney: MERGA.

Callingham, R., & Watson, J. (2004). A developmental scale of mental computation with part-whole numbers, *Mathematics Education Research Journal, 16*(2), 69-86.

Caney, A. (2004). Perception of mental computation practice: Reports for middle school teachers and students. In I. Putt, R. Faragher, & M. McLean (Eds.), *Mathematics education for the third millennium: Towards 2010* (Proceedings of the 27th annual conference of the Mathematics Education Group of Australasia, Townsville, pp. 159-166). Sydney: MERGA.

Carpenter, T. P., & Lehrer, R. (1999). Teaching and learning mathematics with understanding. In E. Fennema, & T. A. Romberg (Eds.), *Mathematics classrooms that promote understanding* (pp. 19-32). Mahwah, N.J.: Lawrence Erlbaum Associates.

Caswell, R. (2005). The value of play to enhance mathematical learning in the middle years of schooling. In P. Clarkson, A. Downton, D. Gronn, M. Horne, A. McDonough, R. Pierce, & A. Roche (Eds.), *Building connections: Research, theory and practice* (Proceedings of the 28th annual conference of the Mathematics Education Research Group of Australasia, Melbourne, pp. 217-225). Sydney: MERGA.

Caswell, R., & Nisbet, S. (2005). Enhancing mathematical understanding through self-assessment and self-regulation: The value of meta-awareness. In P. Clarkson, A. Downton, D. Gronn, M. Horne, A. McDonough, R. Pierce, & A. Roche (Eds.), *Building connections: Research, theory and practice* (Proceedings of the 28th annual conference of the Mathematics Education Research Group of Australasia, Melbourne, pp. 209-216). Sydney: MERGA.

Centre for Applied Educational Research, The University of Melbourne. (2002). *Middle Years Research and Development (MYRAD) Project executive summary*. Melbourne: State Government Victoria, Department of Education, Employment and Training.

Chick, H. L., & Baker, M. K. (2005a). Investigating teachers' responses to student misconceptions. In H. L. Chick, & J. L. Vincent (Eds.), *Proceedings of the 29th annual conference of the International Group for the Psychology of Mathematics Education* (pp. 249-256). Melbourne: University of Melbourne.

Chick, H., & Baker, M. (2005b). Teaching elementary probability: Not leaving it to chance. In P. Clarkson, A. Downton, D. Gronn, M. Horne, A. McDonough, R. Pierce, & A. Roche (Eds.), *Building connections: Research, theory and practice* (Proceedings of the 28th annual conference of the Mathematics Education Research Group of Australasia, Melbourne, pp. 233-240). Sydney: MERGA.

Chinnappan, M. (2005). Children's mappings of part-whole construct of fractions. In P. Clarkson, A. Downton, D. Gronn, M. Horne, A. McDonough, R. Pierce, & A. Roche (Eds.), *Building connections: Research, theory and practice* (Proceedings of the 28th annual conference of the Mathematics Education Research Group of Australasia, Melbourne, pp. 241-248). Sydney: MERGA.

Clarke, D. M., Roche, A., & Mitchell, A. (2007). Year six fraction understanding: A part of the whole story. In J. Watson & K. Beswick (Eds.), *Mathematics: Essential research, essential practice* (Proceedings of the 30th annual conference of the Mathematics Education Research Group of Australasia, Hobart, Vol. 1, pp. 207-216). Adelaide: MERGA.

Clarke, D. M., Sukemik, M., Roche, A., & Mitchell, A. (2006). Assessing student understanding of fractions using task-based interviews. In J. Novotná, H. Moraová, M. Krátká, & N. Stehliková (Eds.), *Proceedings of the 30th annual conference of the International Group of Psychology of mathematics Education* (Vol. 2, pp. 337-344). Prague: Charles University.

Davies, N., & Walker, K. (2005). Learning to notice: One aspect of teachers' content knowledge in the numeracy classroom. In P. Clarkson, A. Downton, D. Gronn, M. Horne, A. McDonough, R. Pierce, & A. Roche (Eds.), *Building connections: Research, theory and practice* (Proceedings of the 28th annual conference of the Mathematics Education Research Group of Australasia, Melbourne, pp. 273-280). Sydney: MERGA.

Davies, N., & Walker, K. (2007). Teaching as listening: Another aspect of teachers' content knowledge in the numeracy classroom. In J. Watson & K. Beswick (Eds.), *Mathematics: Essential research, essential practice* (Proceedings of the 30th annual conference of the Mathematics Education Research Group of Australasia, Hobart, Vol. 1, pp. 217-225). Adelaide: MERGA.

Department of Education and Training Victoria (2004). *Middle years matters* [Discussion paper]. Retrieved August 1, 2007 from: http://www.sofweb.vic.edu.au/ mys/pdf/MYMPedagogy.pdf.

Diezmann, C. M., & Lowrie, T. J. (2006). Primary students' knowledge of and errors on number lines. In P. Grootenboer, R. Zevenbergen, & M. Chinnappan (Eds.), *Identities, cultures and learning spaces* (Proceedings of the 29th annual conference of the Mathematics Education Research Group of Australasia, Canberra, pp. 171-178). Adelaide: MERGA.

Diezmann, C. M., Lowrie, T. J., & Kozak, N. (2007). Essential differences between high and low performers' thinking about graphically-oriented numeracy items. In J. Watson & K. Beswick (Eds.), *Mathematics: Essential research, essential practice* (Proceedings of the 30th annual conference of the Mathematics Education Research Group of Australasia, Hobart, Vol. 1, pp. 226-235). Adelaide: MERGA.

Doig, B. (2005). *Research pointers to practice: A review of research to inform middle years mathematics teaching and learning*. Adelaide: Australian Association of Mathematics Teachers.

Doig, B., Groves, S., Tytler, R., & Gough, A. (2005). Primary and secondary mathematics practice: How different is it? In P. Clarkson, A. Downton, D. Gronn, M. Horne, A. McDonough, R. Pierce, & A. Roche (Eds.), *Building connections: Research, theory and practice* (Proceedings of the 28th annual conference of the Mathematics Education Research Group of Australasia, Melbourne, pp. 305-312). Sydney: MERGA.

Doyle, K. (2007). The teacher, the tasks: Their role in students' mathematical literacy. In J. Watson & K. Beswick (Eds.), *Mathematics: Essential research, essential practice* (Proceedings of the 30th annual conference of the Mathematics Education Research Group of Australasia, Hobart, Vol. 1, pp. 246-245). Adelaide: MERGA.

Drake, M. (2007). Informal knowledge and prior learning: Student strategies for identifying and locating numbers on scales. In J. Watson & K. Beswick (Eds.), *Mathematics: Essential research, essential practice* (Proceedings of the 30th annual conference of the Mathematics Education Research Group of Australasia, Hobart, Vol. 1, pp. 255-264). Adelaide: MERGA.

Ellerton, N. F., & Olson, J. (2005). The assessment dilemma: Correct answers with no understanding and incorrect answers with some understanding. In H. S. Dhindsa, I. J. Kyeleve, O. Chukwu, & J. S. H. Q. Perera (Eds.), *Future directions in science, mathematics and technical education.* (Proceedings of the tenth international conference of the Department of Science and Mathematics Education, Universiti Brunei Darussalam, pp. 226-235). Brunei: University Brunei Darussalam.

English, L. (2004). Mathematical modelling in the primary school. In I. Putt, R. Faragher, & M. McLean (Eds.), *Mathematics education for the third millennium: Towards 2010* (Proceedings of the 27th annual conference of the Mathematics Education Group of Australasia, Townsville, pp. 207-214). Sydney: MERGA.

English, L. (2007). Interdisciplinary modelling in the primary mathematics curriculum. In J. Watson & K. Beswick (Eds.), *Mathematics: Essential research, essential practice* (Proceedings of the 30th annual conference of the Mathematics Education Research Group of Australasia, Hobart, Vol. 1, pp. 275-284). Adelaide: MERGA.

English, L., & Doerr, H. (2004). Learning through interacting with students' ways of thinking. In I. Putt, R. Faragher, & M. McLean (Eds.), *Mathematics education for the third millennium: Towards 2010* (Proceedings of the 27th annual conference of the Mathematics Education Group of Australasia, Townsville, pp. 215-222). Sydney: MERGA.

English, L., & Fox, J. L. (2005). Seventh-graders' mathematical modelling on completion of a three–year program. In P. Clarkson, A. Downton, D. Gronn, M. Horne, A. McDonough, R. Pierce, & A. Roche (Eds.), *Building connections: Research, theory and practice* (Proceedings of the 28th annual conference of the Mathematics Education Research Group of Australasia, Melbourne, pp. 321-328). Sydney: MERGA.

Falle, J. (2005a). Take 'em out of the equation: Student understanding of "cancelling". In M. Coupland, J. Anderson, & T. Spencer (Eds.), *Making mathematics vital: Proceedings of the 20th biennial conference of the Australian Association of Mathematics Teachers* (pp. 113-119). Adelaide: Australian Association of Mathematics Teachers.

Falle, J. (2005b). Towards a language-based model of students' early algebraic understandings: Some preliminary findings. In P. Clarkson, A. Downton, D. Gronn, M. Horne, A. McDonough, R. Pierce, & A. Roche (Eds.), *Building connections: Research, theory and practice* (Proceedings of the 28th annual conference of the Mathematics Education Research Group of Australasia, Melbourne, pp. 345-352). Sydney: MERGA.

Falle, J. (2007). Students' tendency to conjoin terms: An inhibition to their development of algebra. In J. Watson & K. Beswick (Eds.), *Mathematics: Essential research, essential practice* (Proceedings of the 30th annual conference of the Mathematics Education Research Group of Australasia, Hobart, Vol. 1, pp. 285-294). Adelaide: MERGA.

Fitzallen, N. (2006). A model of students' statistical thinking and reasoning about graphs in an ICT environment. In P. Grootenboer, R. Zevenbergen, & M. Chinnappan (Eds.), *Identities, cultures and learning spaces* (Proceedings of the 29th annual conference of the Mathematics Education Research Group of Australasia, Canberra, pp. 203-210). Adelaide: MERGA.

Gervasoni, A., Hadden, T., & Turkenburg, K. (2007). Exploring the number knowledge of children to inform the development of a professional learning plan for teachers in the Ballarat diocese as a means of building community capacity. In J. Watson & K. Beswick (Eds.), *Mathematics: Essential research, essential practice* (Proceedings of the 30th annual conference of the Mathematics Education Research Group of Australasia, Hobart, Vol. 1, pp. 305-314). Adelaide: MERGA.

Gould, P. (2005a). Drawing sense out of fractions. In M. Coupland, J. Anderson, & T. Spencer (Eds.), *Making mathematics vital: Proceedings of the 20th biennial conference of the Australian Association of Mathematics Teachers* (pp. 133-138). Adelaide: Australian Association of Mathematics Teachers.

Gould, P. (2005b). Year 6 students' methods of comparing the size of fractions. In P. Clarkson, A. Downton, D. Gronn, M. Horne, A. McDonough, R. Pierce, & A. Roche (Eds.), *Building connections: Research, theory and practice* (Proceedings of the 28th annual conference of the Mathematics Education Research Group of Australasia, Melbourne, pp. 393-400). Sydney: MERGA.

Gould, P., Outhred, L., & Mitchelmore, M. (2006). One-third is three-quarters of one-half. In P. Grootenboer, R. Zevenbergen, & M. Chinnappan (Eds.), *Identities, cultures and learning spaces* (Proceedings of the 29th annual conference of the Mathematics Education Research Group of Australasia, Canberra, pp. 262-269). Adelaide: MERGA.

Handal, B., & Herrington, T. (2005). Getting school maths online: Possibilities and challenges. In M. Coupland, J. Anderson, & T. Spencer (Eds.), *Making mathematics vital: Proceedings of the 20th biennial conference of the Australian Association of Mathematics Teachers* (pp. 146-152). Adelaide: Australian Association of Mathematics Teachers.

Handal, B., Handal, P., & Herrington, T. (2005). Evaluating online mathematics resources: A practical approach for teachers. In M. Coupland, J. Anderson, & T. Spencer (Eds.), *Making mathematics vital: Proceedings of the 20th biennial conference of the Australian Association of Mathematics Teachers* (pp. 153-165). Adelaide: Australian Association of Mathematics Teachers.

Hansen, A., & Pratt, D. (2005). How do we provide tasks for children to explore the definitions of quadrilaterals? In P. Clarkson, A. Downton, D. Gronn, M. Horne, A. McDonough, R. Pierce, & A. Roche (Eds.), *Building connections: Research, theory and practice* (Proceedings of the 28th annual conference of the Mathematics Education Research Group of Australasia, Melbourne, pp. 408-415). Sydney: MERGA.

Hartnett, J. (2007). Categorisation of mental computation strategies to support teaching and to encourage classroom dialogue. In J. Watson & K. Beswick (Eds.), *Mathematics: Essential research, essential practice* (Proceedings of the 30th annual conference of the Mathematics Education Research Group of Australasia, Hobart, Vol. 1, pp. 345-352). Adelaide: MERGA.

Hilton, G. (2007). Students representing mathematical knowledge through digital filmmaking. In J. Watson & K. Beswick (Eds.), *Mathematics: Essential research, essential practice* (Proceedings of the 30th annual conference of the Mathematics Education Research Group of Australasia, Hobart, Vol. 1, pp. 382-391). Adelaide: MERGA.

Hunter, J. (2007). Relational or calculational thinking: Students solving open number equivalence problems. In J. Watson & K. Beswick (Eds.), *Mathematics: Essential research, essential practice* (Proceedings of the 30th annual conference of the Mathematics Education Research Group of Australasia, Hobart, Vol. 1, pp. 421-429). Adelaide: MERGA.

Hunter, R. (2007). Scaffolding small group interactions. In J. Watson & K. Beswick (Eds.), *Mathematics: Essential research, essential practice*, (Proceedings of the 30th annual conference of the Mathematics Education Research Group of Australasia, Hobart, Vol. 1, pp. 430-439). Adelaide: MERGA.

Hurst, C. (2007). Numeracy in action: Students connecting mathematical knowledge to a range of contexts. In J. Watson & K. Beswick (Eds.), *Mathematics: Essential research, essential practice* (Proceedings of the 30th annual conference of the Mathematics Education Research Group of Australasia, Hobart, Vol. 1, pp. 440-449). Adelaide: MERGA

Irwin, K., & Britt, M. (2004). Operating with decimal fractions as a part-whole concept. In I. Putt, R. Faragher, & M. McLean (Eds.), *Mathematics education for the Third Millennium: Towards 2010* (Proceedings of the 27th annual conference of the Mathematics Education Group of Australasia, Townsville, pp. 312-319). Sydney: MERGA.

Lesh, R. (2006). New directions for research on mathematical problem solving. In P. Grootenboer, R. Zevenbergen, & M. Chinnappan (Eds.), *Identities, cultures and learning spaces* (Proceedings of the 29th annual conference of the Mathematics Education Research Group of Australasia, Canberra, pp. 15-33). Adelaide: MERGA.

Lowrie, T. (2004a). Authentic artefacts: Influencing practice and supporting problem solving in the mathematics classroom. In I. Putt, R. Faragher, & M. McLean (Eds.), *Mathematics education for the third millennium: Towards 2010* (Proceedings of the 27th annual conference of the Mathematics Education Group of Australasia, Townsville, pp. 351-358). Sydney: MERGA.

Lowrie, T. (2004b). Authentic problem solving: The influence of cultural artefacts on sense making. In J. Cia., & J. Mamona Downs (Eds.), *Problem solving in mathematics education.* (Refereed proceedings of the Problem Solving Topic Study Group, TSG 18, of the International Congress of Mathematics Education, Copenhagen, Denmark). Retrieved August 1, 2007 from: http://www.icme-10.dk/

McGee, L, Kervin, L., & Chinnappan, M. (2006). Emerging issues in the investigation of the construct of partitive quotient. In P. Grootenboer, R. Zevenbergen, & M. Chinnappan (Eds.), *Identities, cultures and learning spaces* (Proceedings of the 29th annual conference of the Mathematics Education Research Group of Australasia, Canberra, pp. 360-367). Adelaide: MERGA.

Mitchell, A. (2005). Measuring fractions. In P. Clarkson, A. Downton, D. Gronn, M. Horne, A. McDonough, R. Pierce, & A. Roche (Eds.), *Building connections: Research, theory and practice* (Proceedings of the 28th annual conference of the Mathematics Education Research Group of Australasia, Melbourne, pp. 545-552). Sydney: MERGA.

Mitchelmore, M., White, P., & McMaster, H. (2007). Teaching ratio and rates for abstraction. In J. Watson & K. Beswick (Eds.), *Mathematics: Essential research, essential practice* (Proceedings of the 30th annual conference of the Mathematics Education Research Group of Australasia, Hobart, Vol. 2, pp. 503-512). Adelaide: MERGA.

Morony, W., & Stocks, C. (Eds.), (2005). *Quality mathematics in the middle years*. Adelaide: Australian Association of Mathematics Teachers.

Muir, T., & Beswick, K. (2005). Where did I go wrong? Students' success at various stages of the problem solving process. In P. Clarkson, A. Downton, D. Gronn, M. Horne, A. McDonough, R. Pierce, & A. Roche (Eds.), *Building connections: Research, theory and practice* (Proceedings of the 28th annual conference of the Mathematics Education Research Group of Australasia, Melbourne, pp. 561-568). Sydney: MERGA.

Mulligan, J., & Busatto, S. (2004). The numeracy research in NSW primary schools project 2001-2003. In I. Putt, R. Faragher, & M. McLean (Eds.), *Mathematics education for the third millennium: Towards 2010* (Proceedings of the 27th annual conference of the Mathematics Education Group of Australasia, Townsville, pp. 382-389). Sydney: MERGA.

Nataraj, M. S., & Thomas, M. O. J. (2007). Developing the concept of place value. In J. Watson & K. Beswick (Eds.), *Mathematics: Essential research, essential practice* (Proceedings of the 30th annual conference of the Mathematics Education Research Group of Australasia, Hobart, Vol. 2, pp. 522-532). Adelaide: MERGA.

Nisbet, S., & Grimbeek, P. (2004). Primary teachers' beliefs and practices with respect to compulsory numeracy testing. In I. Putt, R. Faragher, & M. McLean (Eds.), *Mathematics education for the third millennium: Towards 2010* (Proceedings of the 27th annual conference of the Mathematics Education Group of Australasia, Townsville, pp. 406-413). Sydney: MERGA.

Norton, S. (2006). Pedagogies for the engagement of girls in the learning of proportional reasoning through technology practice. *Mathematics Education Research Journal, 18*(3), 69-99.

Norton, S., & Irvin, J. (2007a). A concrete approach to teaching symbolic algebra. In J. Watson & K. Beswick (Eds.), *Mathematics: Essential research, essential practice* (Proceedings of the 30th annual conference of the Mathematics Education Research Group of Australasia, Hobart, Vol. 2, pp. 551-560). Adelaide: MERGA.

Norton, S., & Irvin, J. (2007b). Developing positive attitudes towards algebra. In J. Watson & K. Beswick (Eds.), *Mathematics: Essential research, essential practice* (Proceedings of the 30th annual conference of the Mathematics Education Research Group of Australasia, Hobart, Vol. 2, pp. 561-570). Adelaide: MERGA.

Organisation for Economic Co-operation and Development (OECD) (2003). The PISA 2003 assessment framework – mathematics, reading, science and problem solving knowledge and skills. Paris: OECD. Retrieved February 28, 2008 from: http://www.oecd.org/dataoecd/46/14/33694881.pdf

Paterson, A., & Bana, J. (2005). The effect of money as a context on mental computation performance of students in years 3, 5, 7, and 9. In P. Clarkson, A. Downton, D. Gronn, M. Horne, A. McDonough, R. Pierce, & A. Roche (Eds.), *Building connections: Research, theory and practice* (Proceedings of the 28th annual conference of the Mathematics Education Research Group of Australasia, Melbourne, pp. 617-624). Sydney: MERGA.

Pearn, C., & Stephens, M. (2004). Why you have to probe to discover what year 8 students really think about fractions. In I. Putt, R. Faragher, & M. McLean (Eds.), *Mathematics education for the third millennium, Towards 2010* (Proceedings of the 27th annual conference of the Mathematics Education Research Group of Australasia, Townsville, pp. 430-437). Sydney: MERGA.

Pearn, C., & Stephens, M. (2007). Whole number knowledge and number lines help to develop fraction concepts. In J. Watson & K. Beswick (Eds.), *Mathematics: Essential research, essential practice* (Proceedings of the 30th annual conference of the Mathematics Education Research Group of Australasia, Hobart, Vol. 2, pp. 601-610). Adelaide: MERGA.

Pegg, J., Graham, L., & Bellert, A. (2005). The effect of improved automaticity of basic number skills on persistently low-achieving pupils. In H. L. Chick, & J. L. Vincent (Eds.), *Proceedings of the 29th annual conference for the Psychology of Mathematics Education* (Vol. 4, pp. 49-56). Melbourne: University of Melbourne.

Perso, T. (2005). Algebra outcomes for your K-9 class. In M. Coupland, J. Anderson, & T. Spencer (Eds.), *Making mathematics vital: Proceedings of the 20th biennial conference of the Australian*

Association of Mathematics Teachers (pp. 209-216). Adelaide: Australian Association of Mathematics Teachers.

Roche, A., & Clarke, D. (2004). When does successful comparison of decimals reflect conceptual understanding? In I. Putt, R. Faragher, & M. McLean (Eds.), *Mathematics education for the third millennium, Towards 2010.* (Proceedings of the 27th annual conference of the Mathematics Education Research Group of Australasia, Townsville, pp. 486-493). Sydney: MERGA.

Scott, A. (2007). Seeking evidence of thinking and mathematical understandings in students' writing. In J. Watson & K. Beswick (Eds.), *Mathematics: Essential research, essential practice* (Proceedings of the 30th annual conference of the Mathematics Education Research Group of Australasia, Hobart, Vol. 2, pp. 641-650). Adelaide: MERGA.

Shield, M. (2005). Building a methodology for the comparison and evaluation of middle years mathematics textbooks. In P. Clarkson, A. Downton, D. Gronn, M. Horne, A. McDonough, R. Pierce, & A. Roche (Eds.), *Building connections: Research, theory and practice* (Proceedings of the 28th annual conference of the Mathematics Education Research Group of Australasia, Melbourne, pp. 680-687). Sydney: MERGA.

Siemon, D., Izard, J., Breed, M., & Virgona, J. (2006). The derivation of a learning assessment framework for multiplicative thinking. In J. Norotna, H. Moraova, M. Kratka, & N. Stehlikova (Eds.), *Proceedings of the 30th annual conference for the Psychology of Mathematics Education* (Vol. 5, pp.113-120). Prague: PME.

Skalicky, J. (2007). Exploring teachers' numeracy pedagogies and subsequent student learning across five dimensions of numeracy. In J. Watson & K. Beswick (Eds.), *Mathematics: Essential research, essential practice* (Proceedings of the 30th annual conference of the Mathematics Education Research Group of Australasia, Hobart, Vol. 2, pp. 661-670). Sydney: MERGA.

State of Victoria, Department of Education and Training. (2004). Middle years of schooling overview of Victorian research 1998-2004. Retrieved July 07, 2005 from: www.sofweb.vic.edu.au/mys/docs/research/

Steinle, V., & Stacey, K. (2004). A longitudinal study of students' understanding of decimal notation: An overview and refined results. In I. Putt, R. Faragher, & M. McLean (Eds.), *Mathematics education for the third millennium: Towards 2010*, (Proceedings of the 27th annual conference of the Mathematics Education Group of Australasia, Townsville, pp. 541-548). Sydney: MERGA.

Stephens, M. (2005). Relational thinking about numbers as a bridge to algebraic reasoning. In M. Coupland, J. Anderson, & T. Spencer (Eds.), *Making mathematics vital: Proceedings of the 20th biennial conference of the Australian Association of Mathematics Teachers* (pp. 235-242). Adelaide: Australian Association of Mathematics Teachers.

Stephens, M. (2006). Describing and exploring the power of relational thinking. In P. Grootenboer, R. Zevenbergen, & M. Chinnappan (Eds.), *Identities, cultures and learning spaces* (Proceedings of the 29th annual conference of the Mathematics Education Research Group of Australasia, Canberra, pp. 479-486). Adelaide: MERGA.

Stephens, M. (2007). Students' emerging algebraic thinking in the middle school years. In J. Watson & K. Beswick (Eds.), *Mathematics: Essential research, essential practice* (Proceedings of the 30th annual conference of the Mathematics Education Research Group of Australasia, Hobart, Vol. 2, pp. 678-687). Adelaide: MERGA.

Stillman, G., Galbraith, P., Brown, J., & Edwards, I. (2007). A framework for success in implementing mathematical modelling in the secondary classroom. In J. Watson & K. Beswick (Eds.), *Mathematics: Essential research, essential practice*, (Proceedings of the 30th annual conference of the Mathematics Education Research Group of Australasia, Hobart, Vol. 2, pp. 688-697). Adelaide: MERGA.

Sullivan, P., McDonough, A., & Turner Harrison, R. (2004). Students' perceptions of factors contributing to successful participation in mathematics. In M. J. Johnsen Høines, & A. Fuglestad (Eds.), *Proceedings of the 28th annual conference of the International Group for the Psychology of Mathematics Education* (pp. 289-297). Bergen: PME.

Sullivan, P., & McDonough, A. (2007). Eliciting positive student motivation for learning mathematics. In J. Watson & K. Beswick (Eds.), *Mathematics: Essential research, essential practice* (Proceedings of the 30th annual conference of the Mathematics Education Research Group of Australasia, Hobart, Vol. 2, pp. 698-707). Adelaide: MERGA.

Taylor, M., Hawera, N., & Young-Loveridge, J. (2005). Children's views of their teacher's role in helping them learn mathematics. In P. Clarkson, A. Downton, D. Gronn, M. Horne, A. McDonough, R. Pierce, & A. Roche (Eds.), *Building connections: Research, theory and practice* (Proceedings of the 28th annual conference of the Mathematics Education Research Group of Australasia, Melbourne, pp. 728-734). Sydney: MERGA.

Vaiyatvutjamai, P., & Clements, M. A. (2004). Analysing errors made by middle-school students on six linear inequations tasks. In I. P. A. Cheong, H. S. Dhindsa, I. J. Kyeleve, & O. Chukwu (Eds.), *Globalisation trends in science, mathematics and technical education 2004* (Proceedings of the ninth international conference of the Department of Science and Mathematics Education, Universiti Brunei Darussalam, pp. 173-182). Brunei: University Brunei Darussalam.

Vincent, J., & Stacey, K. (2007). Procedural complexity and mathematical solving processes in year 8 mathematics textbook questions. In J. Watson & K. Beswick (Eds.), *Mathematics: Essential research, essential practice* (Proceedings of the 30th annual conference of the Mathematics Education Research Group of Australasia, Hobart, Vol. 2, pp. 735-744). Adelaide: MERGA.

Watson, J. (2005a). Variation and expectation as foundations for the chance and data curriculum. In P. Clarkson, A. Downton, D. Gronn, M. Horne, A. McDonough, R. Pierce, & A. Roche (Eds.), *Building connections: Research, theory and practice* (Proceedings of the 28th annual conference of the Mathematics Education Research Group of Australasia, Melbourne, pp. 35-42). Sydney: MERGA.

Watson, J. (2005b). Lessons from research: Students' understanding of statistical literacy. In M. Coupland, J. Anderson, & T. Spencer (Eds.), *Making mathematics vital: Proceedings of the 20th biennial conference of the Australian Association of Mathematics Teachers* (pp. 253-260). Adelaide: Australian Association of Mathematics Teachers.

Watson, J., Caney, A., & Kelly, B. (2004). Beliefs about chance in the middle years: Longitudinal change. In I. Putt, R. Faragher, & M. McLean (Eds.), *Mathematics education for the third millennium: Towards 2010* (Proceedings of the 27th annual conference of the Mathematics Education Group of Australasia, Townsville, pp. 581-588). Sydney: MERGA.

Watson, J., & Kelly, B. (2005). Cognition and instruction: Reasoning about bias in sampling. *Mathematics Education Research Journal, 17*(1), 24-57.

Watson, J., Kelly, B., & Izzard, J. (2006). A longitudinal study of student understanding of chance and data. *Mathematics Education Research Journal, 18*(2), 40-55.

White, A. L., & Singh, P. (2005). Computation in the primary classroom: Going mental. *Foundations and creativity: The strengths and weaknesses of mathematics education in East Asia. Proceedings of the joint ICMI – EARCOME 3 conferences.* [CD]. Shanghai: ICMI-EARCOME.

White, A. L. (2006). Shaping the future of secondary school mathematics through modelling with the graphics calculator. In H. S. Dhindsa, I. J. Kyeleve, O. Chukwu, & J. S. H. Q. Perera (Eds.), *Shaping the future of science, mathematics and technical education* (Proceedings of the eleventh international conference of the Department of Science and Mathematics Education, Universiti Brunei Darussalam, pp. 139-146). Brunei: University Brunei Darussalam.

White, P., & Mitchelmore, M. (2005). Teaching percentage as a multiplicative relationship. In P. Clarkson, A. Downton, D. Gronn, M. Horne, A. McDonough, R. Pierce, & A. Roche (Eds.), *Building connections: Research, theory and practice* (Proceedings of the 28th annual conference of the Mathematics Education Research Group of Australasia, Melbourne, pp. 783-790). Sydney: MERGA.

White, P., Wilson, S., Faragher, R., & Mitchelmore, M. (2007). Percentages as part whole relationships. In J. Watson & K. Beswick (Eds.), *Mathematics: Essential research, essential practice* (Proceedings of the 30th annual conference of the Mathematics Education Research Association of Australasia, Hobart, Vol. 2, pp. 805-814). Sydney: MERGA.

Wong, M., & Evans, D. (2007). Students' conceptual understanding of equivalent fractions. In J. Watson & K. Beswick (Eds.), *Mathematics: Essential research, essential practice* (Proceedings of the 30th annual conference of the Mathematics Education Research Association of Australasia, Hobart, Vol. 2, pp. 824-833). Sydney: MERGA.

Woodward, J., & Irwin, K. (2005). Language appropriate for the New Zealand numeracy project. In P. Clarkson, A. Downton, D. Gronn, M. Horne, A. McDonough, R. Pierce, & A. Roche (Eds.), *Building connections: Research, theory and practice* (Proceedings of the 28th annual conference of the Mathematics Education Research Group of Australasia, Melbourne, pp. 799-806). Sydney: MERGA.

Zevenbergen, R. (2004). Reconceptualising numeracy in new times. *Curriculum Perspectives, 24*(3), 1-7.

AFFILIATION

Allan L. White
University of Western Sydney, Australia

LEIGH WOOD

UNIVERSITY LEARNERS OF MATHEMATICS

Mathematics learning and teaching at university is in a state of flux and this is evident from the contradictory findings reported. Is lecture attendance up or down? Yes and no. The real issue of the conditions for the effectiveness of lectures has not been discussed. On computing tools, the majority of authors espouse the use of professional software and hardware tools, such as Excel and CAS, rather than teaching-only tools such as graphics calculators. This reflects two trends: the need to assist students in their transition to professional work; and professional software interfaces becoming more student friendly. It is time for our profession to investigate learning in more depth and publish more widely.

Key words: assessment, computer algebra systems (CAS), graduate skills, technology, tertiary, transition

INTRODUCTION

In Australia, around 18,000 people study undergraduate mathematical sciences at university (DEST, 2006) with around 9,000 of these studying at the first year level. These numbers have not changed over the past four years. In New Zealand, there are also large numbers of students studying university level mathematics (Holton, 2005). Not all of these students are taught by academics in mathematics departments; indeed the number of mathematicians in mathematics departments at universities has decreased in the past four years (Bourguignon, Dietrich, & Johnstone, 2006). For this chapter, only peer-reviewed publications have been included. A request for papers was sent to the Australian Mathematical Society as well as the Mathematics Education Research Group of Australasia. An extensive search of local and international journals and conferences was also conducted. The vast majority of publications were from Australia, several from New Zealand and none from Oceania. Many publications were co-authored with contributors from around the globe and the number of contributions co-authored by multidisciplinary teams such as Britton, New, Sharma, and Yardley (2005) has increased. Research on the learning of statistics is not included here, but computing, which has significant overlap with mathematics, was included if it had significant mathematical content.

Around 80 papers were reviewed, with 20 journal articles, two book chapters, three theses and the rest refereed conference proceedings. It is clear that this area of research in Australasia is not being published in international journals except for opportunities offered by conferences. This shows how critical conferences, such as

H. Forgasz et al. (eds.), Research in Mathematics Education in Australasia 2004-2007, 73-97.

the DELTA series of international conferences focussing on the southern hemisphere and the International Conferences on the Teaching of Mathematics at the tertiary level (ICTM), are to the dissemination of research and scholarship of university mathematics learners.

OVERVIEW

When you consider the numbers of students involved and the importance to the national economy of a workforce with high level mathematical skills, it is surprising that so little research is being done on learners of mathematics at university. Most of the research is done by mathematicians who are passionate about their students and mathematics. However, they need to keep up a research profile in their mathematical discipline area and so tend to work on mathematics education research as a sideline. Most do not have formal qualifications in mathematics education and find it difficult to keep up with the educational literature and methodologies.

Many of the papers in this chapter are descriptions of subjects where the lecturer has made a change to the learning experience of their students and undertaken some student evaluation. These are invariably reported as well received by students, which suggests that failures may not being reported! Surely this would be useful so that others do not fall into the same traps.

Much has been written on assessment again in the context of lecturers researching their own classes. In this four year period, the interest in computer-aided assessment has increased as the computing tools have improved to make this more feasible – though is it is still not easy to do well.

One large area of research has been the influence of various factors, such as language background, on how students perform within a mathematics course. There were few papers on specific mathematics subjects and of these more than half were on teaching using Computer Algebra Systems (CAS). As noted in the previous four year review, investigating the opinions and perceptions of lecturers of mathematics is rare (one paper: Houston et al., 2006) and very few people are studying graduates, though many report working towards developing graduate capabilities.

The methodologies used in investigating university learners of mathematics were mainly student evaluations using questionnaires and the analysis of student performance on assignments and examinations. Many papers were written by lecturers reflecting on their experiences of teaching mathematics over the years, such as Sowey (2005) and Easdown (2006b). There were only two controlled interventions; a study comparing the Keller Plan with the lecture/tutorial format in which the Keller Plan was found to be more effective for the engineering students (Freislich, 2006), and D'Arcy-Warmington's (2006) study in which two groups of first year statistics students were compared and it was found that the one that used group work had better scores and adapted better to university.

Many of the studies fall under the action research paradigm even if the authors do not explicitly mention this. A series of papers by Barton and others (Barton,

Chan, King, & Neville-Barton, 2004; Barton, Chan, King, Neville-Barton, & Sneddon, 2005) used pre-tests, observation and then analysis of student performance. Several studies of learning took a phenomenographic approach with in-depth interviews to investigate learning, such as Reid, Wood, Petocz, and Smith (2005). Wood (2006) used discourse analysis to investigate the language needs of graduates. Petocz, Smith, Wood, and Reid (2004) and Petocz and Reid (2006) describe their phenomenographic research methodology for investigating students' development as professionals. In investigating student learning using a computer algebra system, Blyth, Clarke, and Labovic (2006) adapted the video observation methods used in school mathematics education research (Clarke, 2006) to observe learning in a computer environment. Coupland (2004) and FitzSimons (2006) used activity theory as their learning theory. Jacobs (2006) developed teaching materials using a constructivist approach. Most authors did not explicitly espouse a learning theory.

Atweh and Clarkson (2005) discuss "how mathematics education colleagues in various countries experience aspects of globalisation of their discipline" (p. 1). In Australia, education has become the third largest export product, based on the large number of overseas students enrolled in all sectors of Australia's education system with more than half of these enrolled in the tertiary sector. Internationalisation is one of the major sources of income for many Australian universities. Using focus groups they found "[t]here was a clear realization that as mathematics educators in the Australian tertiary sector we are part of the globalisation process whether we like it or not" (p. 6). There are problems around adapting the local curriculum (or not) and cultural blindness within Australian institutions when working on international projects and working with international students.

One article (Holton, 2005) looks to the future of tertiary mathematics education in 2024. Holton suggests it is not possible to predict future technological developments and "it is not clear to me that the fundamental thing that we call mathematics will change" (p. 305). Therefore, he chose to look at what he would like to see. He considers mathematics to have "two aspects: a created side and a creating side" (p. 306), that need to be acknowledged in any vision of the future. Most importantly mathematics "is a living, breathing entity that students can participate in" (p. 306). This creative side needs to be emphasised more than it has been in order to retain students and attract new ones – it is tertiary educators who need to take a lead.

These are wise words and not reflected in many of the papers reviewed in this chapter. Many authors are researching current problems and perceived student deficiencies, without looking to the future needs of their students and, perhaps, deficiencies in their own knowledge of learning in order to engage diverse student groups in the beauty and utility of mathematics.

THEMES

Assessment

University mathematics classes are often large, so marking and feedback on assessment becomes logistically difficult. The papers which focus on assessment involve a variety of mathematical levels and content areas and show that changes in delivery (due to technology) and assessment types (more open-ended, collaborative) are having an influence on the student experience of university mathematics. The papers all show that changes need to have a learning philosophy that is well communicated to staff and students. Staff resistance to new styles of assessment is often strong so changes need to be well planned.

Groen (2006) and Keady, Fitz-Gerald, Gamble, and Sangwin (2006) describe in detail how they implemented computer-aided assessment. Groen found that students showed no preference for online assessment over traditional methods. They were positive, however, about the value of the practice test and Groen found that the open access led to improved marks. For the lecturer, there were efficiencies which included marking, administration and resource use. Both papers noted that due to economies of scale, the more students who use a computer-aided system, the more worthwhile it becomes since the set-up costs are roughly the same regardless of the number of students. Keady et al. recommend that the management goal for these systems should be to "keep the systems simple" (p. 73), and note that the integration of computerised systems and the Managed Learning Environment (MLE) is increasing, so the approach will continue to improve.

Cretchley (2006a) reports on a study on the use of MATLAB for assessment. MATLAB is a numerical computing and programming language which allows matrix manipulation, plotting of functions and data, implementation of algorithms, creation of user interfaces, and interfacing with programs in other languages. Although it specializes in numerical computing, an optional toolbox interfaces with Maple, allowing it to be part of a full computer algebra system. There has been little investigation of such technological use in timed test conditions – though there has been quite a bit of research on students' reactions to learning using such a system. MATLAB was used in an algebra and calculus course and the results revealed "no difference in technology performance or pen-and-paper performance [in mathematics examinations] due to prior use of computers for doing mathematics" (p. 3). There was a strong correlation between pen-and-paper performance and technology performance but technology performance showed little correlation with mathematics confidence, computer confidence, or with attitudes to technology. However, Cretchley points out the surprising result that "[c]omputer confidence correlated negatively with pen-and-paper performance" (p. 5) which is explained as "some computer-confident students neglect to build appropriate pen-and-paper skills, and some who welcome technology for learning underestimate the skills needed" (p. 5).

McDonald and Taylor (2005) used a method of grading assessments rather than scoring because the subject incorporated "open-ended problem-solving assignments" (p. 88). They give details of the assessment regime and conclude that the changing nature of mathematical assessment, for instance in response to the use of open-ended questions, will necessitate changes to assessment but these need to be dealt with to ensure that bias in the final grades does not result. Taylor (2006) undertook a review of assessment techniques in a first year mathematics subject enrolling 700-800 students and concluded that for these students in transition the first assignment should be brief, the second "summative and developmental" (p. 7), while later assessment could include final examinations and essays. Such an assessment regime would ensure early contact with students and a focus on assessment for learning rather than only assessment of learning. An essential factor is continual self-assessment, based on regular small assignments.

D'Souza and Wood (2004) examined the use of collaborative learning in a large undergraduate mathematics subject. Some students believed that being assessed as a group was not fair, but were positive about the experience of group work in helping their transition to university. Using statistical analysis to investigate performance, MacGillivray and Turner (2005) examined a second level course in linear algebra: "the motivation for the analysis in this paper includes the effects of the continuous assessment, the balance of theory and practice, and the predictors of achievement from first level units" (p. 74). They note that the best predictors of exam achievement were a combination of a theoretical task and a practical task and postulate "that both theory and practice contribute to overall learning and understanding in linear algebra" (p. 80).

Penglase (2004) believes that assessment is "one of the most critical factors in the issue of student learning quality … [and] it appears to be students' perceptions of assessment requirements which cause them to choose specific learning strategies" (p. 1). The study reported here investigated the links between assessment and learning in a calculus course, based on the concept that learning approaches can be divided into deep and surface types. These can lead to different levels of achievement (surface approaches can be linked with low course results). It appeared that most students came to value the deeper types of learning that the lecturer had been trying to encourage.

Teaching methods

Varsavsky (2006) suggests that the university environment has changed in the last decade, for instance the development of a "market-driven approach" alongside reduced government funding has led to an increase in competition for students and "a high reliance on the international market" (p. 1). Cretchley (2005) proposes that in a period of change for universities, teaching styles may need to adapt, particularly if this may affect "learner pragmatism" (p. 1) since courses are now more directed towards a profession than general studies. She conjectured that the use of the internet might imply the development of short attention spans, thus potentially reducing the value of lectures or other face-to-face methods.

Cretchley (2005) undertook to test this conjecture for science and engineering students doing mathematics subjects. She found that the reasons given for attending lectures included: gaining understanding; personal learning style; for interaction via questions (their own and others); motivation; and enjoyment. She notes that "[t]he views and behaviour of lecture attendees investigated in these mathematics courses presented little evidence of the kind of 'pragmatism' reported by Dolnicar" (p. 46), and that "the attendees studied here regarded lectures as an efficient, companionable, and convenient way to meet new mathematical material, build understanding of the concepts" (p. 47). In addition, students who attended lectures performed substantially better on all course assessment items than non-attendees. One should note here that better results were correlated with lecture attendance; however, no causal relationship is shown. Cretchley notes that lecture attendance in these courses is high (60-65%) despite other methods being available to gain the information (such as online and in hard copy).

Hubbard (2006) also studied students' opinion about the usefulness of lectures. She compared the results with a similar study done in 1997, to chart whether changes in student circumstances – online materials and increased university costs – have affected student opinion of lectures and tutorials. She points out that "[t]he current challenge to the predominance of lectures is coming from internet-based courses" (p. 1). However, "there has been very little research, specifically in mathematics, on what students actually learn from lectures and why they continue to attend them" (p. 1). There were two main changes from 1997: the frequency of attending to meet friends decreased because of mobile technology, and attending to find when tests and assignments are due also decreased due to the availability of online communication. With regard to the source of learning, lectures and tutorials still ranked very highly, although the author notes that there was a wide spread of opinions. Overall, the results are quite definite that for most students "human contact is still an important part of learning" (p. 3). Furthermore it is clear that the availability of online materials has not made a substantial difference to reasons for attending lectures. Drawing similar inferences to those of Harding, Kaczynski, and Wood (2005), Hubbard's conclusion is that "there is no evidence in this study to suggest that lectures should be abandoned. On the other hand it does suggest that all the time and effort used to produce online material is wasted on many of our students while lectures are still available to them" (p. 5). Wood, Joyce, Petocz, and Rodd (2007), in a paper on optimising learning from lectures, suggest that having different representations available in lectures is central to building a "concept-image" and offer ideas for helping students to make links between different mathematical representations. They also highlight the difficulties faced by participants of different language backgrounds.

In contrast to Cretchley (2005), Varsavsky (2006) found that attendance at lectures has significantly reduced over the last decade and postulates that this phenomenon has a number of causes. She suggests that the university environment has changed in the last decade, for instance regular surveys of student satisfaction with courses and universities are now common, placing pressure on university staff. At the same time, information and communication technologies have changed the way students live and interact with the rest of the world. This, in turn, has influenced university teaching: "teaching staff are expected to offer the best of both the 'real' and the 'virtual' worlds" (p. 1). Varsavsky discusses changes in the student profile, for instance, "first year students have a better sense of purpose for doing their university course, and clearer career goals" (p. 2). They are also accustomed to using online sources to get information, "and more than one half find that these resources help them to learn at their own pace" (p. 2). However, email and online discussion forums are much less popular, particularly with high achievers. The question then is how to effectively teach mathematics in these changing conditions.

Integrating assessment and feedback in order to encourage active rather than passive learning is the focus of a subject described by Easdown (2006b). He points out that "adults unconsciously resist addressing themselves to new learning" (p. 37), so in the university environment this resistance needs to be overcome. He argues that his aim is to establish the basics and then go beyond to develop independent enquiry and the ability to express oneself. The types of education are passive (such as lectures) – imparting knowledge – and active (tutorials). He uses a method in an intermediate level subject beginning with the passive phase and then using a worksheet of problems with the examples and learning tasks designed to lead the students through a series of ideas. There is further repetition of the problem-solving process, balanced by reflective thought. He has found that in the final examination even the less able students will attempt to do the more difficult questions.

D'Souza (2006), using an action research methodology, investigated the use of collaborative learning methods in a large first year engineering mathematics subject. She found that students believed that they adjusted well to university and having collaborative learning helped them find friends and develop informal study groups. Students from Middle Eastern backgrounds stated a preference for working alone, while those from European, Asian and Indian backgrounds preferred to work in groups. Middle Eastern students displayed the highest level of computer anxiety. D'Souza recommends that students be offered extra support when collaborative learning and computer algebra systems are introduced into university learning.

Parnell's (2005) discussion is based on the idea that "some educators ... have recognised the value of writing in the teaching, learning and assessment of mathematics" (p. 95). It is also recognised that there are many factors that influence students' success in mathematics, such as emotional responses that can interfere with the students' ability to learn mathematics. The concept of writing reflections about their learning was in part to encourage students to express those feelings, which could help the teacher and the students overcome anxiety problems. The method was used in first year mathematics, in a subject where the students came from a wide range of backgrounds and often had limited mathematics education. The author concluded that the reflections "give excellent feedback for both the lecturer and the student and give an opportunity to acknowledge the needs of the students and hence increase their mathematics learning" (p. 100). The technique has now been taken up in higher level mathematics subjects.

Higher level programs in mathematics are rarely reported (perhaps rarely studied), though there are more in this four year period than in the previous one.

One paper (Altas, Messing, & Howarth, 2006) is a description of a distance course in the mathematical concepts of cryptography, which included online assessment, and where many students were mature age. The course was "Network Security" but it included the basics of cryptography, such as basic number theory, which can present problems for students who have not studied mathematics for many years and can consequently have a high level of anxiety. To alleviate this anxiety, the authors have found a number of techniques useful: "subject forums, email, telephone consultation, detailed lecture notes, and a one day workshop" (p. 2).

Teaching – technological aids

FitzSimons (2006) considers her work on adult numeracy has application to tertiary issues of learning mathematics and she has developed "a framework … to evaluate the use of new learning technologies for adult numeracy and mathematics education" (p. 4). She also lists questions she considers important to ask before adoption of an activity, or a technology, into the classroom, with an emphasis on group work and creating a learning community. She has used activity theory as a framework, incorporating the notions of "operation, action, and activity", and assessed the use of technology from that standpoint. This "tool use" can be difficult for students to learn particularly if the tool has not been directly designed for educational use (CAS, as an example). Reference is made to Engelbrecht and Harding's (2005a, 2005b) research on web-based instruction, which highlighted problems for some students when there is little face-to-face contact. Many of the following papers are examples of the introduction of technology into mathematics learning for a variety of reasons.

Aminifar, Caladine, Porter and Nelson (2006) describe the development of multimedia resources combining visual and audio components for delivery over the web. This was done as part of a service mathematics course, for mainly engineering students. The transition from school to tertiary mathematics is discussed, and they found that many students have an inadequate grasp of high school mathematics. Students take a skills test and the results show that those who perform poorly on the skills test are more likely to fail the course. Modification of delivery styles has been done to address this problem, including making resources available on a *WebCT* page, giving students the flexibility of accessing them as and when they needed to. The particular approach focuses on helping students work through problems using a format in which a fully worked video solution is combined with an audio track. Preliminary findings show that "by the end of the fourth week use of the resources allowed the weaker students to catch up to the stronger students" (p. 6). This result may reflect the difficulty of diagnostic tests where some students have simply forgotten the material and can relearn quickly. Students who never knew the material in the first place may still be failing.

Barry and Webb (2006) took a multi-disciplinary approach to teaching numerical methods to first year engineering students by running the course jointly with three separate disciplines, with each putting their own professional perspective on the course, such as how they would approach problems. Various difficulties in developing the course are discussed, such as the different approaches in each discipline: "A typical lecture delivery would start with a description of the engineering problem to be solved by the engineer, followed by descriptions of the mathematical background including the relevant MATLAB functions, programming techniques and numerical issues" (p. C220). They note that "[t]he deep learning happens in the computer laboratories where the students tackle the assignments under supervision" (p. C220). The range of approaches and the supported laboratory environment also helped many students overcome a phobia of computers.

Berger and Cretchley (2005) discuss the use of CAS in early undergraduate mathematics at a South African university, with comparisons to the Australian and UK experience. In summary, the study results showed that their students "have similar attitudes to the use of computers in mathematical learning as these overseas [UK and Australian] students" (p. 104). An interesting result was that "the less privileged students … did not generally feel threatened by the introduction of computers into the mathematics course. Indeed, it seems that these less-resourced students are more appreciative of the opportunities afforded by the technology than other students" (p. 104).

Griffin (2005a, b) and Griffin and Gudlaugsdottir (2006) report on an online interactive approach to teaching and learning mathematics and statistics, called the MacQTeX quiz project. It was developed to assist students to come to grips with the mathematical requirements of their courses. The quiz design, presentation and technical details are described in Griffin (2005a), and examples of problems are presented. The design has also incorporated features for staff access for reviewing student progress and the facility for messages to be directed to students, such as directing them to an on-campus service. Griffin (2005b) includes a discussion of the project in terms of formative assessment. Their online quiz approach assisted students as "[s]tudents are given immediate feedback on their work" (2005b, p. 234). Griffin and Gudlaugsdottir (2006) explain the development of the project as being in response to the need to provide flexible learning choices for cohorts of students having a wide range of mathematical backgrounds, often insufficient for their chosen field of study. The approach was useful for identification of students who were experiencing difficulties and for a minimum time delay between students doing the assessment task and feedback from teaching staff. As well as being able to follow individual students, the system can highlight difficulties being encountered by the whole cohort, which can be dealt with by re-designing the topics and quizzes for the whole course.

Mathematics can be a difficult area of learning and one way of overcoming this can be to use a visual approach (Jacobs, 2005), which demonstrates the visual representation of concepts as well as their technical significance. Computers and web-based applications can provide an avenue for visual presentation particularly when allied with interactive tools and animation. This research deals with the development of an online differential equations course with graphics and interaction as important features of the course material and presentation. Online assessment tasks were also developed, in the form of "randomized multiple choice quizzes and randomized worksheets" (p. 765), where students are prompted to repeat question types until they can answer correctly. In a further paper, Jacobs (2006) stated that technological tools can "enhance enjoyment and satisfaction as well as presenting a broad visual overview to improve conceptual understanding" (p. 1). Online assessment can be combined with the teaching mode to benefit the student's understanding. Online modules were created for two courses "using a student-centred, interactive approach that adheres to the fundamental constructivist guidelines …" (p. 2).

Loch and Donovan (2006) and Loch and McDonald (2007) have been investigating a multimodal approach to lecture presentation built on tablet technology. Loch and Donovan report on the approach and student responses to it. The results led the authors to the conclusion that "this technology is a useful tool for the modern mathematics lecture" (p. 4). However, "[s]urvey results and student comments suggest that the lecturer's competency and dexterity with the tablet is a key factor in the successful teaching with this tool" (p. 4). Loch and McDonald (2007) deal with using web-based instruction for distance students in the form of chat software (MSN Messenger). The technology "allows users to directly post and edit mathematical formulae and diagrams while communicating synchronously" (p. 1). There is a discussion of the requirements of distance education, in particular the need for a higher level of interaction. The study was based on two first year courses offered only to distance students over one summer. An online chat tutorial program – which was not included in the assessment – was run throughout the course. Students were surveyed at the end of the course, before the final examination. Despite not being linked to assessment, students were positive about the method and they "even reported that they used the chat outside the tutorial hour" (p. 12).

Tonkes, Loch, and Stace (2005) were concerned with the difficulty of implementing computer software in first year mathematics classes, and they suggest that "student confidence towards mathematics can improve significantly when learning strategies incorporate a software tool" (p. 752). There is a discussion of the problems associated with integrating computers into the required topics. The teaching resources that they developed led to improved attitudes to computation and survey results confirmed that it is the total learning experience which has improved learning outcomes (p. 756). Another discussion of computational tools (Viskic, Cooks, & Petocz, 2006) focuses on the increasing use of computer algebra systems. The result is a system in flux where traditional methods of performing such tasks – for instance, graph sketching – are still being taught alongside the software applications. Examinations exemplify this mix, and the use of CAS has generated concern on the part of university mathematics educators about the design of exam questions. That is, if there are too many questions that can be easily answered by using a CAS then student learning may be compromised. The authors undertook an investigation of exams at the level of secondary school matriculation, tertiary preparation, and first year university service mathematics. They found that a student who could use *Mathematica* in a straightforward way would pass most of these examinations. The authors state that "[t]his result supports the contention …

that pedagogical approaches lag behind the available technology" (p. 5). They suggest that there has been a "slow uptake of CAS into mathematics courses and particularly into assessment" (p. 5).

Mathematics subjects and topics

This section considers particular topics in mathematics, including technology as a topic rather than as an adjunct to learning mathematics. Two papers examine differences between countries and this is a growing trend. It is pleasing to see authors working across national boundaries as this may offer benchmarking on our programs as well as provide new ideas. Hillel and Mueller (2005) consider difficulties with learning linear algebra. They devised an examination-style question that tested different aspects of the same concept and studied students' performance on that questioning in three different settings. The basis for the research was that "[a] very common phenomenon in mathematics is that the same mathematical object (or concept) is interpreted from different perspectives" (p. 57). This flexibility can be problematic for students, particularly with linear algebra as it also has differing modes of description (abstract, algebraic, geometric). The current tendency is to couch a first linear algebra course within the algebraic mode (matrices). They found that "most students of linear algebra are not able to link different but equivalent conceptions of the same underlying object, but that the ability to do so improves somewhat by the time they are in their second course" (p. 62). Stewart and Thomas (2007) were also concerned with the conceptual difficulties posed by linear algebra and found that the use of visual, embodied notions assisted learning, as opposed to a reliance on definitions.

A paper by the UK authors, Hosein, Aczel, and Clow (2006) examines how linear programming (LP) is taught in many disciplines including mathematics. The authors investigated teaching methods and the types of software used, focusing on the LP topics of problem formulation, problem-solving methods and sensitivity analysis. An online questionnaire was sent to lecturers in Australia, New Zealand, the USA and the UK. The lowest response rates were from Australia and New Zealand. The three LP topics were all covered in all courses. The paper gives a useful overview of computing tools and differences in how topics are taught.

Undergraduate mathematics has seen a wide response to technology in the uptake of computer software (Kissane, 2006). Although required professionally, he believes software packages are not necessarily the most appropriate for teaching and learning mathematics, particularly for undergraduates. The author suggests that hand-held technologies may be preferable, partly because, for students, they are portable and inexpensive, and for teachers "there are good prospects that teachers will be able to incorporate it into standard classrooms …" (p. 1). This paper deals with using ClassPad 300, which offers interactive opportunities for learning mathematics that are not easily accessible otherwise. Kissane sums up by noting that the ClassPad 300 meets the aim of student access to "a minimum level of technology" while meeting the needs of teaching and learning mathematics.

Stewart, Thomas, and Hannah (2005) discuss the benefits of using CAS and which areas it can be used in, while suggesting that there are factors which need considering for it to be successful for aiding learning. One of these is "the individual student's instrumentation of the technological tool" (p. 741). There can be difficulties with the transformation of "an artefact into an instrument … [and] the student has to be able to adapt it to the particular mathematical task at hand" (p. 742). This takes place in two stages: "In the first the subject adapts the tool to himself while in the second he adapts himself ... to the tool" (p. 742). The teacher can assist with this adaptation, in particular with the transition to viewing the tool as a conceptual one, not just computational or procedural. Stewart (2005) took the unusual step of introducing TI-89 calculators into first year pure mathematics and argues that students find it useful. Pierce, Herbert, and Giri (2004) suggest that encouraging students to develop effective use of CAS is not easy. This paper reports on a group of undergraduate students who, despite carefully planned lectures and CAS availability for all learning and assessment tasks, failed to capitalize on its advantages. Teachers clearly need to provide assistance with technical difficulties, actively demonstrate CAS's value and reward its strategic use in assessment.

The chapter by Thomas, Monaghan, and Pierce (2004) is based on the discussions of a working group on CAS within an International Commission on Mathematical Instruction (ICMI) study. The major focus is on CAS within schools but the general discussion of CAS is applicable in the tertiary context. It is largely a critical literature review of research on CAS and teaching. They suggest that "CAS have the potential to change the teaching and learning of algebra …" (p. 155). These systems were originally designed for mathematicians, scientists and engineers, and have a range of capabilities: "Of particular interest in this chapter is the impact on teaching mathematics of the symbolic manipulation facility" (p. 156). There is a detailed discussion of the features and advantages of CAS alongside the effect it will have on learning algebra, such as facilitating different ways of understanding and representing concepts as well as varied methods for solving equations. Another area covered in the chapter is that of instrumentation (and see Stewart et al., 2005 above). It is an important consideration as "technological tools may be employed in qualitatively different ways by users and … this could be expected to influence their conceptual development" (p. 172). They note that "both the technical sequence of actions performed on the machine and the conceptual aspects of instrumentation need explicit attention …" (p. 174). It is important to properly integrate CAS with by-hand techniques and "many of the possibilities of CAS rely on students forming a partnership with it in their everyday mathematical work. Through their individual interactions they have to learn to decide what CAS is useful for and what will be better done by hand" (p. 175).

Graduate skills

Mathematics graduates are failing to secure positions at the rate of those from other disciplines, thus it is an important issue to address. The Australian Government graduate destination surveys of 2004 showed that of the mathematics graduates who wanted full-time employment, only 64.4% were in full-time employment while 35.6% were seeking full-time employment. In all the graduate fields surveyed, at 64.4% mathematics had the second-lowest full-time rate for students graduating in 2004, the lowest was Visual and Performing Arts with 56% (Graduate Careers Council of Australia, 2004). FitzSimons (2006) believes it is essential that curriculum and pedagogy are oriented towards the outside world, and that there is some integration with potential work experience. This is a growing area of research, and changes in curriculum are being influenced by the current research which has led to changing policy in universities.

As an example of changes driven by university policy, Franklin (2005) describes the establishment and content of a professional issues and ethics course in a university school of mathematics, founded because of concern across the university that "the concentration of undergraduate education on technical content was not preparing students for the workplace" (p. 98). The course begins with an emphasis on graduate attributes required by employers, such as communications skills, teamwork and ethical behaviour. There are guest speakers from industry to discuss requirements in a workplace: "Students soon to graduate learn something genuine about what they face …" (p. 99). The subject gives an overview of mathematics and its uses. General ethical issues are presented, such as confidentiality and duty of care, with the use of case studies. In support of developing such a course at other universities, Franklin suggests that "[t]hey need a break on something that connects them with what they will do when we let them out" (p. 100).

The research basis for the above course has been established by Wood and Reid (2005a, b, 2006), Wood (2006) and Wood and Smith (2007). These papers and thesis report on research that investigates the transition to work for mathematics graduates, with a particular focus on communication skills. Wood and Reid (2005a) note that "[a]djusting to the workforce can be problematic for students as they discover that what they have learned at university needs to be contextualized for work …" (p. 125). Leaving the university environment can present obstacles that can be remediated within university such as "some training in interpersonal skills and office dynamics" (p. 126), as well as communication skills. The paper describes the experience of communication skills, which are likely to be different for a student and a professional mathematician, a point that has had "insufficient attention paid to [it] within curriculum design at university" (p. 126). The authors recognise that communication within mathematics can be problematic for students, but assert that a greater difficulty lies in determining "how practicing mathematicians use mathematical discourse …" (p. 126) in order to "develop teaching and learning strategies to inculcate students into the discourse of the discipline" (p. 126). Wood and Smith (2007) also found that graduates often play the role of teacher in the workplace in the form of "explaining methods to colleagues, taking new mathematical material and presenting it to colleagues and turning information into an understandable form for their managers" (p. 1).

The studies found that no graduate believed that they had studied mathematical communication at university and graduates did not believe that they had been well prepared for the workforce and the process of getting a job. Communication skills are used as an indicator of the success of the transition to the workplace. The authors recommend that universities should have a compulsory professional development subject for all students (as in Franklin, 2005). Other changes are recommended and it is emphasised that most work should be done within the teaching and learning of the technical subjects themselves due to the discipline-specific nature of the language and communications required.

An example of integrating communication and other graduate attributes into mathematics leaning was presented by Galligan (2004a). The paper reports on the restructuring of a preparatory program for international students – which included some mathematics – based on "more recent theories of transfer and situated cognition" (p. 28). She points out that these may be reflected in expected graduate attributes, such as those developed by the Australian Council of Educational Research: "problem solving, critical thinking, interpersonal understandings and written communication" (p. 29). The university program includes a Mathematics Communication course which is integrated with the other courses within the program in terms of "transfer of knowledge, rich language tasks and application of academic skills …" (p. 29). In conclusion, the author notes that "[b]y providing such an integrated program, students are alerted to many interrelated skills and they transfer skills learned in one course to another" (p. 40).

Petocz and Reid (2006) found that studying mathematics was perceived by recent graduates as having contributed to their ability to solve problems and think logically: "all our respondents put forward the idea that studying maths had helped them in the general area of problem solving" (p. 4). They link the graduates' interpretation of these skills with the graduate views of being a professional, using the notion of the "Professional Entity" previously developed by the authors.

Graduate attributes are also important in the learning and teaching of service mathematics (Mills & Sullivan, 2005). The area of graduate attributes is described (such as for engineers) in regard to how mathematics can contribute to these. They make suggestions as to how professional requirements can be built into teaching, with a few examples (such as how a question is posed in mathematics). They also make it clear that research should inform teaching since these are the functions of universities so "it is important that undergraduate students get a taste of both throughout their training" (p. 84). The situation of the student is also presented, in terms of deciding what is relevant to them, and ensuring that they are asked questions to develop active listening skills: "We suggest that students should begin the process of shouldering this responsibility in their first year …" (p. 86).

Language

Many authors have worked on mathematical communication and the use of writing to develop mathematical skills. In the previous section, researchers show how important mathematical communication is in enabling a smooth transition to the workplace for graduates. Here we consider the few authors who have investigated the impact of English as an Additional Language (EAL) on learners of mathematics. Barton and Neville-Barton (2004), Barton et al. (2004) and Barton et al. (2005) have a long-term project investigating the disadvantages EAL students face in studying mathematics at university. The authors suggest that some students with poor English language take mathematics under the impression that they will not be so disadvantaged, because many perceive it to be relatively language-free. They have investigated the discourse features that cause problems for EAL students by attending lectures and examining texts and course notes, and trying to identify significant mathematical discourse features at this level of mathematics. They found there is a disadvantage of at least 10% in EAL students' performance in mathematics, and that the students were unaware of this disadvantage. The authors showed also that the language skills required for third year mathematics courses are quite different to first year courses. They pointed out an interesting problem about dealing with these EAL students: for international students there would be language testing, but the students in the study were all New Zealand residents and so were not tested (the same is true for Australian residents).

In an investigation of how students change or switch between languages, Galligan (2004b) reports on a study of two Chinese-English bilingual students and their language preference as they solve various mathematics problems. Many students learn mathematics in a second or third language, but research in the area still leaves some issues unclear, in particular how to blend the languages and in what order the students need to learn the language of mathematics, such as should they learn mathematics in their first language before moving to another. She suggests that "[b]ilinguals who have already learnt most of the mathematics content in L1 [first language] and then revisit this mathematics in L2 [second language], may provide some answers" (p. 1). One of the conclusions was that "conceptual knowledge is shared across L1 and L2 and is accessed with either language" (p. 10). Access is determined by efficiency and "students appeared to reduce the word problems to a minimum and to break down the problem – using problem solving heuristics and not language switching as a first strategy to solve the problem. Switching from one language to the other to solve the problem in fact appeared to be avoided by these participants" (p. 11).

Easdown (2006a) is interested in general principles which govern the way we comprehend information. In particular, he is concerned with why many students have difficulties in mathematics courses, which he thinks are related to what we understand by knowledge and its representation using language. Easdown suggests that "the person imparting the information has a 'semantic' conceptual model in his or her head …" (p. 1), while the student's "first typical response is 'syntactic'" (p. 1). He explains semantics as being the meaning or concept behind the language, and syntax as the structure or form. Understanding is something that happens gradually, with syntax overlaying the "seabed" of semantics. He then discusses how choosing the quality and volume of detail are essential in effective mathematics teaching. That is, "[c]ertain details take the recipient deeper towards semantics. Other details can be a distraction towards syntactic oblivion" (p. 4). He invokes the "Halmos Principle" which focuses on positively enhancing learning and is based on the selection of "accessible examples" to illuminate theory; these examples would preferably have "deep or subtle connections which could be appreciated at many levels" (p. 4).

Although not directly related to EAL students, Taylor and McDonald (2007) formally taught and assessed writing skills to develop non-routine mathematical problem-solving skills. Group work was another intrinsic feature of the first year subject in order to enhance learning through cooperation and discussion. Assessment tasks related to all of these skills. At the start of the course, student reaction was negative (particularly towards a workshop style of delivery), but at the end of the course "[m]ost … agreed that the workshops assisted with the development of mathematics communication skills in general and problem solving in particular …" (p. 645). Improved performance on problem solving and communication skills in assignments also occurred. The authors conclude that: "the introduction of generic skills, such as communication skills, need not necessarily over-extend students … but in fact can be an ally in the development of one of the most important mathematical skills, non-routine problem solving" (p. 651).

Understanding performance

There is pressure to lower the failure rates for mathematics, which are often higher than the average over all university courses. This may be due to the differing mathematical backgrounds of students, who have done bridging and high school mathematics at different levels despite comparable educational backgrounds. Carmody, Godfrey, and Wood (2006) and Carmichael and Taylor (2005) explore reasons for failure and ways to reduce this, because it brings a high social and financial cost. Carmody et al. (2006) describe diagnostic testing of beginning students and found a positive correlation between the test and results in the final exam, in that most of those who failed the test then failed the subject. For those who passed, the test was less useful as a predictor of results and it was posited there were probably other factors operating leading to failure. Aminifar et al. (2006) also used a skills test on entry and show that students who fail are more likely to fail their mathematics subject. Jourdan, Cretchley, and Passmore (2007) also developed a preliminary skills test and this paper reports on the differences between cohorts. The conclusion reached was that for the most recent students, "no significant differences were found in overall skills in each of the six areas ... There were differences in some specific skills … but the few that showed improvement remained at a low level" (p. 7), that is they showed some improvement but were still at a low level of mathematical performance. A variation on this theme is a paper by Britton, Daners, and Stewart (2007), who developed a self-assessment test for students to ensure they enrol in the appropriate first year mathematics course to match their skills. They found a positive correlation between results in the test and results in the highest level mathematics subject.

Bridging courses in mathematics for students entering university have become a common method for dealing with the transition from high school to university and many papers reported on the different teaching approaches used to promote improved performance. Parnell and Statham (2007) report on the success of a bridging course in mathematics where students gave positive responses to the emphasis in the course on teaching for understanding. Miller-Reilly (2007) tested three different teaching methods against affective change in relation to achievement. She found that students overall responded well to an emphasis on the relevance of mathematics to a range of contexts, although a significant minority performed worse at the end (generally those in lower achieving groups). She speculated that resource restraints made it not possible to address many of the problems of these students. She found that students' reactions to the three teaching approaches reflected the nature of the approach and that the two approaches which presented mathematics in a new way changed the students' views of the nature of mathematics.

A high failure rate in a bridging course led Carmichael and Taylor (2005) to investigate motivational effects, noting that research on motivation has concentrated on "achievement anxiety and self-efficacy" (p. 714). The former undermines performance while the latter (that is, confidence) can enhance it, although it is further modified by student beliefs about the nature of intelligence (that is, whether it is immutable, or subject to change). They found that "significant correlations occurred between prior academic knowledge, … confidence to solve specific mathematics questions, and performance" (p. 718).

An area of critical interest to university mathematics learning is one of knowledge transfer (Britton, New, Sharma, & Yardley, 2005). This paper focuses on a study about the transfer of skills learnt in mathematics to other scientific contexts. "'Transfer of learning' refers, in general, to the use of skills and knowledge in a context different from that in which the skills and knowledge were initially learned" (p. 1). The topic is of interest to mathematics educators because so much mathematics "is taught … on the basis that it is useful in different contexts" (p. 1). However, "Lecturers in … other disciplines typically complain that students either do not have sufficient mathematics or are unable to apply it in context" (p. 2). The study was conducted by researchers from mathematics, physics, microbiology and computer science. A study instrument was developed that contained mathematical questions relevant to other disciplines, but relied only on mathematical knowledge. The researchers also designed a "transfer rating" scheme referring to the rate of success of using the mathematical concepts in the other disciplines in reference to their overall mathematical skill. Although they acknowledge that their ranking system was flawed, they conclude that: "The combination of these techniques, applied to an appropriately re-designed instrument, would seem to make measurable the notion of transfer of mathematical abilities" (p. 13).

Educators are using technology for learning to enable student competency in the workplace, but also as motivation for learning (Cretchley, 2006b). Nonetheless, "some learners … express anxieties about using computers for learning, and some feel disadvantaged by their lack of experience with technology" (p. 30), which they argue poses a potential problem for these students' learning and hence achievement. The author set out to assess students' confidence in their ability to learn in a computer environment. This study revealed, perhaps surprisingly, that mathematics and computer confidence correlate very poorly. Not unexpected, however, was the finding that there are "consistently moderate correlations between computer confidence and attitudes to the use of technology for learning mathematics" (p. 34). Most surprising of all was that "[t]his study revealed no evidence at all that computer confidence related to achievement on a wide range of course tasks, not even those that specifically required the use of technology" (p. 34). The author suggests that forcing students to use unfamiliar technology could be an unnecessary extra burden and so "[e]nforced use of software and assessment thereof may be more appropriate in later undergraduate courses" (p. 35).

The needs of mature age local and international students are the focus of an article by Leder and Forgasz (2004). They found a number of differences between the groups, for instance local students generally enrolled "to meet their own personal goals" (p. 195), while the international students were following parental pressure. The international students were generally better off financially, but, "[l]oneliness was clearly a problem for some international students" (p. 195). The data reveal the need for "support services that could fruitfully be put in place …" (p. 196), such as promotion of social interaction. For the local students, "identifying and addressing needed prerequisite skills early in their studies may bridge the deficiencies …" (p. 196) perceived by some students, and there needs to be better information dispersal about the financial support schemes that are available.

Taylor and Mander (2007) were interested in the early study experiences of first year students. They differentiated between on-campus and distance education students, and focused on students taking mathematics as a service subject. This study continues the recurring theme of the transition from high school to university, but is distinguished by their investigation of the type, quantity and perceptions of study. Taylor and Mander found that on-campus students studied "significantly less time than distance education students …" (p. 4) with the mean time for both groups being less than the recommended time, except for some distance education students who spent considerably more time than recommended. When asked to select (in a questionnaire) the characteristics which they thought were necessary for success and those which they personally possessed, students displayed a mismatch between the most commonly desirable and actual characteristics. The overall finding was that the students understood the characteristics of an effective student; however, "in those early weeks few students actually carried forward these ideals into actual practice" (p. 9).

In a long-term study to examine the influence of students' conceptions of mathematics and its uses, Reid et al. (2005) used a phenomenographic methodology and Biggs (1999) 3P model to show that students' conceptions of mathematics and their intentions influenced their approach to learning. This was followed by a large international study reported in Petocz et al. (2007) and Wood et al. (2006) which categorised responses from 1200 students in five countries. They found that most students considered mathematics to consist of components, lists of topics or techniques, without structure. Some students held the conception that mathematics was modelled on the real world or had a structure (such as an algebraic structure). The authors present teaching strategies to encourage students to move to higher levels of conceptions of mathematics.

CONCLUSION

Mathematics learning and teaching is in a state of flux and this is evident from the contradictory findings reported in many papers. Is lecture attendance up or down? Yes and no. The real issue of the conditions for the effectiveness of lectures is not discussed. The failure rates in mathematics subjects are higher than in other areas according to some authors and so diagnostic tests and student support are being offered. But, this deficit model of student learning has not really progressed to a discussion of teaching changes to support diverse learners.

On computing tools, the majority of authors espouse the use of professional software and hardware tools, such as Excel, CAS and computers rather than teaching-only tools such as graphics calculators. This reflects two trends: the need to assist students in their transition to professional work; and professional software interfaces becoming more student friendly. There is a distinct split between universities that favour computing tools for mathematics learning and those who work only with pen and paper. Some of the reported dissatisfaction with computing may be due to lecturer reluctance, inadequate student training in the use of the software or poor implementation. Several authors reported problems with implementation. The software is becoming *more* student friendly but there is room for improvement. Some tools, such as the internet or use of computer software, are claimed to achieve good results *per se*, without considering the nature of the learning underpinning the tool. A software tool may be successful because it forces a certain style of learning that is effective, not because it is software.

On the research front, many authors are failing to place their findings in the literature of the field, and to discuss why their findings are the same or different to others. Many papers describe students' different learning style preferences, mathematical knowledge and aptitude, but they tend to be piecemeal. No attempt seems to have yet been made to integrate this information into a single coherent theory of how best to address the needs of all mathematics students. This is surely the key issue to be addressed. Equally it is possible that diversity in student learning experiences at different universities reflects the diversity in student learning.

There have been no Australian Research Council Grants in the area. Perhaps the Carrick Institute grants will supply sufficient resources to help mathematicians develop a more cohesive picture of learning in mathematics at university.

Publishing opportunities occur particularly in refereed conference publications and journals associated with conferences.

Perhaps it is time for our discipline to publish more widely and in more depth.

REFERENCES

Altas, I., Messing, J., & Howarth, J. (2006). I haven't done maths for 20 years – How can I learn it now? *Proceedings of the 3rd International Conference on the Teaching of Mathematics at the Undergraduate Level.* (CD). Istanbul, Turkey: Turkish Mathematical Society.

Aminifar, E., Caladine, R., Porter, A., & Nelson, M.I. (2006). Online solutions to mathematical problems: Combining video, audio and stills on the web. In T. Reeves & S. Yamashita (Eds.), *Proceedings of the World Conference on E-Learning in Corporate, Government, Healthcare, and Higher Education.* Retrieved April 10, 2007 from: http://www.editlib.org/index.cfm?fuseaction=Reader.ViewAbstract&paper_id=23650

Atweh, B., & Clarkson, P. (2005). Conceptions and tensions in globalisation and their effects on mathematics educators. In P. Clarkson, A. Downton, D. Gronn, M. Horne, A. McDonough, R. Pierce, & A. Roche (Eds.), *Building connections: Theory, research and practice* (Proceedings of the 28th annual conference of the Mathematics Education Research Group of Australasia, Melbourne, pp. 105-112). Sydney: MERGA.

Barry, S. I., & Webb, T. (2006). Multi-disciplinary approach to teaching numerical methods to engineers using Matlab. In A. Stacey, B. Blyth, & A. J. Roberts (Eds.), *Proceedings of the 7th biennial Engineering Mathematics and Applications Conference 2005,* (*ANZIAM Journal, 47,* C216-C230). Australian Mathematics Society. Retrieved February 29, 2008 from: http://anziamj.austms.org.au/V47EMAC2005/Barry/Barry.pdf

Barton, B., Chan, R., King, C., & Neville-Barton, P. (2004). The mathematical discourse of advanced undergraduate mathematics. In I. Putt, R. Faragher, & M. McLean (Eds.), *Mathematics education for the third millennium: Towards 2010* (Proceedings of the 27th annual conference of the Mathematics Education Research Group of Australasia, Townsville, pp. 79-86). Sydney: MERGA.

Barton, B., Chan, R., King, C., Neville-Barton, P., & Sneddon, J. (2005). EAL undergraduates learning mathematics. *International Journal of Mathematical Education in Science and Technology, 36*(7), 721-729.

Barton, B., & Neville-Barton, P. (2004). *Undergraduate mathematics learning in English by speakers of other languages.* Paper presented at International Congress of Mathematics Education (ICME-10), Copenhagen. Retrieved April 16, 2007 from: http://www.icme-organisers.dk/tsg25/plenary/barton.doc

Berger, M., & Cretchley, P. (2005). Technology and mathematics learning at university level: A South African perspective. *African Journal of Research in SMT Education, 9*(2), 97-108.

Biggs, J. (1999). *Teaching for quality learning at university. Society for research in higher education.* UK: Open University Press.

Blyth, W. F., Clarke, D., & Labovic, A. (2006). Video analysis to understand e-learning of vector calculus. In A. Stacey, B. Blyth, & A. J. Roberts (Eds.), *Proceedings of the 7th Biennial Engineering Mathematics and Applications Conference 2005* (ANZIAM Journal, *47*, C185-C199). Australian Mathematics Society. Retrieved February 28, 2008 from: http://anziamj.austms.org.au/V47EMAC2005/Blyth/home.html

Bourguignon, J. P., Dietrich, B., & Johnstone, I. M. (2006). Mathematics and statistics: Critical skills for Australia's future. *National strategic review of mathematical sciences in Australia.* Retrieved January 14, 2007 from: http://www.review.ms.unimelb.edu.au/MSReviewShort.pdf

Britton, S., Daners, D., & Stewart, M. (2007). A self-assessment test for incoming students. *International Journal of Mathematical Education in Science and Technology. 38*(7) 861-868.

Britton, S., New, P. B., Sharma, M. D., & Yardley, D. (2005). A case study of the transfer of mathematics skills by university students. *International Journal of Mathematical Education in Science and Technology, 36*(1) 1-13.

Carmichael, C., & Taylor, J. A. (2005). Analysis of student beliefs in a tertiary preparatory mathematics course. *International Journal of Mathematical Education in Science and Technology, 36*(7), 713-719.

Carmody, G., Godfrey, S., & Wood, L. (2006). Diagnostic tests in a first year mathematics subject. *Assessment in science teaching and learning. Symposium proceedings* (pp. 24-30). Sydney NSW: UniServe Science.

Clarke, D. J. (2006). The LPS research design. In D. J. Clarke, C. Keitel, & Y. Shimizu (Eds.), *Mathematics classrooms in twelve countries: The insider's perspective* (pp. 15-37). Rotterdam: Sense Publishers.

Coupland, M. (2004). *Learning with new tools.* Unpublished PhD thesis, University of Wollongong.

Coupland, M., & Crawford, K. (2006). Many dimensions: The complex picture of student encounters with a computer algebra system. In P. Grootenboer, R. Zevenbergen, & M. Chinnappan (Eds.), *Identities, cultures and learning spaces* (Proceedings of the 29th annual conference of the Mathematics Education Research Group of Australasia, Canberra, pp. 155-162). Adelaide: MERGA.

Cretchley, P. (2005) Mathematics and dumping lectures? Another perspective on the shift towards learner pragmatism. In M. Bulmer, H. MacGillivray, & C. Varsavsky (Eds.), *Proceedings of Kingfisher Delta '05* (Fifth southern hemisphere conference on undergraduate mathematics and statistics teaching and learning, pp. 42-48). Brisbane: University of Queensland.

Cretchley, P. (2006a). Students using computer software for mathematics assessment: A comparison of technology and pen-and-paper performance. *Proceedings of the 3rd international conference on the teaching of mathematics at the undergraduate level.* (CD). Istanbul, Turkey: Turkish Mathematical Society.

Cretchley, P. (2006b). Does computer confidence relate to levels of achievement in ICT-enriched learning models? *Education and Information Technologies, 11*(4), 29-39.

D'Arcy-Warmington, A. (2006). Stand and deliver: A complex interaction to mathematical learning. *Proceedings of the 3rd international conference on the teaching of mathematics at the undergraduate level.* (CD). Istanbul, Turkey: Turkish Mathematical Society.

D'Souza, S. M., & Wood, L. N. (2004). The need for higher education reform – problems faced with assessing collaborative team based work. In A. Rogerson (Ed.), *The future of mathematics education, international conference* (pp. 24-28). Ciechocinek Poland.

D'Souza, S. M. (2006). *An examination into the effects of incorporating collaborative learning methods in a core first-year mathematics subject.* Unpublished PhD dissertation. UTS, Sydney, Australia.

Easdown, D. (2006a). Teaching mathematics: The gulf between semantics (meaning) and syntax (form). *Proceedings of the 3rd international conference on the teaching of mathematics at the undergraduate level.* (CD). Istanbul, Turkey: Turkish Mathematical Society.

Easdown, D. (2006b). Integrating assessment and feedback to overcome barriers to learning at the passive/active interface in mathematics courses. *Assessment in science teaching and learning, Symponium proceedings* (pp. 37-42). Sydney NSW: UniServe Science.

Engelbrecht J., & Harding A. (2005a). Teaching undergraduate mathematics on the internet. Part 1: Technologies and taxonomy. *Educational Studies in Mathematics, 58*(2), 235-252.

Engelbrecht J., & Harding A. (2005b). Teaching undergraduate mathematics on the internet. Part 2: Attributes and possibilities. *Educational Studies in Mathematics, 58*(2), 253-276.

FitzSimons, G. E. (2006). An activity theory perspective on technology mediated mathematics education for undergraduates. *Proceedings of the 3rd international conference on the teaching of mathematics at the undergraduate level.* (CD). Istanbul, Turkey: Turkish Mathematical Society.

Franklin, J. (2005). A "Professional issues and ethics in mathematics" course. *Australian Mathematical Society Gazette, 32*(2), 98-100.

Freislich, M. R. (2006). The effects of Keller Plan instruction on the achievement and attitudes of tertiary mathematics students. *Proceedings of the 3rd international conference on the teaching of mathematics at the undergraduate level.* (CD). Istanbul, Turkey: Turkish Mathematical Society.

Galligan, L. (2004a). Preparing international students for university: Mathematics as part of an integrated program. *Australian Senior Mathematics Journal, 18*(1), 28-41.

Galligan, L. (2004b). The role of language-switching in bilingual students' processing of mathematics. Paper presented at the International Congress of Mathematics Education (ICME-10), Copenhagen. Retrieved April 16, 2007 from: http://www.icme-organisers.dk/tsg25/subgroups/galligan.doc

Graduate Careers Council of Australia. (2004). GradStats. Retrieved February 29, 2008 from: http://www.graduatecareers.com.au/content/download/2148/8520/file/Gstats2004.pdf

Griffin F. (2005a). The MacQTeX quiz project. In F. Olviero & R. Sutherland (Eds.), *Visions of mathematics education: Embedding technology in learning. Proceedings of the 7th international conference on technology in mathematics teaching,* (pp. 242-249). Bristol, UK.: Graduate School of Education, University of Bristol.

Griffin F. (2005b). Practising for perfection. In F. Olviero & R. Sutherland (Eds.), *Visions of mathematics education: Embedding technology in learning. Proceedings of the 7th international conference on technology in mathematics teaching,* (pp. 233-241). Bristol, UK: Graduate School of Education, University of Bristol.

Griffin F., & Gudlaugsdottir, S. (2006). Using online randomised quizzes to boost student performance in mathematics and operations research. *Proceedings of the 7th international conference on information technology based higher education and training.* Retrieved February 28, 2008 from: http://ieeexplore.ieee.org/Xplore/login.jsp?url=/iel5/4141586/4084515/04141663.pdf?tp=&isnumber=&arnumber=4141663

Groen, L. (2006). Enhancing learning and measuring learning outcomes in mathematics using online assessment. In *Assessment in science teaching and learning. Symposium proceedings* (pp. 56-61). Sydney NSW: UniServe Science.

Harding, A., Kaczynski, D., & Wood, L. N. (2005) Evaluation of blended learning: analysis of qualitative data. *Blended learning in science teaching and learning. Symposium proceedings* (pp. 56-62). Sydney NSW: UniServe Science.

Hillel. J., & Mueller, U. A. (2005). Understanding multiple interpretations of the notion of solution in linear algebra. In M. Bulmer, H. MacGillivray, & C. Varsavsky (Eds.), *Proceedings of Kingfisher Delta '05* (Fifth southern hemisphere conference on undergraduate mathematics and statistics teaching and learning, pp. 57-64). Brisbane: University of Queensland.

Holton, D. (2005). Tertiary mathematics education for 2024. *International Journal of Mathematical Education in Science and Technology, 36*(2-3), 305-316.

Hosein, A., Aczel, J., & Clow, D. (2006). The teaching of linear programming in different disciplines and in different countries. *Proceedings of the 3rd international conference on the teaching of mathematics at the undergraduate level.* (CD). Istanbul, Turkey: Turkish Mathematical Society.

Houston, S. K., Wood, L. N., Engelbrect, J., Harding, A., Kagesten, O., Holton, D., & Barton, W. (2006). Professional development of mathematics academics. *Proceedings of the 3rd international conference on the teaching of mathematics at the undergraduate level.* (CD). Istanbul, Turkey: Turkish Mathematical Society.

Hubbard, R. (2006). Have students' attitudes to mathematics lectures changed now that they have access to online materials? *Proceedings of the 3rd international conference on the teaching of mathematics at the undergraduate level.* (CD). Istanbul, Turkey: Turkish Mathematical Society.

Jacobs, K. L. (2005). Investigation of interactive online visual tools for the learning of mathematics. *International Journal of Mathematical Education in Science and Technology, 36*(7), 761-769.

Jacobs, K. (2006). Candy-coated mathematics, a more digestible approach. *Proceedings of the 3rd international conference on the teaching of mathematics at the undergraduate level.* (CD). Istanbul, Turkey: Turkish Mathematical Society.

Jourdan, N., Cretchley, P., & Passmore, T. (2007). Secondary-tertiary transition: What mathematics skills can and should we expect this decade. In J. Watson & K. Beswick (Eds.), *Mathematics:*

Essential research, essential practice (Proceedings of the 30th annual conference of the Mathematics Education Research Group of Australasia, Hobart, pp. 463-472). Adelaide: MERGA.

Keady, G., Fitz-Gerald, G., Gamble, G., & Sangwin, C. (2006). Computer-aided assessment in mathematical sciences. *Assessment in science teaching and learning. Symposium proceedings* (pp. 69-73). Sydney NSW: UniServe Science.

Kissane, B. (2006). Using the *ClassPad 300* for teaching and learning early undergraduate mathematics. *Proceedings of the 3rd international conference on the teaching of mathematics at the undergraduate level.* (CD). Istanbul, Turkey: Turkish Mathematical Society.

Leder, G. C., & Forgasz, H. J. (2004). Australian and international mature students. *Higher Education Research & Development, 23*(2), 183-198.

Loch, B., & Donovan, D. (2006). Progressive teaching of mathematics with tablet technology. *e-Journal of Instructional Science and Technology, 9,* 2. Retrieved April 10, 2007 from: http://eprints.usq.edu.au/archive/00001716/

Loch, B. I., & McDonald, C. (2007). Synchronous chat and electronic ink for distance support in mathematics. *Innovate: Journal of Online Education, 3,* 3. Retrieved April 11, 2007 from: http://www.innovateonline.info/index.php?view=article&id=374

McDonald, C., & Taylor, J. A. (2005). Preliminary investigation into aggregating grades in a first year course with mixed assessment types. In M. Bulmer, H. MacGillivray, & C. Varsavsky (Eds.), *Proceedings of Kingfisher Delta '05* (Fifth southern hemisphere conference on undergraduate mathematics and statistics teaching and learning, pp. 88-94). Brisbane: University of Queensland.

MacGillivray, H., & Turner, I. (2005). Components of learning and assessment in linear algebra. In M. Bulmer, H. MacGillivray, & C. Varsavsky (Eds.), *Proceedings of Kingfisher Delta '05* (Fifth southern hemisphere conference on undergraduate mathematics and statistics teaching and learning, pp. 74-81). Brisbane: University of Queensland.

Miller-Reilly, B. M. (2007). Three different teaching approaches in pre-calculus bridging mathematics. *International Journal of Mathematical Education in Science and Technology, 38*(7), 891-906.

Mills, T. M., & Sullivan, P. (2005). Mathematics – at your service. In H. K. Ko & D. Arganbright (Eds.), *Proceedings of KAIST international symposium on enhancing university mathematics teaching* (pp. 81-95). Korea: KAIST.

Parnell, S. (2005). Do student reflections increase undergraduate mathematics learning? In M. Bulmer, H. MacGillivray, & C. Varsavsky (Eds.), *Proceedings of Kingfisher Delta '05* (Fifth southern hemisphere conference on undergraduate mathematics and statistics teaching and learning, pp. 95-103). Brisbane: University of Queensland.

Parnell, S., & Statham, M. (2007). An effective preparation for tertiary mathematics. *International Journal of Mathematical Education in Science and Technology. 38*(7), 869-880.

Penglase, M. (2004). Learning approaches in university calculus: The effects of an innovative assessment program. In I. Putt, R. Faragher, & M. McLean (Eds.), *Mathematics education for the third millennium: Towards 2010* (Proceedings of the 27th annual conference of the Mathematics Education Research Group of Australasia, Townsville, pp. 446-453). Sydney: MERGA.

Petocz, P., & Reid, A. (2006). *The contribution of mathematics to graduates' professional working life.* Paper presented at the annual conference of the Australian Association for Research in Education. Retrieved July 16, 2007 from: http://www.aare.edu.au/05pap/pet05141.pdf

Petocz, P., Reid, A., Wood, L., Smith, G., Mather, G., Harding, A., Engelbrecht, J., Houston, K., Hillel, J., & Perrett, G. (2007). Undergraduate students' conceptions of mathematics: An international study. *International Journal of Science and Mathematics Education, 3*(5), 439-459. Retrieved April 16, 2007 from: http://www.springerlink.com/content/w74365258080538q/

Petocz, P., Smith, G. H., Wood, L. N., & Reid, A. (2004). On becoming a mathematician: An international perspective for future professionals in the mathematical sciences. Paper presented at the International Congress on Mathematics Education (ICME10), Copenhagen. Retrieved April 16, 2007 from: http://www.icme-organisers.dk/tsg28/ICME10_TSG28_Petocz_etal.doc

Pierce, R, Herbert, S., & Giri, J. (2004). CAS: Student engagement requires unambiguous advantages. In I. Putt, R. Faragher, & M. McLean (Eds), *Mathematics education for the third millennium:*

Towards 2010 (Proceedings of the 27th annual conference of the Mathematics Education Research Group of Australasia, Townsville, pp. 462-469). Sydney: MERGA.

Reid, A., Wood, L. N., Petocz, P., & Smith, G. (2005). Intention, approach and outcome: University mathematics students' conceptions of learning mathematics. *International Journal of Science and Mathematics Education, 3*, 567-586.

Sowey, E. (2005). From a logical point of view: An illuminating perspective in teaching statistical inference. *International Journal of Mathematical Education in Science and Technology, 36*(7), 801-811.

Stewart, S (2005). Concerns relating to the CAS use at University level. In P. Clarkson, A. Downton, D. Gronn, M. Horne, A. McDonough, R. Pierce, & A. Roche (Eds.), *Building connections: Theory, research and practice* (Proceedings of the 28th annual conference of the Mathematics Education Research Group of Australasia, Melbourne, pp. 704-710). Sydney: MERGA.

Stewart, S., & Thomas, M. O. J. (2007). Embodied, symbolic and formal thinking in linear algebra. *International Journal of Mathematical Education in Science and Technology. 38*(7), 927-938.

Stewart, S., Thomas, M. O. J., & Hannah, J. (2005). Towards student instrumentation of computer-based algebra systems in university courses. *International Journal of Mathematical Education in Science and Technology, 36*(7), 741-749.

Taylor, J. A. (2006). Assessment: A tool for development and engagement in the first year of university study. Paper presented at the *9th Pacific Rim in Higher Education (FYHE) conference: Engaging students,* Griffith University and Queensland University of Technology, Gold Coast. Retrieved August 20, 2007 from http://eprints.usq.edu.au/archive/00000994/01/FYE_2006_Assessment.pdf

Taylor, J. A., & McDonald, C. (2007). Writing in groups as a tool for non-routine problem solving in first year university mathematics. *International Journal of Mathematical Education in Science and Technology, 38*(5), 639-655.

Taylor, J. A., & Mander, A. (2007). Studying at university: Early perceptions and experiences of first year service mathematics students. *Studies in Learning, Evaluation, Innovation and Development. 4*(3), 29-43.

Thomas, M. O. J., Monaghan, J., & Pierce, R. (2004). Computer algebra systems and algebra: Curriculum, assessment, teaching, and learning. In K. Stacey, H. Chick, & M. Kendal, M. (Eds.). *The future of the teaching and learning of algebra: The 12th ICMI study* (pp. 154-186). Boston: Kluwer Academic Publishers.

Tonkes, E. J., Loch, B. I., & Stace, A. W. (2005). An innovative learning model for computation in first year mathematics. *International Journal of Mathematical Education in Science and Technology, 36*(7), 751-759.

Varsavsky, C. (2006). Managing student expectations and improving student engagement in a first year calculus course. *Proceedings of the 3rd international conference on the teaching of mathematics at the undergraduate level.* (CD). Istanbul, Turkey: Turkish Mathematical Society.

Viskic, D., Cooks, L., & Petocz, P. (2006). Examining *Mathematica*: An analysis of the performance of CAS on examinations in mathematics. *Proceedings of the 3rd international conference on the teaching of mathematics at the undergraduate level.* (CD). Istanbul, Turkey: Turkish Mathematical Society.

Wood, L. N. (2006). Graduate voices: The nexus between learning and work. Unpublished PhD dissertation, Macquarie University, Australia.

Wood, L. N., & Harding, A. (2007). Can you show you are a good lecturer? *International Journal of Mathematical Education in Science and Technology, 38*(7), 939-947.

Wood, L. N., Joyce, S, Petocz, P., & Rodd, M. (2007). Learning in lectures: Multiple representations. *International Journal of Mathematical Education in Science and Technology*, 38(7), 907-915.

Wood., L., & Reid, A. (2005a). Mathematics communication for graduates. In M. Bulmer, H. MacGillivray, & C. Varsavsky, (Eds.), *Proceedings of Kingfisher Delta '05* (Fifth southern hemisphere conference on undergraduate mathematics and statistics teaching and learning, pp. 125-131). Brisbane: University of Queensland.

Wood, L., & Reid, A. (2005b). *Graduates' initial experiences of work.* Refereed paper presented at the annual conference of the Australian Association for Research in Education. Retrieved February 25, 2008 from: http://www.aare.edu.au/05pap/woo05147.pdf

Wood L., & Reid, A. (2006). Conversations with graduates: reflections on learning mathematics. *Proceedings of the 3rd international conference on the teaching of mathematics at the undergraduate level.* (CD). Istanbul, Turkey: Turkish Mathematical Society.

Wood, L., Smith, G., Mather, G., Harding, A., Engelbrecht, J., Houston, K., Perrett, G., Hillel, J., Petocz, P., & Reid, A. (2006). Student voices: Implications for teaching mathematics. *Proceedings of the 3rd international conference on the teaching of mathematics at the undergraduate level.* (CD). Istanbul, Turkey: Turkish Mathematical Society.

Wood, L. N., & Smith, N. F. (2007). Graduate attributes: Teaching as learning. *International Journal of Mathematical Education in Science and Technology. 38*(6), 715-727.

AFFILIATION

Leigh Wood
Macquarie University, Australia

LINDA GALLIGAN & JANET A. TAYLOR

ADULTS RETURNING TO STUDY MATHEMATICS

This chapter focuses on adults learning mathematics in two distinct sections: adults learning mathematics in the workplace and community; and adults engaged in further studies (e.g., university or vocational studies) including bridging mathematics. In this review, research available on adults in basic numeracy education through to those gaining access to or supported at university study involving mathematics in a variety of forms is investigated. Consideration is also given to relevant definitions of *adult*, *mathematics*, and *numeracy*, and to the relationship between research into learning and research into teaching practices or curriculum design. It is concluded that the area of adults learning mathematics is still under-theorised and under-researched at a time when 21st century needs demand a population with increasing numeracy-based skills and knowledge.

Key words: adult, bridging, numeracy, university, workplace

INTRODUCTION

This overview of research on adults returning to study mathematics by necessity involves the intersection of two research domains; adult learning and mathematics education. This intersection leads to the distinct regime of adults learning mathematics, a young and growing domain which has been described as under-theorised (FitzSimons & Godden, 2001) and research scant. In 2001 the Australian Council for Adult Literacy (ACAL), wrote:

> Ten years ago Australia was at the forefront of adult literacy and numeracy teaching, learning and research. Now, we do not even have a national policy and we are going backwards (ACAL, 2001, cited in Coben, 2006, p. 84).

So it is not surprising that no previous Mathematics Education Research Group of Australasia (MERGA) review of adults learning mathematics has specifically been conducted, although components have been incorporated into other sections of previous reviews. However, since 2002 there has been a resurgence in research, especially within the vocational education sector, often under the banner of literacy research.

With these developments in mind, in this review mathematics is viewed within a broad conception, not limited to specialised mathematics (FitzSimons, Coben, &

H. Forgasz et al. (eds.), Research in Mathematics Education in Australasia 2004-2007, 99-118.

O'Donoghue, 2003). It incorporates research available on adults in basic numeracy education through to those gaining access to, or studying at, university level. Consideration is given to definitions associated with mathematics, numeracy, and adult which help define the parameters of the chapter. In the body of the chapter, the focus is on research on adults learning and using mathematics in the workplace and community, in further studies, such as the Vocational Education and Training (VET) sector or bridging or enabling mathematics programs at the university level. Within this literature there is a strong relationship between research into learning and research into teaching practice or curriculum design. Hence, research into the pedagogies for adults returning to study mathematics in all its forms will be included.

DEFINITIONS

Adult

In commencing this review it became apparent that there was some confusion about what constituted *an adult*. A precise definition has proved elusive and has been described as "nebulus and revolves around the student rather than the level of mathematics" (FitzSimons & Godden, 2001, p. 14). Even using the legal definition of all persons over 18 years of age is problematic. Thus for the purpose of this review we focus on adults who are returning to study mathematics or those who have never studied mathematics. So that means that a 16 year old who left school at 15 years and is now in the work force is classified as an adult learner, while a 17 or 18 year old who has entered university directly from school is not so classified.

Mathematics and numeracy

In this chapter we focus on the definition of adults returning to mathematics as clarified above and the notion of numeracy that surrounds them. We are aware of various terms often used synonymously with numeracy (critical numeracy, quantitative literacy, financial literacy, document literacy, mathematical literacy, democratic numeracy, functional mathematics) but will not attempt to define them here.

The nature of numeracy has been touched on in previous MERGA reviews. In a chapter on language factors, Ellerton, Clarkson, and Clements (2000) devoted some time to the definition of numeracy. At the time it appeared that too little was known about how mathematics was used in aspects of life. In the years between 2000 and 2003 the MERGA literature was silent about numeracy, while in other arenas its definition was hotly debated. Concerns about it being politicised and subsumed by literacy dominated. In recent years, the debate has reduced with the National Centre for Vocational Education Research (NCVER) (2003) concluding that:

> Attempts to propose any single definition of literacy and numeracy are relatively futile when social, cultural and technological changes shape our understandings and alter the way we think about literacy and numeracy. As our concepts change with the times, so do approaches to developing these skills. (p. 1)

This is reflected in FitzSimons (2006) who quoted van Groenestijn (2002) to emphasise that numeracy is essentially "dynamic and contextually bound to time and place" (para. 4). In a suite of research-based reports on literacy and numeracy from the NCVER (see section on workplace and community), *literacy* was used as a *catch-all* term for "English language, numeracy and information technology literacy" (Wickert & McGuirk, 2005, p. 11). Hence in this review, literature in which the term literacy was used was only included if, upon closer inspection, some aspect of numeracy was found.

In recent literature a change in the concept of numeracy has been chronicled (Kemp, 2005; Miller-Reilly, 2006). Between 2004 and 2007 there appeared to be a group of core definitions with a number of common characteristics. An international group of mathematics educators (Adults Learning Maths: A Research Forum), with a substantial Australian and New Zealand membership, has extensively debated the topic, with Johnston and McGuire (2005) indicating that numeracy involved:

> managing a situation of solving a problem in a real context by responding to information about mathematical ideas that is represented in a range of ways and requires activation of a range of enabling knowledge, behaviours and processes. (p. 128)

FitzSimons (2005b) used Coben's (2003) definition which expands the Johnston-McGuire definition to include a sense of confidence when describing numerate behaviour on the job:

> To be numerate means to be competent, confident, and comfortable with one's own judgements on whether to use mathematics in a particular situation and if so, what mathematics to use, how to do it, what degree of accuracy is appropriate, and what the answer means in relation to the context. (FitzSimons, 2005b, p. 27)

The most recently used definitions include mathematics, its context, and the notion of being able to use the mathematics for decision making or problem solving. This is reflected in a definition developed by the Australian Association of Mathematics Teachers (AAMT). Here numeracy is defined as the ability "to use mathematics effectively to meet the demands of life at home, in paid work, and for participation in community and civic life" (AAMT, 1998, p. 3). In a review by Hartley and Horne (2005) it was noted that:

> many Citizens Advice Bureau (CAB) enquirers have well developed numeracy and literacy skills but are unable to identify the optimum financial decision or strategy based on the information available to them. (p. 22)

The CAB enquirers, in fact, do not have well developed numeracy skills, according to the recent numeracy definitions, as they could not solve the problem of finding the optimal financial decision.

The term academic numeracy, modified from Yasukawa and Johnston (1994), was used by Galligan and Taylor (2005) to clarify the skills necessary for success in the university context as:

> a critical awareness which allows the student to situate, interpret, critique, use and perhaps even create mathematics in context, in this case the academic context. It is more than being able to manipulate numbers or being able to succeed at mathematics. (p. 87)

Neill (2001), attempting to clarify the notion of numeracy, suggested that there were two threads running through many of the forty definitions he examined: *location* (home, work, society etc) and *context* (specific problems or situations). At the time, few definitions included both of these components. It should be noted that the three definitions quoted in the review (FitzSimons, 2005b; Galligan & Taylor, 2005; Johnson & McGuire, 2005) all have the concepts of *location* and *context.*

FitzSimons (2005d) brought a new dimension to the numeracy debate situating it in an international setting and focusing on Bernstein's concepts of vertical and horizontal discourse. While vertical discourse centres mainly on school mathematics, horizontal discourse is closely linked to numeracy as it is related to on-going practices, is affective, has specific immediate goals, and is highly relevant. She emphasised that these discourses are different with different practices, and that vertical discourses will not guarantee numerate activity.

FitzSimons and Coben (in press) referred to Maguire and O'Donoghue's (2002) framework of adult numeracy as a continuum from formative (phase 1: basic skills) to mathematical (phase 2: mathematics in the context of everyday life) to integrative (phase 3: mathematics integrated with the cultural, social, personal, and emotional). At phase 3, FitzSimons and Coben argue, adults can become *knowledge producers* as well as *knowledge consumers*. Thus adults become "technologically, socially, personally, and/or democratically numerate" (Condelli, 2006, p. 7).

Finally, Zevenbergen (2004) placed the concept of numeracy in the context of the workforce. She defined numeracy as "the application of, and disposition towards using contextually appropriate mathematics to solve everyday problems" (p. 100) and visited the notion of multiple numeracies. While there has been debate over multi-literacies in these *new times*, until then no such debate had emerged in mathematics and numeracy. She posed the question of whether there are different forms of numeracy in these changing times and concluded that we may need to re-theorise knowledge, skills and dispositions, to re-conceptualise definitions of numeracy relevant to workforce needs.

The next section is divided into two parts. The first part reviews adult numeracy; a series of studies commissioned by NCVER; and other research focussing on workplace and community. In the second part of the review, research on adults in further studies is reported on.

WORKPLACE, COMMUNITY AND THE ADULT LEARNER OF MATHEMATICS

Reviews of adult numeracy

Five contributions were found in which Australasian research on aspects of adult literacy that included numeracy was reviewed. The first of these was from New Zealand, three were from Australia and one jointly researched with an Irish author.

Benseman, Sutton, and Lander's (2005a) review, commissioned by the New Zealand Ministry of Education, located 300 studies (world wide) in which the effectiveness of adult literacy, numeracy, and language teaching was investigated. Within the 105-page report, only six pages were devoted to numeracy, reinforcing the notion of the under-researched nature of adult numeracy. The authors were led to conclude:

> the research evidence on numeracy does not provide guidance on the time and the nature and extent of teaching required for adults to make significant progress… our understanding of adult numeracy is so limited that at this point rather than evaluating the impact of what does take place it may be more appropriate to undertake insight studies that consider the range and nature of adult mathematical understandings, how that understanding develops and what interventions might be effective. (Benseman et al., 2005a, p. 84)

Hartley and Horne (2005) reported on an Australian NCVER project in which the social and economic benefits of improved adult literacy were investigated. The report focussed on business, health and financial literacy (including research projects by the ANZ Bank in 2003 and the Commonwealth Bank in 2005), family literacy and crime, together with brief reviews of research related to two specific population groups (older people and Australian Aborigines). Some useful frameworks for measuring numeracy directly were also discussed (Hartley & Horne, 2005, 2006). While the report did not explore detailed benefits and costs of numeracy, it recognised numeracy as important area, and that further research is required.

McKenna and Fitzpatrick (2004) conducted an extensive audit of policies and practices of adult literacy and adult basic skills in Canada, Ireland, New Zealand, USA, UK and Australia. Their report focussed on four areas of literacy: "'lingering basics', 'new basics', 'elite literacies' and 'foreign language literacy'" (p. 10). Numeracy was implicitly embedded in three of the areas. While the authors concluded that Australia was successful in integrating literacy within VET, national policies in that sector had not been renewed since 1996. Outside the VET sector there appeared to be a lack of understanding of the literacy provision.

Perkins (2005) reported on the National Reporting System (NRS) which sets levels of language, literacy and numeracy in Australia. The author concluded that although this system was generally accepted by practitioners, it needed to be updated.

Johnston and McGuire (2005) aimed ambitiously to "provide a snapshot of international adult numeracy policy, pedagogy and provision over the last 20 years with case studies of interesting and significant initiatives" (p. 8). The seven chapters included policy, financial and lifelong numeracy, and international comparisons. Chapters on research, professional development and implications were not available at the time of this review. The study so far has helped to expand understandings of numeracy internationally and has clarified issues of the links between numeracy, democracy and social justice.

Workplace and Community Projects

In New Zealand a *Learning for Living* project has produced a number of general research based reports commissioned by the Ministry of Education into aspects of adult literacy. This includes research into assessment practices (New Zealand Council for Educational Research [NZCER], 2006) and an observational study on adult literacy, numeracy and language teachers (Benseman, Sutton, & Lander, 2005b). These reports, while having an overall focus on literacy, numeracy and language, had little specific numeracy research. However, a research project on the development of a community of practice for numeracy teachers has been completed by Gill Thomas but was not available in time for this publication.

In 2002, 2003, and 2004, the NCVER in Australia managed adult literacy projects producing 25 publications between 2004 and 2007. Of these one general (Wickert & McGuirk, 2005) and six specific reports included research on numeracy.

The Wickert and McGuirk (2005) study was a preliminary investigation into building literacy capabilities in communities and the workplace. They used the term 'literacy' as a general term including English language, numeracy, and IT literacy, and interviewed coordinators of initiatives from five worksites and four community sites. Only in one example (a women's community health centre) were numerate activities identified. These involved using street directories and timetables. While acknowledging this as a small study, the authors concluded that there was potential to engage agencies outside education and the training sector (e.g., industry partnerships) to improve literacy competencies.

The numeracy demands of the chemical spraying industry have been reported in numerous studies. FitzSimons and others examined the numeracy in this industry (FitzSimons, 2005b; FitzSimons, 2005c; FitzSimons, Mlcek, Hull, & Wright, 2006; FitzSimons & Wedege, 2004) using work shadowing, observation, semi-structured interviews, and the collection of artefacts. They concluded that all the workplaces investigated required a wide variety of mathematical skills based on Mayer's key competencies, cited by FitzSimons et al. (2006) to be: collecting, analysing and organising information; communicating ideas and information; planning and organising activities; working with others and in teams; using mathematical ideas and techniques; solving problems; and using technology collecting. FitzSimons also identified Bishop's (1998) 'pan-cultural' activities (counting, designing, explaining, locating, measuring and playing), plus underlying skills of algebra (spreadsheet) calculations, estimations, geometric thinking, storage, retrieval display, and interpretation of data (Bishop, cited in FitzSimons, 2005b). This research importantly recognised the complexity of the workplace, where workers require a strong general mathematics education. It was purported that the research should assist vocational teachers to address the broad range of concepts beyond "simplistic rational number and measurement skills" (FitzSimons, 2005b, p. 38). Following this work, and underpinned by activity theory, was FitzSimons' (2005c) project on 'An evaluative framework for new learning technologies'. In these numerous works FitzSimons argues that while learning numeracy in the workforce is informal, it is filled with meaning, and the relation between such things as rules of the community, the division of labour, and knowledge is important.

Cumming and Wilson (2005) investigated the literacy and numeracy demands of alternative dispute resolution processes in Australia. After surveying six alternative dispute resolution organisations, 51 individual dispute resolution practitioners (response rate of 22%) and analysing numerous documents, they suggested that dispute resolution required a high level of numeracy (level 5, National Reporting System [NRS]). The numeracy skills identified included financial, map reading (for property settlement), interpreting data, and calculating percentages. In one instance, the poor numeracy skills of lawyers were noted. The results indicated that this level of numeracy demand may prevent Australians from participating in effective dispute resolution processes, and suggested a need for specific training of practitioners. This research appeared to be the first of its kind, as even though dispute resolution involved a "considerable range of financial considerations and mathematical procedures involved in legal cases" (Cumming & Wilson, 2005, p. 17), no previous international research had involved numeracy considerations. The authors expressed a concern that limited literacy and numeracy skills could "act as a barrier to participation in the justice system in Australia" (Cumming & Wilson, 2005, p. 17).

Hayes, Golding, and Harvey (2006) investigated the literacy and numeracy skills of public safety volunteers. In this study of over 300 participants from twenty fire brigades and State Emergency Services (SES) units in small and remote Australian communities, the authors described the numeracy demands of the profession and the participants' perceptions of their personal competencies. Questions on skills produced the following results on the importance of skills to the fire brigade and a participant's personal ratings on certain skills: reading a map (50% rated their personal skills as high and 82% rated this as highly important to the brigade); reading dials on equipment (51% personal skills high and 79% highly important to the brigade); doing basic calculations (44% high personal skills high and 43% highly important to the brigade). The authors concluded that the development of skills is best embedded into training that is of direct interest to the volunteers. The providers should have strong links with the volunteer organisations. Informal training was also concluded to be significant for the development of numeracy skills and the importance of the volunteer with local knowledge should not be overlooked. They recommended the use of mentors to deal with the wide range of learners.

A project by Gleeson (2005) sought to explore the benefits of further training for adults with low levels of numeracy. Statistical modelling was applied to the 1975 cohort of the Longitudinal Survey of Australian Youth (LSAY) and the 1979 cohort of the American National Longitudinal Survey of Youth (NLSY79). Analysis of the LSAY cohort indicated that individuals with low numeracy skills were less likely to receive further education, but gained positive and significant economic returns when they did. Continuing with formal schooling was positive and significant for individuals with very low numeracy skills.

A study by Balatti, Black, and Falk (2006) was concerned with the changes in students' connections with people and the social capital outcomes experienced by students undertaking numeracy or literacy courses were examined. This qualitative study involved interviews with 57 students and 18 teachers in the states of Northern Territory, Queensland, and New South Wales in Australia. Data were coded for the presence of social capital based on the Australian Bureau of Statistics (ABS) framework (ABS, 2004), and evidence of course participation exerting socio-cultural impacts. The findings showed that the outcomes were highly valued by students and teachers. Almost 80% gained social capital, that is, they found changes in network structures and transactions. The course impacted on their lives not by improved literacy and numeracy, but by social capital outcomes.

Marr and Hagston (2007) investigated how numeracy skills are learned in the workplace and how these skills are related to the workers' school learning experiences. They conducted semi-structured interviews with ten key stakeholders and in three worksites (an aged care facility, an automotive engineering manufacturer, and a sheet metal engineering factory). They confirmed the necessity for numeracy skills now and in the future; found workers competent but under-confident in numeracy related skills; found workers had a lack of understanding of the link between school mathematics and their paid work; and suggested training be developed and delivered by a team with adult numeracy expertise and local industry knowledge. They also found the term 'numeracy' to be problematic as it "conveys a narrow picture of basic number calculations" (Marr & Hagston, 2007, p. 8) and suggested it be extracted from the LLN (language, literacy and numeracy) acronym at the policy level, and that the term be unpacked to explicitly extend the scope and breadth of numeracy from policy statements down to training packages.

Zevenbergen (2004, 2005) and Zevenbergen and Zevenbergen (2004) reported on the results of a three year project in which numeracy practices of young people in the Australian workplace were examined. It was suggested that there was little research on 'multi-numeracies' and the following question was posed: "Are there new ways of working mathematically brought about by the use of technology?" (Zevenbergen, 2004, p. 99). In the study, the ways young people (22 years or less) perceived and worked mathematically in the workplace, compared to more senior people (older than 40), were examined. Workers involved included: a bricklayer, boat-builder, hairdresser, motor mechanic, retail assistant, baker, builder, sign writer, painter, printer, room service employee, laundry attendant, lab technician, receptionist, short order cook, and chef. In total 880 participants were surveyed, work-shadowed, or interviewed using stimulus recall. Results suggested younger and older people have a "different orientations towards mathematics in the workplace" (p. 107). Senior people see the value of mental calculations, whereas younger people see this as a menial task best undertaken using technology. The workers' tasks then revolve around more the important skills of problem solving and estimation. Zevenbergen (2005) suggested there was evidence of dematheматization of the workplace in many sites.

In a large Australian study, the Commonwealth Bank Foundation (2005) investigated the links between financial literacy and individual outcomes. This study initially included a telephone survey of 5000 individuals aged between 16 and 65, who were asked questions about such things as how to choose the best way to reduce credit card interest (82% correct) or if they understood how to control mobile phone bills (18% correct). In follow-up stages, the micro and macro economic effects of improving financial literacy were investigated using hypothetical scenarios and modelling. The authors concluded that if the financial literacy of the 10% of the Australian population with the lowest scores was increased over a ten-year period, Australia's GDP would increase by $6 billion per year and 16,000 new jobs would be created.

In a case study of the numeracy skills of adults with Downs Syndrome, Faragher and Brown (2005) attempted to identify links between numeracy and the quality of life model (personal contexts, variability, life-span perspectives, values, choices and personal control, perception, and self image). Results demonstrated numeracy competencies in six areas of the quality of life model. Questions as to whether Downs Syndrome adults should learn the mathematics associated with aspects of life (e.g., knowing the bowling score and then knowing they lost); how to prepare a numeracy development plan and how to tailor it, were posed.

While there were a number of conference papers and journal articles on adult numeracy in the period 2004 to 2007, many of them did not report on actual research, or the research was framed loosely. In particular, Inglis (2005), in a plenary in the Adults Learning Maths Conference in Goteborg, highlighted numeracy demands in plumbing. Research however was not the focus of this paper. Similarly, Lee (2006) reported on investigations of numeracy in hairdressing and beauty therapy that were conducted through workshops in a learning centre at an institute of technology in New Zealand. Feedback from the workshops indicated improved student attitudes towards mathematics, and confidence with numeracy associated with tills, pricing guns, and diluting solutions. In other papers only descriptions of numeracy in particular contexts were provided including, for example, Faine (2006) on saving house water, and Porter (2006) on statistics in marine studies.

In 1996, Australia and New Zealand were part of an international survey of aspects of literacy. In the second half of 2006 this was followed by a second wave of data collection under the auspices of the Adult Literacy and Life Skills Survey (ALLS). The latter survey, of about 9000 persons in Australia and 7000 in New Zealand, targeted more aspects of numeracy than did the 1996 survey. One of the objectives was to assess the skills of adults in four areas: prose, document literacy, numeracy, and problem solving and analytical reasoning. It also aimed to identify students at risk with low level skills, and to evaluate assistance programs. The results of the survey will be published in 2008 (ABS, 2007; Ministry of Education, 2006).

The studies on workplace and community reported in this section suggest a surge in Australian and New Zealand government led research. In particular the research by NCVER to 2007 indicated increased interest in the area. Whether this research will have an impact on teaching practices and curriculum design is unclear. However, funding for the adult literacy and numeracy research through NCVER ceased from March 2007 and the impact of such cuts on further research is uncertain.

ADULTS IN FURTHER STUDIES

Adults in further studies can be enrolled in a range of different types of courses. Although we acknowledge that overlap exists between the contents of this chapter and the review of tertiary learners of mathematics elsewhere in this book, we have restricted the review to students enrolled in two types of programs. The first type refers to courses that lead to university entrance of some form. These are routinely called bridging, enabling, or occasionally developmental courses. Here this will primarily involve students who can be called second chance students. This means that they are adults returning to study at the tertiary level of education with limited backgrounds in mathematics (Miller-Reilly, 2006). The second type includes courses, programs and strategies that are designed to support students whilst enrolled in university studies. These can take the form of consultations in learning centres or interventions embedded within traditional university courses. Care is taken within these two categories not to intersect with research available on the traditional learning and teaching of mathematics in undergraduate courses and is thus focussed on mature students in university study rather than recent school leavers.

Although there are a significant number of courses across Australasia within the VET or Technical and Further Education (TAFE) sectors, the paucity of research on learning and teaching within these programs precludes their inclusion in this section, with the exception of one article. In this, Cooper, Baturo, Ewing, Duus, and Moore (2007) used an action-research methodology to investigate how mathematically-underachieving Torres Strait Islander students could meet the requirements of a block-laying course.

Learning and teaching within bridging mathematics

Bridging mathematics has been a focus within mathematics education since the early 1990s. Wood (2004), in the previous MERGA four-yearly review, included a brief section on research in this area. In a recent review of research, Taylor and Galligan (2006b) indicated that there have been four broad areas of research within the bridging mathematics community, namely: evaluation of specific courses, tests and other methods of establishing student needs, ways of overcoming mathematics anxiety, and what are the important cognitive differences between bridging mathematics and traditional mathematics teaching. In the years 2004-2007, the types of research undertaken within the bridging mathematics sector changed little.

Miller-Reilly (2006), in a significant long term analysis of two large bridging courses and one individual bridging program in New Zealand, used quantitative and qualitative measures to compare students' reactions. These multiple strands of evidence provided a complex overall picture of three largely successful teaching approaches. A one-to-one supervised course focused on understanding fear of mathematics and early mathematics experiences. The course empowered the student who came to believe that mathematics was a creative and enjoyable process. A second course (100 students) focused on the mathematization of realistic situations. Here, students came to regard mathematics as useful, interesting, and relevant to real life. The third course (100 students) was a carefully structured re-introduction of mathematics. The students appreciated the course and were pleased that they could now do mathematics that they could not do in school. Students in all programs were highly motivated, mature, and had not seen formal mathematics for some years. One surprising result from the study was that if students were unsuccessful they were, in fact, worse off than before, and often confused. A significant component of the study was the focus placed on dealing with students' mathematics anxiety or fear. Quantitative measures and qualitative descriptors indicated a decrease in mathematics anxiety throughout the duration of the three programs, and in the larger courses this correlated with achievement. Beliefs about mathematics in general, however, did not necessarily change, although students in the larger courses did see the practical nature of mathematics.

The importance of attitudes and beliefs about mathematics was taken up by a number of other researchers within the bridging mathematics community. In an analysis of the written reflections of mature students within a preparatory course, Viskic and Petocz (2006) found that although their students' mathematical abilities varied, many showed a maturing understanding of mathematics and its place in their own self development and in various aspects of their lives, including their working life, intellectual life, and civic life. In particular, they could categorise students' beliefs into 'component', 'models', 'life' and 'techniques' previously defined by Reid, Petocz, Smith, Wood, and Dortins (2003). Overall the students' reflections on projects they had to complete indicated that although initially surprised by the project's connection with mathematics, there was growth in their awareness of the usefulness of mathematics.

In a quantitative approach, Klinger (2006) investigated attitudes, self-efficacy beliefs, and mathematics anxiety of a diverse group of pre-tertiary adult learners using a self-designed survey and pre- and post-testing. He found that although students initially shared the negative views reported in the wider population, these views changed significantly by the conclusion of the topic(s). He concluded that adults' perceptions of, and capacity to engage with, mathematical content was strongly influenced by early learning experiences, but like authors before, confirmed that these could be changed. In a later study, Klinger (in press) repeated the analysis with commencing arts, humanities, and science students and found the arts/humanities students to have significantly lower mathematical self-efficacy and poorer attitudes than the science students. The results were strongly linked with gender, with females scoring lower than males. He concluded that these kinds of results indicated that undergraduate support programs would have to address such poor attitudes to enable students to succeed and develop mathematical literacy necessary for their discipline.

In two similar quantitative studies underpinned by Bandura's social cognitive theory (Carmichael, Dunn, & Taylor, 2006; Carmichael & Taylor, 2005), motivation (particularly self-efficacy) and other factors such as prior knowledge were examined as measures of prediction of success in a university preparatory mathematics course. Results suggested that students' assessment of their confidence was based, in part, on their current levels of knowledge and skills, and that this assessment influenced their current performance. Females and those who had been away from study for a long period of time had lower entering levels of confidence and self-efficacy than males and recent school leavers, but final performance in the course was not significantly different.

Taylor and Galligan (2006a) used conversation theory framework to design a multimedia package to simultaneously develop students' mathematical skills while improving mathematics confidence. The evaluation of this program indicated the effectiveness of this approach. It is apparent through the work discussed above that although attitudes and beliefs about mathematics are important for students enrolled in bridging programs, the programs can change students' attitudes and beliefs about mathematics as well as their achievement. An alternative to the above is to design programs that can assist in self-selection. Egea (2004) has designed and evaluated an online self assessment tool that assists students to choose the appropriate level of bridging mathematics program. It has been specifically designed with mathematics anxious students in mind, and participating students indicated that the test was 'enjoyable', 'fun', 'easy', and even 'relaxing', confirming that its design had addressed both the cognitive and affective domains.

Bridging or foundation programs are widely available across Australasia. However, few research results have been published on the learning and teaching within these programs. Galligan (2004b) described one such program for on-campus international students, highlighting the added difficulties for students whose first language was not English. Significant difficulties are associated with English language proficiency and the role that language plays in bi-lingual students' processing of mathematics (Galligan, 2004a). Galligan (2005, 2007) reported further on the difficulties associated with extending such programs to offshore delivery. Using an ethnographic approach, she investigated potential conflicts in a mathematics foundation classroom in Hong Kong, and analysed the data using Valsiner's Zone theory. Although the emerging themes of teaching, student, curriculum issues, outside influences, and use of English, would not be surprising to anyone working in the field, the methodology used in the study allowed for greater understanding to be developed between teacher, student and program provider. Birney (2006) briefly described what numeracy based activities worked with Melanesian adults living in Papua New Guinea. One such activity was learning the decimal place system. Here the students learnt by building on their traditional language and the concept of number embedded in that language.

Mathematics support within the university sector

For some years, universities across Australia have supported the mathematical learning and development of undergraduate students. They do this through a variety of methods and strategies. This type of support is often offered to all students, not just students who are enrolled in traditional mathematics courses. Traditionally the support is distinct from the teaching of such courses, but increasingly such support is integrated into the teaching (e.g., Taylor, Mander, & McDonald, 2004). Research on the teaching and learning within this integrated support is rare, but growing. In this section the focus is on research on support courses rather than studies associated with the teaching of undergraduate mathematics courses, which are reviewed by Wood elsewhere in this book. In most instances, research on support courses focuses on mature aged university students rather than recent school leavers. Leder and Forgasz (2004), in their research on the daily challenges of mature students of mathematics, recommended that identification of required prerequisite skills early in students' studies may bridge perceived deficiencies. Research in that area is growing, as typified by the work of Egea, Dekkers, and Flanders (2004) and Wilson and MacGillivray (2007). Carmody and Wood (2006) describe a peer-tutoring system in which third year students are used to support the development of the mathematics and statistics skills of first year students. Benefits were reported not only to the first year students but also to the peer tutors in terms of increasing their own understanding of mathematics.

Although support programs focus on a range of disciplines, current research was apparent only within nursing. Two studies were identified in the area of drug calculations. Behrend, Clark, Hall, and Hill (2006) designed a pencil and paper test to diagnose numeracy skills and used the results to develop and evaluate a comprehensive one-stop shop to manage the safe administration of drugs. Gillies (2004) compared two methods of teaching drug calculation: the traditional formula method, and building on students' existing mathematical problem solving skills. Using a pre-/post-test model, performance in drug calculations was better when using the application of formulas than when allowing students to employ a problem solving strategy of their choice from a suite of strategies explored in classes. However, the groups who were taught a variety of problems solving strategies demonstrated increased confidence, and better long-term recall and transfer of skills.

More general learning skills were researched in two other instances. Taylor and McDonald (2007) investigated the use of writing to improve non-routine problem solving and mathematical communication among students enrolled in a service mathematical course enhanced by mathematics support staff. Results indicated that both communication and problem solving improved after introducing a writing heuristic into group problem solving sessions. Kemp (2005), in an extensive study of critical numeracy at the tertiary level, initially developed a qualitative instrument to measure the ability of students to read and interpret tables of data, and then used the instrument to diagnose and thus address students' table reading behaviour. Through a workshop-based intervention, she investigated the efficacy of teaching table reading and interpreting skills. Overall Kemp (2005) concluded that table reading and interpreting skills can be taught and learned, and that students should not be assumed to be able to perform these tasks without instruction. Finally Winn and Keuskamp (2007) leant heavily on the work of Kemp (2005) and a pattern-based approach to provide a strategy for developing problem solving skills for university students requiring support for the development of academic numeracy skills. Their "getting started" strategy should provide a useful device for staff to support such students.

CONCLUSION

Adults studying in formal or informal numeracy programs learn more than mathematics skills. They encounter rich embedded numeracy tasks, whether they are involved in spraying chemicals, analysing options to buying a mobile phone, or studying/preparing for university courses with embedded mathematics (e.g., drug calculations).

Research on numeracy in the workplace suggests that as the workforce is becoming demathematized, numeracy is becoming more hidden in complex tasks. Teachers, it appears, are ill equipped to deal with these complexities and have few resources at their disposal. Strong themes related to pedagogy emerged from this section of the review. For example, there are consistent, strong calls for the development of initial training and professional development of teachers in numeracy (Benseman et al., 2005a). Another example is that no mechanism currently exists to accurately quantify the language, literacy, and numeracy teaching workforce in Australia (Mackay, Burgoyne, Warwick, & Cipollone, 2006). Also, there is an urgent need for mandatory specialised preparation and/or professional development of teachers and trainers involved in adult numeracy, even if this is not their principal role (FitzSimons, 2005a), although there does appear to be a move in the direction of "improving the skills and strategies of foundation skills tutors" (Benseman & Sutton, 2007, p. 26).

In an NCVER report on the professional development needs of the literacies workforce (Mackay et al., 2006), it was found that the sector had high levels of casualization (i.e. the employment of sessional or temporary staff), with limited access to resources and learning materials. FitzSimons (2005b) says that such numeracy teaching requires a "fundamentally different curriculum and pedagogy from that of school mathematics" (p. 39), a sentiment reflected by McKenna and Fitzpatrick (2004). Marr (2006) described a model of cooperative collaboration between herself, a researcher and a group of seven practicing adult numeracy teachers which appeared effective in developing a holistic adult numeracy assessment resource.

Research on adults returning to study mathematics in the tertiary sector has some similarities to those occurring in the workplace and community, although there is little to no discussion of teaching practices. The focus is often on the students and what they bring with them, rather than the pedagogical practices they experience, although there are a growing number of such studies. Taylor and Galligan (2006b) listed four key unanswered questions within bridging or enabling mathematics, including:

> How is success defined in bridging mathematics activities? What are the numeracy demands on entry to 'non-mathematical' university study? Are successful bridging mathematics students, successful university students? Is there more than mathematics? (connection of mathematics skills with other academic skills, e.g., language and learning skills). (p. 18)

With the exception of some key research projects, most of the research explored in this review primarily focused on the descriptive. Only in some recent cases in adult numeracy literature for the workforce and the community have more sophisticated methodologies been adopted in the research (e.g., work-shadowing, stimulated recall). In only a few papers was the discussion situated within theoretical frameworks, such as Bernstein's horizontal and vertical discourse, activity theory, conversation theory or Valsiner's zone theory. This may be a result of the high teaching demands placed on many of the authors highlighted in this review which leave little room for research and publication. The resurgence of government funding for workplace and community research associated with literacy assisted in the development of numeracy research, but in Australia one of the major sources for this funding has now been cut. Funding is also not easily accessible in the tertiary sector and this has resulted in the stagnation of research in this area, after a flurry of activity in the early 1990s. Clearly, adults returning to study mathematics is still an under-theorized research domain, with little activity in recent years to move the domain to the next level of research. Zevenbergen (2004) takes this criticism even further in her calls to re-theorize our understandings of knowledge, skills, and dispositions to apply to 21st century needs.

REFERENCES

Australian Association of Mathematics Teachers (AAMT). (1998). Policy on Numeracy Education in Schools. Adelaide: AAMT, Retrieved December 12, 2007 from: www.aamt.edu.au/content/download/724/19518/file/numpol.pdf

Australian Bureau of Statistics (ABS). (2004). *Measuring social capital: An Australian framework and indicators, information paper* 1378.0. Retrieved August 8, 2007 from: http://www.abs.gov.au.

Australian Bureau of Statistics (ABS). (2007). *Adult literacy and life skills survey*. Retrieved April 27, 2007 from: http://www.connected.gov.au/case_studies/adult_literacy_and_lifeskills_survey

Balatti, J., Black, S., & Falk, I. (2006). *Reframing adult literacy and numeracy course outcomes: A social capital perspective*. Adelaide: NCVER.

Behrend, M., Clark, R., Hall, C., & Hill, P. (2006). Online numeracy resources in the context of safe administration of medicines (SAM). In M. Horne & B. Marr (Eds.), *Connecting voices:*

Practitioners, researchers & learners in adult mathematics and numeracy (Proceedings of the 12th International Conference on Adults Learning Mathematics (ALM), pp. 61-68). Melbourne: ACU National.

Benseman, J., & Sutton, A. (2007). *OECD/CERI formative assessment project background report: New Zealand.* Retrieved January 7, 2008 from: http://www.oecd.org/dataoecd/53/58/38205154.pdf

Benseman, J., Sutton, A., & Lander, J. (2005a). *Working in the light of evidence: A literature review of the best available evidence about effective adult literacy, numeracy and language teaching.* Auckland: The University of Auckland & University Services Ltd. Retrieved April 17, 2007 from: http://www.minedu.govt.nz/web/downloadable/dl10603_v1/fl-literature-review-final-document-jan-2005-unis.doc

Benseman, J., Sutton, A., & Lander, J. (2005b). *Pedagogy in practice: An observational study of literacy, numeracy and language teachers.* Auckland: The University of Auckland & University Services Limited. Retrieved January 07, 2007 from: http://www.minedu.govt.nz/index.cfm?layout=index&indexid=11194&indexparentid=1028

Birney, T. (2006). Adult learning and oral culture. In M. Horne & B. Marr (Eds.), *Connecting voices: Practitioners, researchers & learners in adult mathematics and numeracy* (Proceedings of the 12th International Conference on Adults Learning Mathematics (ALM), pp. 69-75). Melbourne: ACU National.

Carmichael, C., & Taylor, J. A. (2005). The analysis of student beliefs in a tertiary preparatory mathematics course. *International Journal of Mathematical Education in Science and Technology, 36*(7), 713-719.

Carmichael, C., Dunn, P., & Taylor, J. A. (2006). Identifying at-risk students: Is it possible in a tertiary preparation course for adults? In P. Grootenboer, R. Zevenbergen, & M. Chinnappan (Eds.), *Identities, cultures and learning spaces* (Proceedings of the 29th annual conference of the Mathematics Research Group of Australasia, Canberra, pp. 107-114). Adelaide: MERGA.

Carmody, G., & Wood, L. (2006). Bridging the gap for first year students with the use of peer tutoring. In M. Horne & B. Marr (Eds.), *Connecting voices: Practitioners, researchers & learners in adult mathematics and numeracy* (Proceedings of the 12th International Conference on Adults Learning Mathematics (ALM), pp. 76-81). Melbourne: ACU National.

Coben, D. (2006). Surfing the adult numeracy wave: What can we learn from each other in the UK and Australia? In M. Horne & B. Marr (Eds.), *Connecting voices: Practitioners, researchers & learners in adult mathematics and numeracy* (Proceedings of the 12th International Conference on Adults Learning Mathematics (ALM), pp. 82-87). Melbourne: ACU National.

Commonwealth Bank Foundation. (2005). *Improving financial literacy in Australia: Benefits for the individual and the nation.* Melbourne: Monash University.

Condelli, L. (2006). *A review of the literature in adult numeracy: Research and conceptual issues.* Washington, DC: The American Institutes for Research. (ERIC Document Reproduction Service No. ED495456).

Cooper, T. J., Baturo, A. R., Ewing, B., Duus, E., & Moore, K. (2007). Mathematics education and Torres Straight Islander blocklaying students: The power of vocational context and structural understanding. In J. H. Woo, H. C. Lew, K. S. Park, & D. Y. Seo, (Eds.), *Proceedings of the 31st annual conference the Psychology of Mathematics Education* (Vol 2, pp. 177-184). Seoul, South Korea: Seoul National University.

Cumming, J. J., & Wilson, J. M. (2005). *Literacy and numeracy and alternative dispute resolution.* Adelaide: NCVER.

Egea, K. (2004). A method to advise students of their mathematical currency. *Australian Journal of Adult Learning, 44*(3), 389-412.

Egea, K., Dekkers, A., & Flanders, M. (2004). *An undergraduate course design to address the limitations in mathematical knowledge of entry level students in Information Technology.* Paper presented at the Building Foundations 2004 National Conference of Enabling Educators, Newcastle, NSW. Retrieved December 17, 2007 from: http://www.pco.com.au/Foundations04/

Ellerton, N. F., Clarkson, P. C., & Clements, M. A. (2000). Language factors in mathematics teaching and learning. In K. Owens & J. Mousley (Eds.), *Research in mathematics education in Australasia 1996-1999* (pp. 29-95). Sydney: MERGA.

Faine, H. (2006). Numeracy down the drain: Adult education explores house water expert. In M. Horne & B. Marr (Eds.), *Connecting voices: Practitioners, researchers & learners in adult mathematics and numeracy* (Proceedings of the 12th International Conference on Adults Learning Mathematics (ALM), pp. 104-109). Melbourne: ACU National.

Faragher, R., & Brown, R. I. (2005). Numeracy for adults with Downs syndrome: It's a matter of quality of life. *Journal of Intellectual Disability Research, 49*(10), 761-765.

FitzSimons, G. E. (2005a). Adult numeracy in Australian workplaces. In L. Lindberg (Ed.), *"Bildning" and/or training* (Proceedings of the 11th international conference on Adults Learning Mathematics (ALM), pp. 79-85). Kungälv, Sweden: Goteborg University.

FitzSimons, G. E. (2005b). Numeracy and Australian workplaces: Findings and implications. *Australian Senior Mathematics Journal, 19*(2), 27-40.

FitzSimons, G. E. (2005c). Technology mediated post-compulsory mathematics: An activity theory approach. *International Journal of Mathematical Education in Science and Technology, 36*(7), 769-777.

FitzSimons, G. E. (2005d). Can adult numeracy be taught? A Bernsteinian analysis. In M. Goos, C. Kanes, & R. Brown (Eds.), *Mathematics education and society* (Proceedings of the 4th International Mathematics Education and Society Conference, pp. 155-165). Brisbane: Griffith University, Centre for Learning Research.

FitzSimons, G. E. (2006). *Towards a definition of numeracy: A work in progress.* Retrieved August 9, 2007 from: http://www.statvoks.no/emma/index_numeracy_gail_middle.html

FitzSimons, G. E., & Coben, D. (in press). Adult numeracy for work and life: Curriculum and teaching implications of recent research. In R. Mclean & D. Wilson (Eds.), *International handbook of technical and vocational education and training*. Dordrecht, The Netherlands: Springer-Verlag.

FitzSimons, G. E., Coben, D., & O'Donoghue, J. (2003). Lifelong mathematics education. In A. J. Bishop, M. A. Clements, C. Keitel, J. Kilpatrick, & F. K. S. Leung (Eds.), *International handbook of mathematics education* (pp. 105-144). Dordrecht, The Netherlands: Kluwer Academic Publishers.

FitzSimons, G., & Godden, G. (2001). Review of research on adults learning mathematics. In D. Coben, J. O'Donoghue, & G. FitzSimons (Eds.), *Perspective on adults learning maths* (pp. 13-45). Dordrecht, Netherlands: Kluwer.

FitzSimons, G., Mlcek, S., Hull, O., & Wright, C. (2006). *Learning numeracy on the job: A case study of chemical handling and spraying*. Adelaide: NCVER.

FitzSimons, G., & Wedege, T. (2004). *Developing numeracy in the workplace: The context of chemical spraying*. Paper presented at Workplace learning from the learner's perspective. Copenhagen.

Galligan, L. (2004a). The role of language switching in bi-lingual students' processing of mathematics. *Paper presented at the 10th International Congress on Mathematics Education*, Copenhagen, Denmark. Retrieved October 12, 2004 from: http://www.icme-organisers.dk/tsg25/subgroups/galligan.doc.

Galligan, L. (2004b). Preparing international students for university: Mathematics as part of an integrated program. *Australian Senior Mathematics Journal, 18*(1), 28-41.

Galligan, L. (2005). Conflicts in offshore learning environments of a university preparatory mathematics course. In H. L. Chick & J. L. Vincent (Eds.), *Proceedings of the 29th annual conference for the Psychology of Mathematics Education* (Vol. 3, pp. 25-32). Melbourne: University of Melbourne.

Galligan, L. (2007). *Shared understanding of students' approach to learning in transnational programs.* (Working paper No. 18). AIS St Helens, Auckland. Retrieved 1 February 2008 from http://www.crie.org.nz/research_paper/research.html.

Galligan, L., & Taylor, J. A. (2005). Investigating academic numeracy in non-mathematics courses at university. In L. Lindberg (Ed.), *"Bildning" and/or training* (Proceedings of the 11th International

Conference on Adults Learning Mathematics (ALM), pp. 86-95) Kungälv, Sweden: Goteborg University.

Gillies, R. (2004). Numeracy for nurses: The case for traditional versus non-traditional methods for teaching drug calculation. In I. Putt, R. Faragher, & M. McLean (Eds.), *Mathematics education for the third millennium: Towards 2010* (Proceedings of the 27th annual conference of the Mathematics Education Research Group of Australasia, Townsville, Vol. 1, pp. 255-262). Adelaide: MERGA.

Gleeson, L. (2005). *Economic returns to education and training for adults with low numeracy skills*. Adelaide: NCVER.

Hartley, R., & Horne, J. (2005). *Social and economic benefits of improved adult literacy: Towards a better understanding: Support document*. Adelaide: NCVER. Retrieved December 17, 2007 from http://www.ncver.edu.au/research/proj/nr4L06s.doc

Hartley, R., & Horne, J. (2006). Researching literacy and numeracy costs and benefits: What is possible? In M. Horne & B. Marr (Eds.), *Connecting voices: Practitioners, researchers & learners in adult mathematics and numeracy* (Proceedings of the 12th International Conference on Adults Learning Mathematics (ALM), pp. 136-142). Melbourne: ACU National.

Hayes, C., Golding, B., & Harvey, J. (2006). Finding the fire: Numeracy and literacy for public safety volunteers. In M. Horne & B. Marr (Eds.), *Connecting voices: Practitioners, researchers & learners in adult mathematics and numeracy* (Proceedings of the 12th International Conference on Adults Learning Mathematics (ALM), pp. 143-149). Melbourne: ACU National.

Inglis, M. (2005). "I said I taught the horse how to sing; I didn't say it learnt": Issues underlying appropriate training for vocational uses of mathematics. In L. Lindberg (Ed.), *'Bildning' and /or training* (Proceedings of the 11th International Conference on Adults Learning Mathematics (ALM), pp. 28-41). Kungälv, Sweden: Goteborg University.

Johnston, B., & Maguire, T. (2005). *Adult numeracy: Policy and practice in global contexts of lifelong learning*. Unpublished manuscript. Retrieved June 26, 2006 from: http://www.staff.vu.edu.au/alnarc/revamp/publications/05bettyjohns.pdf

Kemp, M. (2005). *Developing critical numeracy at the tertiary level*. Unpublished PhD Thesis, Murdoch University, Australia. Retrieved June 22, 2006 from: http://wwwlib.murdoch.edu.au/adt/browse/view/adt-MU20060831.171947

Klinger, C. (2006). Challenging negative attitudes, low self-efficacy beliefs, and math-anxiety in pre-tertiary adult learners. In M. Horne & B. Marr (Eds.), *Connecting voices: Practitioners, researchers & learners in adult mathematics and numeracy* (Proceedings of the 12th International Conference on Adults Learning Mathematics (ALM), pp. 164-171). Melbourne: ACU National.

Klinger, C. (in press). On mathematics attitudes, self-efficacy beliefs, and math-anxiety in commencing undergraduate students. In *Crossing boarders – Research, reflection and practice* (Proceedings of the 13th International Conference on Adults Learning Mathematics (ALM)). Belfast, N. Ireland.

Leder, G., & Forgasz, H. (2004). Australian and international mature students: The daily challenges. *Higher Education Research & Development, 23*(2), 183-198.

Lee, C. (2006). Mathematics in hair and beauty: Staying in business. In M. Horne & B. Marr (Eds.), *Connecting voices: Practitioners, researchers & learners in adult mathematics and numeracy* (Proceedings of the 12th International Conference on Adults Learning Mathematics (ALM), pp. 188-194). Melbourne: ACU National.

Mackay, S., Burgoyne, U., Warwick, D., & Cipollone, J. (2006). *Current and future professional development needs of the language, literacy and numeracy workforce*. Adelaide: NCVER.

Marr, B. (2006). Creative conversations: Connecting voices of teachers and researchers to rethink assessment. In M. Horne & B. Marr (Eds.), *Connecting voices: Practitioners, researchers & learners in adult mathematics and numeracy* (Proceedings of the 12th International Conference on Adults Learning Mathematics (ALM), pp. 213-219). Melbourne: ACU National.

Marr, B., & Hagston, J. (2007). *Thinking beyond numbers: Learning numeracy for the future workplace*. Adelaide: NCVER. Retrieved January 8, 2008 from: http://www.ncver.edu.au/publications/1795.html

McKenna, R., & Fitzpatrick, L. (2004). *Building sustainable adult literacy provision.* Adelaide: NCVER.

Miller-Reilly, B. (2006). *Affective change in adult students in second chance mathematics courses: Three different teaching approaches.* Unpublished PhD thesis, University of Auckland, New Zealand.

Ministry of Education (2006). *The adult literacy and life skills (ALL) survey.* Retrieved January 08, 2008 from: http://www.educationcounts.govt.nz/themes/research/all/all_2006

National Centre for Vocational Education Research (NCVER). (2003). *NCVER synthesises recent adult literacy research.* Retrieved April 27, 2007 from: http://www.ncver.edu.au/newsevents/insight/issue19/adlit.html

Neill, W. A. (2001). *The essentials of numeracy.* Paper presented at the New Zealand Association of Researchers in Education conference, Christchurch, New Zealand. Retrieved January 12, 2007 from: http://www.nzcer.org.nz/pdfs/10604.pdf

New Zealand Council for Educational Research (NZCER). (2006). *Assessment for foundation learning. Final report.* Wellington, New Zealand: New Zealand Council for Educational Research. Retrieved January 7, 2008 from: http://www.minedu.govt.nz/web/downloadable/dl11359_v1/assessment-study.pdf

Perkins, K. (2005). *Reframe, rename, revitalise. Future directions for the language, literacy and numeracy national reporting system.* Adelaide: NCVER.

Porter, S. (2006). Statistics of marine studies: The issues in teaching statistics to students of mixed ages and backgrounds. In M. Horne & B. Marr (Eds.), *Connecting voices: Practitioners, researchers & learners in adult mathematics and numeracy* (Proceedings of the 12th International Conference on Adults Learning Mathematics (ALM), pp. 226-231). Melbourne: ACU National.

Reid, A. Petocz, P., Smith, G. H., Wood, L. N., & Dortins, E. (2003). Maths students' conceptions of mathematics. *New Zealand Journal of Mathematics, 32*(supplementary issue), 163-172.

Taylor, J. A., & Galligan, L. (2006a). Mathematics for maths anxious tertiary students: Integrating the cognitive and affective domains using interactive multimedia. *Literacy and Numeracy Studies, 15*(1), 23-42.

Taylor, J. A., & Galligan, L. (2006b). Research into research on adults in bridging mathematics: The past, the present and the future. In M. Horne & B. Marr (Eds.), *Connecting voices: Practitioners, researchers & learners in adult mathematics and numeracy* (Proceedings of the 12th International Conference on Adults Learning Mathematics (ALM), pp. 11-19). Melbourne: ACU National.

Taylor, J. A., & McDonald, C. (2007). Writing in groups as tool for non-routine problem solving in first year university mathematics. *International Journal of Mathematics Education in Science and Technology, 38*(5), 639-655.

Taylor, J. A., Mander, D., & McDonald, C. (2004). Transition to engineering mathematics: Issues and solutions. In C. Snook & D. Thorpe (Eds.), *Creating flexible learning environments* (Proceedings of the 15th Australasian Conference for the Australasian Association for Engineering Education, pp. 164-172). Toowoomba: USQ.

Viskic, D., & Petocz, P. (2006). Adult students' views of mathematics: Reflections on projects 6. *Adults Learning Mathematics – An International Journal, 1*(2), 6-15. Retrieved on December 17, 2007 from: http://www.alm-online.net/images/ALM/journals/almij_volume1_2_feb2006.pdf

Wickert, R., & McGuirck, J. (2005). *Integrating literacies: Using partnerships to build literacy capabilities in communities.* Adelaide: NCVER.

Wilson, T., & MacGillivray, H. (2007). Counting on the basics: Mathematical skills among tertiary entrants. *International Journal of Mathematical Education in Science & Technology, 38*(1), 19-41.

Winn, T., & Keuskamp, D. (2007). Communicating concepts of academic numeracy through a pattern-based approach. *Journal of Academic Language and Learning, 1*(1), 110-112. Retrieved January 9, 2008 from: http://www.aall.org.au/journal/ojs/index.php/jall/article/view/26

Wood, L. (2004). University mathematics learning and teaching. In B. Perry, G. Anthony, & C. Diezmann (Eds.), *Research in Mathematics education in Australasia 2000-2003* (pp. 219-242). Flaxton: Post Press.

Yasukawa, K., & Johnston, B. (1994). A numeracy manifesto for engineers, primary teachers, historians … a civil society – Can we call it theory? *Proceedings of the Australian Bridging Mathematics Network Conference* (pp. 191-199). Sydney: University of Sydney.

Zevenbergen, R. (2004). Technologizing numeracy: Intergenerational differences in working mathematically in new times. *Educational Studies in Mathematics*, *56*(1), 97-117.

Zevenbergen, R. (2005). Prioritising the voice of the researched: Using photographs to elicit mathematical thinking of participants. In P. Clarkson, A. Downton, D. Gronn, M. Horne, A. McDonough, R. Pierce, & A. Roche (Eds.), *Building connections: Research, theory and practice* (Proceedings of the 28th annual conference of the Mathematics Education Group of Australasia, Melbourne, Vol. 2, pp. 815-822). Sydney: MERGA.

Zevenbergen, R., & Zevenbergen, K. (2004). *The numeracies of boat building*. Paper presented at the 10th International Congress on Mathematics Education, Copenhagen, Denmark. Retrieved December 17, 2006 from: http://www.icme-organisers.dk/tsg07/ZEVENBERGEN.pdf

AFFILIATIONS

Linda Galligan
The University of Southern Queensland, Australia

Janet A. Taylor
The University of Southern Queensland, Australia

TAMSIN MEANEY, COLLEEN MCMURCHY-PILKINGTON, & TONY TRINICK

MATHEMATICS EDUCATION AND INDIGENOUS STUDENTS

Indigenous students are often classified as disadvantaged learners of mathematics and this is manifested in significant under-achievement. In this chapter we have explored four types of interactions – societal, community, teacher, and student – that we feel contribute to the positioning of Indigenous students as knowers, doers, and learners of mathematics. Using the ideas of Sfard and Prusak (2005) in regard to how stories become students' identities, we show how the interactions reported in the research contribute to Indigenous students being considered as disadvantaged. Alternative stories to those of disadvantage involve active mathematics program development with Indigenous communities.

Key words: Aboriginal, community involvement, disadvantage, ethnomathematics, identity, interactions, language, Māori

INTRODUCTION

In the Australasian region, Indigenous students are a significant proportion of students in mathematics classrooms. In the chapter on social justice and socio-cultural perspectives, in the last four-yearly MERGA review, Atweh, Meaney, McMurchy-Pilkington, Neyland, and Trinick (2004) argued that "Indigenous students are amongst the most disadvantaged in mathematics education" (p. 33). This is compounded by a history in which "Indigenous people's cultures were regarded as primitive and inferior. They were ignored and suppressed in formal schooling and the Western environment" (Paraide, 2005, p. 104). Consequently, there continues to be discussion about how Indigenous students can retain their indigenous identity as well as participating effectively in the mainstream society. McKinley (2004) stated:

> Māori are not interested in producing scientists or mathematicians who just happen to be Māori. The issue of who we are, our language and our knowledge are inextricably tied together and need to be considered together to inform any change. (p. 1)

H. Forgasz et al. (eds.), Research in Mathematics Education in Australasia 2004-2007, 119-139.

We propose that it is the interaction with and around Indigenous students that contribute to their positioning as disadvantaged knowers, doers and learners of mathematics. Therefore, we consider how four types of interactions – societal, community, teacher, and student – affect Indigenous students' current and future achievement in mathematics.

For this chapter, we have used Meaney's (2002) definition of Indigenous students as being those who belong to communities who originally controlled the land and developed distinctive cultures before the arrival of Europeans, but who are presently attending educational institutions which closely resemble those of industrial countries. Consequently, we have reviewed book chapters and articles with foci on Māori students in New Zealand and Aboriginal and Torres Strait Islander students in Australia, as well as Papua New Guinean and Pacific Islander students. We have not included research on students who come from Indigenous cultures but reside in other countries, such as Pacific Islander students living in New Zealand. Readers interested in research on these students could begin their investigation with Anthony and Walshaw (2007).

Since the last MERGA review of research (Perry, Anthony, & Diezmann, 2004), there has been a significant increase in the amount of research in which the issues around Indigenous students learning mathematics have been considered. This increase has been so large that we have not been able to include all reports. Concern for Indigenous students learning mathematics was also raised at two recent international conferences, the Fourth Mathematics Education in Society conference and The Third International Conference on Ethnomathematics. Both of these conferences were held in the Australasia region and had presentations that were relevant to Indigenous students in that region. As well, in New Zealand, the publication of the *Effective Pedagogy in Mathematics/Pāngarau: Best Evidence Synthesis Iteration* (BES) (Anthony & Walshaw, 2007) provided research evidence on "instructional practices that make a difference for diverse groups of learners" with a focus on the "mathematical development of Māori and Pasifika students" (p. xix).

Although we use the term Indigenous to describe a range of students, ideas that are applicable to one community will not be appropriate for another, even if they speak the same first language. Quite often the studies reviewed in this chapter are small, qualitative ones whose applicability for Indigenous students in the Australasia region as a whole is questionable. We are concerned about Rennie's (2006) caution that "myths are often perpetuated by overgeneralising about groups of people" (p. 124). Nevertheless, we are aware that the accumulation of these research stories does contribute to Indigenous and non-Indigenous educators' and others' perceptions of what 'an Indigenous mathematics student' is like.

BEING DISADVANTAGED MATHEMATICS LEARNERS

Disadvantage in mathematics education is manifested in the under-achievement of Indigenous students. Large international studies such as PISA (De Bortoli & Cresswell, 2004; Thomson & De Bortoli, 2008) and smaller studies such as that undertaken by the Queensland Studies Authority (2004) continue to report that Indigenous students in Australia fail to achieve at the same levels as their non-Indigenous peers. The latest PISA study showed that Indigenous students in Australia performed in the 2006 mathematical literacy assessments on average at levels two and a half school years behind their non-Indigenous peers (Thomson & De Bortoli, 2008). In New Zealand, results from the National Education Monitoring Project showed that Māori students in mainstream schools did not perform as well as their European background peers (Flockton, Crooks, Smith, & Smith, 2006). As well, Forbes (2005) found that "Maori students were markedly less likely than non-Maori students to continue in mathematics in all years" of senior high school and that "the accumulated mathematics attainment of Maori students, as a group, was less than two-thirds that of non-Maori students" (p. 3). In Samoa, Afamasaga-Fuata'i (2005) also deplored the under-achievement of students in mathematics "at all levels" (p. 147).

Various researchers hypothesise about the contribution of a broad range of factors to the under-achievement of Indigenous students in mathematics. As stated by Amosa and Cooper (2006) "[t]he question of 'what matters most' in terms of impact on student achievement has long been an issue of political and academic focus in Australia, as elsewhere" (p. 1). Nevertheless, there is a realisation that the teaching/learning process is extremely complex and it is a combination of factors that contribute to Indigenous students' under-achievement (Amosa & Cooper, 2006).

In order to make sense of the complexity of influences on under-achievement, we have drawn on the work of Sfard and Prusak (2005). They proposed that reifying, significant and endorsable stories about a person become his/her identity. Reifying stories are those that suggest that a certain characteristic of a student is an unchangeable component of their personality. In the articles reviewed in this chapter, the authors were considered to be providing reifying stories if they describe Indigenous students as *having* certain characteristics, such as "Indigenous students are disadvantaged". A story is significant "if any change in it is likely to affect the storyteller's feelings about the identified person" (Sfard & Prusak, 2005, pp. 16-17). Endorsable stories are those which the story teller repeats frequently as a true reflection of the situation and the participants within it. By reporting their research findings in academic journals or books, we presume that the authors believe the stories that they tell to be both significant and endorsable. These stories highlight the importance of who the authors are. Although non-Indigenous researchers still proliferate, there has also been an increase in the number of Indigenous researchers in the last few years.

Sfard and Prusak (2005) proposed that there are two kinds of stories: actual, that describe the person's current identity; and designated, that describe their future possibilities. Learning occurs when students attempt to close the gap between these two identities. In this chapter, we will consider the impact of four different types of interactions on the stories that are told about the current and future possibilities for the mathematics learning of Indigenous students as a whole, rather than as individuals. These stories suggest reasons for Indigenous students to be considered disadvantaged and also offer some hope of alternative futures for these students.

Figure 1 illustrates how we perceive the four types of interactions as influencing the positioning of the identity of Indigenous students as knowers, doers and learners of mathematics. The black arrows show how different interactions are connected to each other, whilst the white arrows refer to the effect on Indigenous students. The following discussion looks at each of the influences but in doing so we realise that we highlight some features whilst ignoring what may be equally important interpretations.

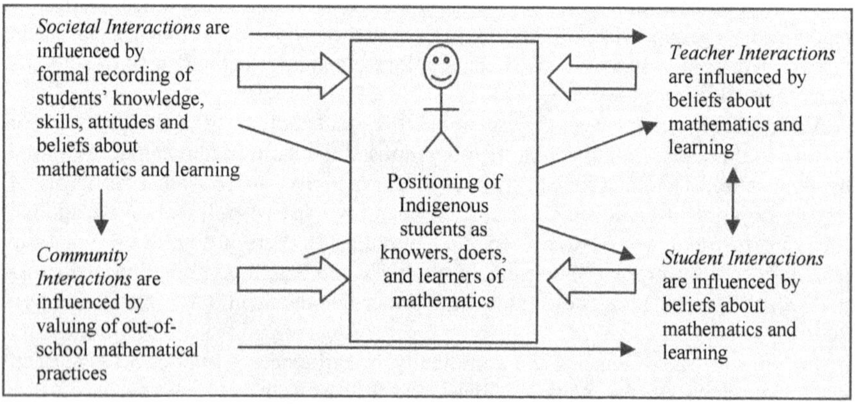

Figure 1: The influence of different kinds of interactions on the positioning of Indigenous students

Societal influences on Indigenous students' achievement in mathematics

Meaney (2004, p. 187) described societal relationships as those that position:

> different groups within a society as a consequence of the development of cultural norms over long periods of time… [and which] allow people to operate without having to consider the implications of every action or thought because their beliefs are ingrained as true representations of how the world is.

Although helpful in being able to effectively operate in society, these relationships are also difficult to identify and thus change. For example, a discussion on the under-achievement of Indigenous students requires an acceptance at the societal level of what it means to achieve in mathematics, as well as the most appropriate ways to identify achievement.

In the PISA 2003 study, mathematical literacy was measured by having students solve problems using pencil and paper (De Bortoli & Cresswell, 2004). However, as Young-Loveridge (2005) pointed out there are a number of ways of assessing students and these will give quite different pictures of how they are doing. She showed that Māori boys' performance would appear to improve dramatically if they were assessed through diagnostic interviews, as used in the New Zealand Numeracy Project, rather than with paper and pencil tests. Given that many Indigenous students are not learning in their first language or dialect, the impact of different assessment types may well affect Indigenous students' perceived performance. The choice of language of instruction is often a societal rather than a community decision. In Samoa, the teaching of mathematics switches from the vernacular language to English at the upper primary level. However, many teachers continued to code-switch between the two languages and felt that many students did poorly in Year 8 examinations because of having to read the questions in English (Afamasaga-Fuata'i, 2005).

Another issue is what constitutes mathematics and who gets to decide this. It is often the beliefs of certain sections of the society that determine which knowledge and practices are acceptable, while those of disadvantaged peoples are often rendered invisible (Meaney, 2004). There has been a tendency for 'whites' to think of themselves as 'normal', ideal, and morally neutral, so that when they engage in education or research to benefit the 'other' there is an emphasis on making 'them' more like 'us'. In discussing the question of what constitutes mathematics, Alekseev, Barton, and Knijnik (2006) wrote:

> when the world view is held by a small group, especially one that has been colonised or is in interaction with a larger group, then comparative debate is likely to be difficult and invidious. The dominant paradigms become the measure against which other conventions are judged, and when some of the criteria are aspects like generality or width of application, then conventions based in smaller communities will not count. It should be noted that the effect of dominant conventions operates within mathematics as well as externally. (p. 568)

They went on to state that it is only those within an Indigenous group who can fully understand their own world view. One attempt to reconsider what mathematics is has been through looking at the ethnomathematical practices within a culture. Ethnomathematics is part of the consideration of how societal views about mathematics can affect the positioning of Indigenous students as mathematics learners. In the next section, ethnomathematics is juxtaposed with ideas about the mathematics that should be taught in schools.

Ethnomathematics

Ethnomathematics is a study of mathematics in cultural practices. Usually, it is seen as contributing to the relationship between mathematical practices carried out in indigenous community settings and those done in school (Paraide, 2005). The use of ethnomathematics, or otherwise, within mathematics classrooms is often determined by societal influences.

In relationship to his own work, Matang (2006) stated that:

> the term ethnomathematics will be used to express the relationship between all culturally-embedded mathematical ideas and activities that can be used in any educational setting aimed at enriching deeper understanding of the formal concepts of mathematics taught in schools. (p. 2)

This is similar to Adams' (2004) definition which she directly linked to school mathematics – "ethnomathematical approaches are intended to make *school* mathematics more relevant and meaningful to learners, and to promote the overall quality of education" (p. 49). However, the problems identified by Alekseev et al. (2006) would suggest that, although Matang (2006) and Adams (2004) were primarily dealing with educators who themselves are from the local communities, their approaches may still face difficulties.

Concerns were raised by Barton (2004) about the merging of mathematics with mathematical practices. He felt that ethnomathematics provides opportunities to reconsider how aspects of Western mathematics are perceived, but labelling cultural activities as 'mathematics' was problematic unless certain conditions were met. He stated that "[t]he knowledge should be systematised, should be formalised and should relate to quantity, relationships, or space. It must also be sufficiently abstracted to be removable from its practice" (p. 23). These conditions would enable practitioners to be able to discuss their ideas as mathematics. Societal interactions, about what mathematics is, would have an impact on how ethnomathematical practices are discussed.

Meaney and Fairhall (2003) described the conditions that would foster the sharing of ideas about mathematics and cultural practices between teachers and community members. However, when these conditions are not implemented, then mainstream beliefs about mathematics and learning can limit any usefulness that might have been gained from using ethnomathematical practices as learning experiences. In their article on card games that are played by children in Aboriginal communities, Baturo, Norton, and Cooper (2004) described the inherent mathematical understandings in them. The assumption was that making links to the out-of-school mathematical activities will improve in-school mathematics achievement. However, as Brenner (1998) suggested, "mathematical content and problem solving processes used by children are quite different in and out of school, and that success in everyday mathematical activities does not necessarily translate to mathematical competence in school" (p. 124). Baturo et al.'s (2004) paper did not reference previous research on card games played in Aboriginal communities in the 1970s by Davidson (1979). This earlier research suggested that rather than using arithmetic skills to add up combinations in card games such as DY as suggested by Baturo et al. (2004), Aboriginal people in fact use visual pattern recognition. Another issue around card games is whether their prevalence in indigenous communities means that communities would support their use in mathematics lessons. In the mid 1990s a unit of work on card games was developed in the Northern Territory (Roberts, 1996). The steering committee for this project which was made up of a number of Aboriginal people, refused to allow it to be released generally because of concern that it would promote gambling. Gambling resulted in some families (and thus some children) going hungry.

Teachers who wanted to use the card game unit had to consult their communities and then to request it from the Department of Education. It is essential that if out-of-school practices are to be used in the mathematics classrooms, then appropriate consultation needs to be done with the community in which the activity originated or was to be taught.

Owens and Kaleva (2007) were aware of the differences between mathematical practices in the home as opposed to those in the school. They investigated the informal measurements used by teacher education students in Papua New Guinea. They wanted to understand how these could be linked to Western mathematical ideas on measurement, and also to discover the cultural contexts in which measurement was undertaken. The importance of visualisation was recognised and that:

> … [m]easurement was linked to purpose. The data indicated that the cultural activity – bride price, canoe making, garden building or hole making – influenced the comparison and measurement technique. The unit size, if used, was appropriate for the kind of length. (Owens & Kaleva, 2007, p. 576)

These ideas were seen as being valuable not just in teaching measurement to Papua New Guinean students but to other Australasian students as well. However, discussing ethnomathematical practices using the mathematics register will result in some features of the practice being made invisible whilst others will be highlighted. The situation is further complicated if the language of instruction is not the students' first langauge.

Esmonde and Saxe (2004) studied six mathematical lessons given to grade 3 students in Oksapmin, Papua New Guinea. They found that although the Tok Ples (the language of the village and is the term used to describe any of the indigenous languages) counting system had been traditionally connected to pointing to body parts, at no time in the classroom did the students refer to body parts. It also seemed that most students only had the number terms to ten in Tok Ples. Students in grades 7 and 8 were unable to count in Tok Ples at all, but instead were proficient in English and Tok Pisin (the common Creole language of Papua New Guinea). The teachers who did use it in the grade 3 classes tended to use it to make explicit links to the Hindu-Arabic numeral system. Esmonde and Saxe (2004) suggested that using vernacular languages and Indigenous counting systems in community schools may revive the use of the Tok Ples counting system, but only by altering its structure so that it more closely matched the base ten of the Hindu-Arabic system. This may contribute to the continuing change in Oksapmin mathematical practices. The use of traditional Indigenous mathematical practices needs to be included in mathematics classrooms with caution.

In a paper for the Third International Ethnomathematics Conference, Meaney, Fairhall, and Trinick (2006) considered the role of language in ethnomathematics. When the mathematical practices are discussed in a different language to the one in which the practice is done, different associations could be connected to the practice. This can mean that Western mathematical ideas that are discussed in an indigenous language could be distorted in a similar way to having an Indigenous mathematical practice discussed in an international language, such as English. They strongly recommended that there be discussion about the mathematics register so that its use would add, rather than subtract, meaning from an ethnomathematical practice.

Societal views on mathematics and ethnomathematical practices have a direct influence on teachers' willingness to adopt alternative pedagogical approaches. This is discussed further in the section on teacher interactions.

COMMUNITY INTERACTIONS

Stories told about the communities in which Indigenous students live also have an impact on how they are positioned as knowers, doers, and learners of mathematics. Even in mainstream communities, parental involvement in their children's education is often considered problematic by schools (Cutler, 2000). Differences between Indigenous communities and Western schooling are often seen by the wider society and teachers as deficits in the communities (Cooper, Baturo, Warren, & Doig, 2004) and what these communities may have to offer to the teaching of mathematics therefore ignored (Rennie, 2006). However, Indigenous communities, like other communities, also have stories to tell about the mathematics education of their students.

Indigenous communities are far from passive recipients of Western Education. In describing the provision of a mobile pre-school pilot project for remote Aboriginal communities, Goos, Jolly, and Kostogriz (2005) stated that "like Aboriginal people elsewhere in Australia, the Yolgnu want to hear what this education has to say, but they want to decide for themselves how to use it" (p. 182). Matang (2005) suggested that more research is needed if Papua New Guinea parents are to accept that teaching in vernacular languages will not result in the lowering of academic standards. Withdrawal of their involvement is one way that a community can show its unwillingness to participate (Meaney, 2004).

If effective mathematics education programs are to be implemented, then it is essential that non-Indigenous educators communicate with communities and actively listen to what they are told. Meaney (2004) suggested that an outsider may use dominating behaviours that can limit the community's chances of controlling the negotiation of meaning. Instead, the outsider's role needs to be negotiated through a mutual sharing of knowledge and requirements. As Goos et al. (2005) commented, "essential relationships cannot be mandated from outside nor built up overnight, but depend on trust and mutual respect which can only be gained over time" (p. 184).

An on-going research project into the building of community capacity demonstrates how Aboriginal and non-Aboriginal communities in rural and urban areas can work together with schools to enhance understanding and respect. This is done by groups working alongside each other to develop the mathematics knowledge of the primary and secondary students in their communities (Howard & Perry, 2005, 2006; Howard, Perry, & Butcher, 2006). Priority is given to Aboriginal voices so that cultural relevance can be to the fore whilst working towards strategies that will support the educational potential of Aboriginal learners. Howard, Perry, and Butcher (2006) identified five challenges in capacity building which included mutual respect, shared power and leadership, evidence-based information, building shared understandings, and strengthening personal and collective efficacy. They saw these challenges as inter-related and complex. Nevertheless, they believed that this project could be a model for other communities to engage in capacity building, with Aboriginal communities taking a lead role so that "well meaning groups of people [who] strived to improve the lot of Aboriginal people without Aboriginal people … can change their approaches" (Howard, Perry, & Butcher, 2006, p. 11).

TEACHER INTERACTIONS

Teachers have a significant influence on the learning of mathematics by students, Indigenous or non-Indigenous (Howard & Perry, 2006). In regard to Indigenous students, the stories that are told about their teachers in the research reports can be divided into several categories. These are stories about: the teachers' working conditions; teachers' beliefs about their students and their communities; teachers' inadequate knowledge of mathematics, pedagogy, and contexts for teaching Indigenous students; and attempts, sometimes successful, to improve teachers' knowledge in these areas. There were also a few stories about successful mathematics teachers of Indigenous students.

Working conditions for teachers

Some stories revolved around the working conditions of teachers that are believed to contribute to Indigenous students' under-achievement in mathematics. For example, in Australia, casual teachers were more likely to be found in classes with high numbers of Indigenous students (Amosa & Cooper, 2006). The retention of skilled mathematics teachers is problematic in many remote areas of Australia (Clark, Aldous, Davies, Barnes, & White, 2006). Consequently, teachers with no mathematics background and limited professional development are conscripted into taking mathematics classes with high numbers of Indigenous students. Cooper, Baturo, Warren, and Grant (2006) stated that the current incentive, to encourage first-year-out teachers to remote schools, of gaining enough points to move to Brisbane after two years was "not supporting Indigenous learning" (p. 374). Instead, it produced a continuous turnover of inexperienced staff.

Teachers' beliefs about Indigenous students and their communities

Teachers also tell stories about the Indigenous students that they teach and this has an impact on the positioning of these students as knowers, doers, and learners of mathematics. Howard and Perry (2006) found that although there was a large number of Aboriginal students in one school, their presence was not obvious in comments made by their teachers about mathematics teaching. It is unlikely that appropriate activities that support "the children's cultural and learning needs" (p. 299) will be provided if teachers take little notice of these students.

As well, teachers who are not from the community can hold deficit views about the Indigenous students they are teaching (see Cooper et al., 2004). An unfavourable comparison with their expectations of European students contributes to a lack of understanding of the knowledge and skills that Indigenous students bring to the mathematics classroom. Rennie (2006) documented a number of out-of-school activities that seven Indigenous students engaged in that she felt had literacy and numeracy practices embedded within them. However, these students' high school teachers failed to recognise these activities and assumed that these students were lacking in all areas of numeracy.

In a small study in Australia, Warren, Young, and de Vries (2007) described the strategies that two experienced non-Indigenous teachers used in teaching mathematics to Indigenous primary school students. These strategies were affected by the teachers' beliefs about the students and the languages that they brought to the classroom. The teachers believed that both Aboriginal English and Creole lacked the ability to make refined comparisons of attributes and therefore there was a need to encourage students to use what was termed "explicit language of mathematics" (p. 333). In fact what was being recommended was the teaching of Standard Australian English so that students could describe comparisons in Western ways. As Aboriginal English and Creole are based on Aboriginal languages and concepts, it is likely that these language describe comparisons in Aboriginal or Torres Strait Islander ways, rather than in Western ways. For example, Rudder (1983) found that in the Yolgnu culture in North-East Arnhemland, comparisons were two dimensional. A person would be viewed as tall or short depending upon what they wanted to do. This is similar to what Owens and Kaleva (2007) found in regard to community measurement systems in Papua New Guinea. Therefore, being able to compare the heights of different people was only necessary in regard to whether or not they could do what was needed. For the teachers in Warren et al.'s (2007) study, understanding alternative conceptions of comparison could help them see that this is a difference rather than a deficit in the students' languages. It also provides the teachers with a broader, non-Western, understanding of measurement that could change how they introduce Western concepts.

The knowledge of teachers of Indigenous students

In the last four years, concern has been raised about the mathematical and pedagogical knowledge of the teachers of Indigenous students. Hawera (2004) reported on Māori pre-service teachers' concern about their own level of mathematical knowledge. Samoan pre-service primary and secondary students are also reported as having significant gaps or misconceptions in their mathematical understanding (Afamasaga-Fuata'i, Meyer, & Falo, 2007; Afamasaga-Fuata'i, Meyer, Falo, & Sufia, 2006). One of the outcomes of an intervention study to support the learning of numeracy by Indigenous students in remote communities in the Northern Territory of Australia was a recognition that teachers, non-Indigenous and Indigenous, needed a mentoring program to improve their teaching of numeracy (Commonwealth of Australia, 2005). Although not specifically discussing mathematics teaching, Amosa and Cooper (2006) also stated that "the findings regarding the quality of teaching experience in schools with high enrolments of Aboriginal and Torres Strait Islander descent suggest that there is a dire need for quality of pedagogy to be enhanced in such schools" (p. 28).

To be a teacher of mathematics to Indigenous students requires a deep understanding of mathematics, plus a knowledge of the students (Howard & Perry, 2006). This knowledge includes an understanding of appropriate contexts and pedagogy for them. The stories that are told about teachers and the impact that these have on Indigenous students' achievements in mathematics are unlikely to improve unless teachers perceive a need to change and have the ability to access professional development and resources (Howard & Perry, 2006). Clark et al. (2006) discussed how the difficulties of finding relief teachers and the cost of travel meant that teachers in remote schools in Australia rarely attended professional development.

Teachers are unlikely to change their teaching practices without significant support (Matang & Owens, 2004). Owens and Matang (2004) describe the current mathematics curriculum used in community schools. This curriculum encourages the use of "vernacular words, events and representations to teach mathematics" (p. 5). Paraide (2005) suggested that little has been done in Papua New Guinea to support the integration of Indigenous knowledge into formal schooling. She went on to state that "[curriculum] documents provide insufficient examples and background information of Indigenous mathematical knowledge for teachers to use as resources, so that they can build on it when they are teaching in their various locations in Papua New Guinea" (p. 8).

In New Zealand, Anderson, Averill, Easton, and Smith (2005) described attempts to incorporate the principles of the Treaty of Waitangi into a pre-service mathematics education program. The Treaty of Waitangi describes the conditions for Māori and non-Māori to live together in New Zealand. The pre-service students were positive about including cultural activities in their mathematics programs but were "less confident in describing mathematical links to Treaty principles" (p. 87). It would seem that understanding the deep ideas about the purposes for incorporating Indigenous perspectives into mathematics education is not a simple process and takes time to develop.

One intervention program, undertaken in the last four years, was conducted with Torres Strait Islander students and their teacher of block laying by Cooper, Baturo, Ewing, Duus, and Moore (2007). They worked together to design and implement a program to cover the mathematics needed in order to gain a Certificate 1 in General Construction. The students strongly supported the integration of practical skills of block laying with the mathematics learning at work sites because they felt that it provided them with a stronger understanding of the mathematics. The teacher changed his teaching approach to mathematics as a result of the intervention of the researchers.

In the pre-service area, Hawera (2006) described a program of mathematics investigations that was used to increase understanding of processes and strategies of Māori pre-service teachers. By having teachers engage more with mathematics, she anticipated that there would be a positive impact on Māori students' achievement in mathematics.

Te Poutama Tau (numeracy learning in the medium of te reo Māori) is a professional development program which began in 2002 to support New Zealand teachers working in Māori-medium contexts (Trinick, 2005, 2006). Based on the number framework developed for New Zealand schools, facilitators work with Māori-medium schools and their teachers. Results show relatively consistent increases each year in students' achievement in numeracy. However, non-Māori students have improved more. Consequently, the gap between Māori and non-Māori students' achievement has increased rather than declined. Improvements for students in Maori medium are evident in some areas, such as counting forward, and number-word-sequence but not in others. Improvements in performance seem to be more difficult to achieve at higher levels. This suggests that teachers' mathematical content knowledge is still lacking in a number of areas and professional development needs to be ongoing (Trinick, 2006). As well, te reo Māori (the Māori language) proficiency of Māori students as rated by their teachers dropped slightly in 2005, which may have future significance for the learning of mathematics in te reo Māori (Trinick & Stevenson, 2005). This is an area that will be the subject of ongoing research.

As many teachers in Māori-medium education are second language learners of te reo Māori themselves, the New Zealand Ministry of Education instigated Whakapiki Pāngarau (McMurchy-Pilkington, 2005). This full time program for six months supports teachers to improve their mathematical, pedagogical, and curriculum knowledge. Additionally, the teachers strengthen their understanding about the complexity of the Māori language and the linguistic structures necessary to teach mathematics in the medium of Māori. The program affirms the importance of making connections between research, theory, and practice. Unlike their pre-service experiences, the teachers declared the readings now had greater meaning for them.

Evaluations of the Early and Advanced Numeracy Projects in English-medium schools show that learners in schools in low socio-economic areas and Māori and Pasifika students have improved less than students from other groups (Higgins, 2005). However, the results were not consistent. Evaluations revealed that some English-medium schools with a high Māori population and located in low socio-economic areas, have Māori learners whose achievement in numeracy is above that for Māori as a whole group. Further research needs to be carried out to ascertain what might be behind these results.

As well as numeracy and language knowledge, teachers also need to know about students' expectations of how learning occurs. Meaney (2005) provided an illustration of an example of an exchange with an Aboriginal student where the student was obviously able to do the mathematics but did not feel obliged to talk about it with her. Consequently Meaney (2005) recommended that when there are differences in beliefs about learning knowledge, then misunderstandings can occur:

> Knowledge of different possible linguistic choices is valuable but so is an understanding of the importance of discussion in mathematics classrooms. As a teacher, it was essential that I learnt about the differences in how the beliefs about knowledge had an impact on our development of a text of mathematics. (p. 132)

In a project to support the teaching of numeracy to Indigenous students in the Northern Territory, there was a discussion about students' learned helplessness (Commonwealth of Australia, 2005). This is a situation in which students who are perceived as failing are provided with so much teacher support that the task is completed, but with little learning by students.

Successful teachers of mathematics to Indigenous students

As well as recognising that there is a need to generally improve teachers' teaching of mathematics to Indigenous students, there have also been a few descriptions of successful teachers of Indigenous students. Zevenbergen, Mousley, and Sullivan (2004) described a teacher who was able to use her knowledge of the culture and dispositions to support Indigenous students gain access to the "mathematical language and ideas embedded within the lesson" (p. 404). Higgins (2005) reported on the classroom pedagogy of one teacher of numeracy to Māori learners in an English-medium school. The Māori teacher viewed the class as a collective of interconnected groups rather than a collection of instructional groups. The 'lead group' became the tool of instruction for the less expert groups. Thus, the teaching role became shared amongst the teacher and the students as the teacher "appropriated the contributions of those who are knowledgeable" (Higgins, 2005, p. 77). This pedagogy can be viewed as building the capacity of the student groups to become the knower and the expert in the mathematics class.

At the university level, Clark (2006) described a tutorial program for Māori and Pasifika students enrolled in a first year statistics course at Victoria University, Wellington. This program was set up at the request of community elders who had read about a similar program used in California. Knowledge about Māori students' previous learning experiences was used to make students comfortable with tutorials, before offering other ways of working that are perceived as being helpful for learning at this level of education. The program has resulted in a larger proportion of these students passing and moving onto other statistics courses. Latu (2004) reported on the similar success of a program set up in the mathematics department at the University of Auckland for Māori and Pasifika from low socio-economic areas.

Matang and Owens (2004), Matang (2005) and Matang (2006) looked at counting systems in several Papua New Guinean languages. In particular, they described how numbers are constructed in Kâte, Matang's first language, by being either related to twos, tens, or twenties. Being able to recognise the mathematical ideas underlying these counting systems can help students to solve simple addition and subtraction problems more easily. Owens and Matang (2004) discussed the work they did with one class of junior school children and the impact of counting strategies such as counting forward, counting backwards, and skip counting in Tok Ples, Tok Pisin and in English. They found that children were able to count quickly, even though the Tok Ples of this village used more syllables in their number words as word for small numbers are combined to form larger numbers. For example, 6 – *ligizani luga hamo* – is the equivalent of 5 – *ligizani luga* – and 1 – *hamo*. Owens and Matang (2004) suggested that the number of syllables did not affect students' learning of counting. They also found that the students were able to give 5 + 2 + 1 as a combination that added to 8 more quickly that 4 + 4. They attributed this to the inherent counting system in Tok Ples. They also found that children who were familiar with English or Tok Pisin, both with number systems based around 10, used these languages to solve the problems and then translated back into Tok Ples. Unfortunately, no explicit data were provided to support these results.

For non-Indigenous people working with Indigenous students, it is important not to consider just what needs to be taught in order for the students to gain Western mathematics. There is also a need to better understand these students' own concepts from the mathematical practices that they engage with. These concepts and the ways that language are used to describe them will not only broaden the teachers' understandings of mathematics but can help them to provide more appropriate teaching of Western mathematics that is not based on deficit models of the students and their home knowledge.

STUDENT INTERACTIONS

In stories about Indigenous students particular characteristics that appear to be related to their ability to achieve in mathematics are described. For example, in Australia, Indigenous students' irregular attendance at school has been well documented (De Bortoli & Cresswell, 2004). Teachers feel that students' attendance has a huge impact on "continuity of learning experiences" (Clark et al., 2006), and therefore students' likelihood to achieve. However, the causes for irregular attendance are not understood and often remain undiscussed or simply labelled as being a community problem.

In the research reports, there were references to characteristics of Indigenous students that had an impact on their learning. These included not feeling comfortable about being publicly asked questions (Clark, 2006; Zevenbergen et al., 2004) and being visual learners (Cooper, Baturo, & Warren, 2005). Watego (2005) cautions about the use of generalisations: "stereotypes have a negative influence on young people. Indigenous children, particularly in urban schools, are continually presented with pictures of themselves that don't always look like them" (p. 769). Similar cautions about the problems of homogenising Indigenous students were raised by McKinley, Stewart, and Richards (2004).

De Bortoli and Creswell (2004) found that factors such as family structure, educational attainment of parents, and home education resources correlated with the mathematics performance of Indigenous students. However, they noted that Indigenous students were affected less by socio-economic background than other groups of students. This was also supported by research by Marks (2006). The implication is that it is important to raise the achievement of all Indigenous students, not just those from lower socio-economic backgrounds (De Bortoli & Cresswell, 2004). However, the contribution of socio-economic and socio-cultural factors on under-achievement was rejected by Rowe (2003) who claimed that what mattered most was quality teachers and teaching.

Indigenous students also had lower educational aspirations (De Bortoli & Cresswell, 2004) and this correlated with their achievement in mathematical literacy. The reasons for these lower aspirations were not discussed and this would seem to be an area that needs further investigation. There does seem to be some agreement that teachers need to raise their expectations of Indigenous students (Clark, 2006; Cooper et al., 2006).

A number of stories were also told through research reports about possibilities for changing Indigenous students' achievement in mathematics. Matang (2005) used the results of a study of 52 primary school children whose first langauge was Kâte, to suggest that learning basic counting and arithmetic skills in a vernacular language does not lower students' understanding. However, some of the children came from schools in which Tok Pisin and English were the languages of instruction, so the relationship to Kâte is not straight forward. Nevertheless , his results suggest that the if students learnt mathematical ideas or skills well, then they could perform them in any language.

Many students who attend kura kaupapa Māori (Māori immersion schools) in New Zealand do not have te reo Māori as their first language. However, they would not necessarily have academic English either because their schooling had not been in English. This situation raises a number of complex linguistic issues. One of these issues is the students' ability to access the specific terms and grammatical expressions of the mathematics register. Using interviews with ten Māori students who had moved from kura kaupapa Māori into secondary schools, some Māori-medium whilst others were English-medium, Herewini (2006) found that the amount of new vocabulary that students had to grapple with increased dramatically. Therefore, teachers needed to provide explicit language teaching to support students' learning of this vocabulary. In both English-medium and Māori-medium classrooms, there was a need to do this using second language acquisition strategies.

Meaney, Fairhall, and Trinick (2007) described the modelling and scaffolding practices that teachers used in a kura kaupapa Māori. Although many of the strategies could be found in English-medium classrooms, there were some that seemed to be specific to Māori culture and language. For example, compared to English, te reo Māori has a larger number of logical connectives that could be used to connect ideas together but also to support the learning process. There was also a huge amount of student interaction both with the teacher and between themselves.

Students were expected to take responsibility for each other in a tuakana-teina, or older-younger sibling relationship. Consequently, there was little reticence by students to take up a teacher role or to be tutored by another student.

In Australia, Matthews, Cooper, and Baturo (2007) described an intervention program in which Indigenous students were introduced to algebraic representations through the use of symbols to represent elements within a story. By using the students' understanding of traditional story telling, it was possible to move them more easily into formal mathematical ideas and representations. Using strengths within the culture of Indigenous students means that their mathematics learning builds on positive experiences and moves away from deficit descriptions of students' mathematical experiences. However, this does require non-Indigenous educators to listen and learn so that they can work with the community in the development of such programs.

Watego (2005) argued that contextualisation needs to consider ideas of cultural safety. It is through seeing what and how meaning is ascribed to mathematical activities that an Indigenous context is provided.

Apart from the study by Herewini (2006), Howard and Perry (2005) were the only ones to report on Indigenous students' own beliefs about how they learnt, "when learning mathematics, the Aboriginal children believed in the importance of listening, watching and working out the mathematics on paper" (p. 158).

The process by which Indigenous students form identities as knowers, doers, and learners of mathematics is extremely complex. Students do this, not just through stories told about the "Indigenous student", but also ones about the "society", the "Indigenous community" and the "mathematics teacher of Indigenous students". Consequently, as Howard and Perry (2005) state, "the ideas of disadvantage and deficit linked to Aboriginal children's learning need to be replaced with beliefs about their competencies" (p. 159).

CONCLUSION

In the last four years, a considerable amount of research has been conducted in regard to Indigenous students in mathematics classrooms. This research presents a very complex situation that is continually evolving. Using the ideas of Sfard and Prusak (2005) on identity, we have considered Indigenous students' current and future positioning as knowers, doers, and learners of mathematics as being influenced by four types of interactions.

The stories presented in research reports showed how perceptions of Indigenous students, in regard to mathematics, by society, communities, teachers, and the students themselves can contribute to them being positioned as disadvantaged. However, some stories also highlighted possibilities for changing this position to one of being successful mathematics achievers. These stories consistently referred to the need to work with students' communities to develop appropriate mathematics programs. Communities need to understand school mathematics so that they can actively discuss with teachers what cultural practices can be shared in the classroom. Teachers and researchers need to learn to listen to students and their communities in order to better understand what students bring into the classroom. This listening needs to include developing an awareness of different perceptions of mathematics and learning. There is also a need to be aware of previous research in this area. Teachers and researchers also have an obligation to share their own understandings about mathematics and learning so that a true collaboration with Indigenous students and their communities can be achieved. It is vitally important that collaborative research between Indigenous and non-Indigenous researchers is done with care. Otherwise, presumptions about what is being learnt and how it is learnt can be at the very least misleading and at worse exploitative of the students who are supposed to gain from the research being conducted (Bishop & Glynn, 1992).

REFERENCES

Adams, S. (2004). Ethnomathematical ideas in the curriculum. *Mathematics Education Research Journal, 16*(2), 49-68.

Afamasaga-Fuata'i, K. (2005). Mathematics education in Sāmoa: From past and current situations to future directions. *The Journal of Sāmoan Studies, 1*, 147-165.

Afamasaga-Fuata'i, K., Meyer, P., Falo, N., & Sufia, P. (2006). *Future teachers' developing numeracy and mathematical competence as assessed by two diagnostic tests.* Paper presented at the annual conference of the Australian Association of Research in Education, Adelaide. Retrieved July 31, 2007 from: http://www.aare.edu.au/06pap/afa06011.pdf.

Afamasaga-Fuata'i, K., Meyer, P., & Falo, N. (2007). Primary student teachers' diagnosed mathematical competence in semester one of their studies. In J. Watson & K. Beswick (Eds.), *Mathematics: Essential research, essential practice* (Proceedings of the 30th annual conference of the Mathematics Education Research Group of Australasia, Hobart, Vol. 1, pp. 83-92). Sydney: MERGA.

Alekseev, V., Barton, B., & Knijnik, G. (2006). Other conventions in mathematics and mathematics education. In F. Leung, K. Graf, & F. Lopez-Real (Eds.), *Mathematics education in different cultural traditions: A comparative study of East Asia and the West. The 13th ICMI Study* (pp. 567-580). New York: Springer Science and Business Media.

Amosa, W., & Cooper S. (2006). *Connections among school contexts, teaching contexts and the quality of teaching.* Refereed paper presented at the annual conference of the Australian Association of Research in Education, Adelaide. Retrieved April 24, 2007 from: http://www.aare.edu.au/06pap/code06.htm (amo06388).

Anderson, D., Averill, R., Easton, H., & Smith, D. (2005). Use of a cultural metaphor in pre-service mathematics teacher education. In P. Clarkson, A. Downton, D. Gronn, M. Horne, A. McDonough, R. Pierce, & A. Roche (Eds.), *Building connections: Theory, research and practice.* (Proceedings of the 28th annual conference of the Mathematics Education Research Group of Australasia, Melbourne, Vol. 1, pp. 81-88). Sydney: MERGA.

Anthony, G., & Walshaw, M. (2007). *Best evidence synthesis: Characteristics of pedagogical approaches that facilitate learning for diverse learners in early childhood and schooling in Pāngarau/Mathematics.* Wellington: Ministry of Education.

Atweh, B., Meaney, T., McMurchy-Pilkington, C., Neyland, J., & Trinick, T. (2004). Social justice and socio-cultural perspectives in mathematics education. In B. Perry, G. Anthony, & C. Diezmann (Eds.), *Research in mathematics education in Australasia: 2000-2003* (pp. 29-52). Flaxton, Qld.: Post Pressed.

Barton, B. (2004). Mathematics and mathematical practices: Where to draw the line? *For The Learning of Mathematics, 24*(1), 22-24.

Baturo, A. R., Norton, S., & Cooper, T. J. (2004). The mathematics of Indigenous card games: Implications for mathematics teaching and learning. In I. Putt, R. Faragher, & M. McLean (Eds.), *Mathematics education for the third millennium: Towards 2010*. (Proceedings of the 27th annual conference of the Mathematics Education Research Group of Australasia, Townsville, Vol. 1, pp. 87-94). Sydney: MERGA.

Bishop, R., & Glynn, T. (1992). He kanohi kitea: Conducting and evaluating research. *New Zealand Journal of Education Studies, 27*(2), 125-135.

Brenner, M. E. (1998). Meaning and money. *Educational Studies in Mathematics, 36*(2), 123-155.

Clark, M. (2006) A first year statistics programme for Indigenous and migrant students arrived at by co-operating with local communities and the students themselves. In A. Rossman & B. Chance (Eds.), *Proceedings of ICOTS-7*. Bahia, Brazil: ICOTS. Retrieved February 17, 2008 from: http://www.stat.auckland.ac.nz/~iase/publications/17/1B1_CLAR.pdf

Clark, J., Aldous, C., Davies, J., Barnes, A., & White, B. (2006). *Science, technology and mathematics teaching and learning in rural South Australia: Problems and prospects.* Refereed paper presented at the annual conference of the Australian Association of Research in Education, Adelaide. Retrieved April 24, 2007 from: http://www.aare.edu.au/06pap/code06.htm (bar06568).

Commonwealth of Australia (2005). *Supporting Indigenous students' achievement in numeracy*. Canberra: Author.

Cooper, T. J., Baturo, A. R., Warren, E., & Doig, S. (2004). Young "white" teachers' perceptions of Aboriginal and non-Aboriginal students in remote communities. In M. J. Høines & A. B. Fuglestad (Eds.), *Proceedings of the 28th annual conference of the International Group for the Psychology of Mathematics Education* (Vol. 2, pp. 239-246). Bergen, Norway: Bergen University College.

Cooper, T. J., Baturo, A. R., & Warren, E. (2005). Indigenous and non-Indigenous teaching relationships in three mathematics classrooms in remote Queensland. In H. L. Chick & J. L. Vincent (Eds.), *Proceedings of the 29th annual conference of the International Group for the Psychology of Mathematics Education* (Vol. 2, pp. 265-272). Melbourne: University of Melbourne.

Cooper, T, J., Baturo, A. R., Warren, E., & Grant, E. J. (2006). Aim high – beat yourself: Effective mathematics teaching in a remote Indigenous community. In J. Novotná, H. Moraová, M. Krátká, & N. Stehliková (Eds.), *Proceedings of the 30th annual conference of the International Group for the Psychology of Mathematics Education* (Vol. 2, pp. 369-376). Prague, Czech Republic: PME.

Cooper, T., Baturo, A., Ewing, B., Duus, E., & Moore, K. (2007). Mathematics education and Torres Strait Islander blocklaying students: The power of vocational context and structural understanding. In J-H. Woo, H-C Lew, K-S. Park, & D-Y. Seo (Eds.), *Proceedings of the 31st annual conference of the International Group for the Psychology of Mathematics Education* (Vol. 2, pp. 177-184). Seoul, Korea: The Korean Society of Educational Studies in Mathematics.

Cutler, W. (2000). *Parents and schools, the 150-year struggle for control in American education*. Chicago: University of Chicago Press.

Davidson, G. R. (1979). An ethnographic psychology of Aboriginal cognitive ability. *Oceania, 49*(4), 270-94.

De Bortoli, L., & Cresswell, J. (2004). *Australia's Indigenous students in PISA 2000: Results from an international study: ACER Research Monograph No. 59*. Melbourne: Australian Council for Education Research.

Esmonde, I., & Saxe, G. B. (2004). 'Cultural mathematics' in the Oksapmin curriculum: Continuities and discontinuities. In Y. Kafai, W. A. Sandoval, N. Enyedy, A. S. Nixon, & F. Herrera (Eds.), *Proceedings of the International Conference of the Learning Sciences 2004* (pp. 174-181). Mahwah, NJ: Lawrence Erlbaum Associates.

Flockton, L., Crooks, T., Smith, J., & Smith, L. F. (2006). *National education monitoring report 37. Mathematics: Assessment results 2005*. Dunedin, New Zealand: University of Otago, Education Assessment Research Unit.

Forbes, S. (2005). *Potential uses of longitudinal analyses to investigate education outcomes*. Paper presented at the IASE sessions at the 55th session of the International Statistical Institute, Sydney,

April, 2005. Retrieved April 24, 2007 from: www.stat.auckland.ac.nz/~iase/publications/13/ Forbes.pdf

Goos, M., Jolly, L., & Kostogriz, A. (2005). Home, school and community partnerships for numeracy education in a remote Indigenous community. In M. Goos, C. Kanes, & R. Brown (Eds.), *Proceedings of the 4th International Mathematics Education and Society Conference* (pp. 176-186). Gold Coast: Griffith University.

Hawera. N. (2004). Addressing the needs of mathematically anxious preservice primary teachers. In I. Putt, R. Faragher, & M. McLean (Eds.), *Mathematics education for the third millennium: Towards 2010* (Proceedings of the 27th annual conference of the Mathematics Education Research Group of Australasia, Townsville, Vol. 1, pp. 287-294). Sydney: MERGA.

Hawera. N. (2006). Maori preservice primary teachers' responses to mathematics investigations. In P. Grootenboer, R. Zevenbergen, & M. Chinnappan (Eds.), *Identities, cultures and learning spaces* (Proceedings of the 29th annual conference of the Mathematics Education Group of Australasia, Canberra, Vol. 1, pp. 286-292). Sydney: MERGA.

Herewini, L. (2006). Māori students' perspective on their mathematical journey through Māori medium. Unpublished Masters thesis, Waikato University, Hamilton, New Zealand.

Higgins, J. (2005). Effective teaching strategies for Maori students in an English-medium numeracy classroom. In J. Higgins, K. C. Irwin, G. Thomas, T. Trinick, & J. Young-Loveridge (Eds.), *Findings from the New Zealand Numeracy Development Project 2004* (pp. 74-78). Wellington, New Zealand: Ministry of Education.

Howard, P., & Perry, B. (2005). Learning mathematics: Perspectives of Australian Aboriginal children and their teachers. In H. L. Chick & J. L. Vincent (Eds.), *Proceedings of the 29th annual conference of the International Group for the Psychology of Mathematics Education* (Vol. 3, pp. 153-160). Melbourne: University of Melbourne.

Howard, P., & Perry, B (2006). Issues in teaching mathematics to Aboriginal students. In P. Grootenboer, R. Zevenbergen, & M. Chinnappan (Eds.), *Identities, cultures and learning spaces* (Proceedings of the 29th annual conference of the Mathematics Education Research Group of Australasia, Canberra, Vol. 1, pp. 293-300). Sydney: MERGA.

Howard, P., Perry, B., & Butcher, J. (2006). *Shared ownership and community capacity building.* Refereed paper presented at the Australian Association of Research in Education conference, Adelaide. Retrieved February 17, 2008 from: http://www.aare.edu.au/06pap/how06148.pdf.

Latu, V. F. (2004). There is more to learning mathematics than mathematics: Attributional beliefs and motivation of Maori and Pacific Island students. In I. Putt, R. Faragher, & M. McLean (Eds.), *Mathematics education for the third millennium: Towards 2010* (Proceedings of the 27th annual conference of the Mathematics Education Research Group of Australasia, Townsville, Vol. 2, pp. 343-350). Sydney: MERGA.

Marks, G. N. (2006). Influences on and the consequences of low achievement. *The Australian Educational Researcher, 33*(1), 95-115.

Matang, R., & Owens, K. (2004). *Rich transitions from Indigenous counting systems to English arithmetic strategies: Implications for mathematics education in Papua New Guinea.* Paper presented at Working Group 15, International Congress on Mathematics Education, Copenhagen. Retrieved February 18, 2008 from: http://www.icme-organisers.dk/dg15/DG15_RM&KO_final_ed.pdf

Matang, R. A. (2005). Formalising the role of Indigenous counting systems in teaching the formal English arithmetic strategies through local vernaculars: An example from Papua New Guinea. In P. Clarkson, A. Downton, D. Gronn, M. Horne, A. McDonough, R. Pierce, & A. Roche (Eds.), *Building connections: Theory, research and practice.* (Proceedings of the 28th annual conference of the Mathematics Education Research Group of Australasia, Melbourne, Vol. 2, pp. 505-512). Sydney: MERGA.

Matang, R. A. (2006). *Linking ethnomathematics, situated cognition, social constructivism and mathematics education: An example from Papua New Guinea.* Paper presented at the 3rd

International Conference of Ethnomathematics. Auckland. Retrieved August 25, 2005 from: www.math.auckland.ac.nz/~poisard/ICEm3/ICEm3.html.

Matthews, C., Cooper, T. J., & Baturo, A. R. (2007). Creating your own symbols: Beginning algebraic thinking with Indigenous students. In J-H. Woo, H-C Lew, K-S. Park, & D-Y. Seo (Eds.), *Proceedings of the 32nd annual conference of the International Group for the Psychology of Mathematics Education* (Vol. 3, pp. 249-256). Seoul, Korea: The Korean Society of Educational Studies in Mathematics.

McKinley, E. (2004). *Māori science and mathematics education: Issues of knowledge, language and identity.* Paper presented at American Educational Research Association annual meeting. Retrieved 18 February, 2008 from: http://convention.allacademic.com/aera2004/view_paper_info.html? pub_id=8449&part_id1=909000

McKinley, E., Stewart, G., & Richards, P. (2004). Māori students in science and mathematics: Junior programmes in secondary schools. *SET, 3,* 9-13.

McMurchy-Pilkington, C. (2005). A mathematics teacher educator's perspective of building connections between research, theory, and practice. In P. Clarkson, A. Downton, D. Gronn, M. Horne, A. McDonough, R. Pierce, & A. Roche (Eds.), *Building connections: Theory, research and practice.* (Proceedings of the 28th annual conference of the Mathematics Education Research Group of Australasia, Melbourne, Vol. 1, pp. 46-49). Sydney: MERGA.

Meaney, T. (2002). Symbiosis or cultural clash? Indigenous students learning mathematics. *Journal of Intercultural Studies*, *23*(2), 167-187.

Meaney, T. (2004). So what's power got to do with it? In M. Walshaw (Ed.), *Mathematics education within the postmodern* (pp. 181-199). Greenwich, CT: Information Age Publishing.

Meaney, T. (2005). Mathematics as text. In A. Chronaki & I. M. Christiansen (Eds.), *Challenging perspectives in mathematics classroom communication* (pp. 109-141). Westport, CT: Information Age.

Meaney, T., & Fairhall, U. (2003). Tensions and possibilities: Indigenous parents doing mathematics education curriculum development. In L. Bragg, C. Campbell, G. Herbert, & J. Mousley (Eds.), *Mathematics education research: Innovation, networking, opportunities.* (Proceedings of the 26th annual conference of the Mathematics Education Research Group of Australasia, Geelong, pp. 507-514). Melbourne: Deakin University.

Meaney, T., Fairhall, U., & Trinick, T. (2006). The role of language in ethnomathematics: Does the language make it ethnomathematics? *Proceedings of the 3rd International Conference of Ethnomathematics.* Auckland. Retrieved August 26, 2006 from: www.math.auckland.ac.nz/~poisard/ICEm3/ICEm3.html.

Meaney, T., Fairhall, U., & Trinick, T. (2007). Acquiring the mathematics register in te reo Māori. In J. Watson & K. Beswick (Eds.), *Mathematics: Essential research, essential practice* (Proceedings of the 30th Mathematics Education Research Group of Australasia, Hobart, Vol 2, pp. 493-502). Adelaide: MERGA.

Owens, K., & Kaleva, W. (2007). Changing our perspective on measurement: A cultural case study. In J. Watson & K. Beswick (Eds.), *Mathematics: Essential research, essential practice* (Proceedings of the 30th annual conference of the Mathematics Education Research Group of Australasia, Hobart, Vol 2, pp. 571-580). Adelaide: MERGA.

Owens, K., & Matang, R. (2004). *Multilingual elementary school teaching in Papua New Guinea.* Paper presented to the Multilingual Discussion Group, satellite meeting of the *International Group for the Psychology of Mathematics Education*, Bergen, Norway.

Paraide, P. (2005). *The value of Indigenous mathematical knowledge in formal learning.* Paper presented at the conference on sustainable curriculum development: The PNG experience. Retrieved May 8, 2007 from: http://www.pngcurriculumreform.ac.pg/research/evaluation.htm

Perry, B., Anthony, G., & Diezmann, C. (Eds.). (2004). *Research in mathematics education in Australasia 2000-2003.* Flaxton, Qld.: Post Pressed.

Queensland Studies Authority. (2004). *Overview of statewide student performance in aspects of literacy and numeracy: Report to the Minister for Education and Minister for the Arts.* Brisbane, Qld.:

Queensland Studies Authority.

Rennie, J. (2006). Meeting kids at the school gate: The literacy and numeracy practices of a remote Indigenous community. *The Australian Educational Researcher, 33*(3), 123-142.

Roberts, T. (1996). *Implementing the common curriculum in Aboriginal schools: Mathematics support materials for stage 4* (Unit of Work on Card Games). Darwin: Northern Territory Department of Education.

Rowe, K. (2003). *The importance of teacher quality as a key determinant of students' experiences and outcomes of schooling.* A context and discussion paper prepared on behalf of the Interim Committee for the NSW Institute of Teachers. Retrieved December 21, 2007 from: http://nswteachers.nsw.edu.au/library/Rowe.html.

Rudder, J. (1983). *Qualitative thinking: An examination of the classification systems, evaluative systems and cognitive structures of the Yolnu people of North-East Arnhem Land.* Unpublished Masters thesis, Australian National University, Canberra.

Sfard, A., & Prusak, A. (2005). Telling identities: In search of an analytical tool for investigating learning as a culturally shaped activity. *Educational Researcher, 34*(4), 14-22.

Thomson, S., & De Bortelli, L. (2008). *Exploring scientific literacy: How Australia measures up.* Melbourne: Australian Council for Education Research.

Trinick, T. (2005). Te poutama tau: A case study of two schools. In J. Higgins, K. C. Irwin, G. Thomas, T. Trinick, & J. Young-Loveridge (Eds.), *Findings from the New Zealand Numeracy Development Project 2004* (pp. 80-88). Wellington: Ministry of Education.

Trinick, T. (2006). Te poutama tau – A case study of two schools. In F. Ell, J. Higgins, K. C. Irwin, G. Thomas, T. Trinick, & J. Young-Loveridge (Eds.), *Findings from the New Zealand Numeracy Development Project 2005* (pp. 103-114). Wellington: Ministry of Education.

Trinick, T., & Stephenson, B. (2005). An evaluation of te poutama tau 2004. In J. Higgins, K. C. Irwin, G. Thomas, T. Trinick, & J. Young-Loveridge (Eds.), *Findings from the New Zealand Numeracy Development Project 2004* (pp. 56-65). Wellington: Ministry of Education.

Warren, E., Young, J., & de Vries, E. (2007). Australian Indigenous students: The role of oral language and representations in the negotiation of mathematical understanding. In J. Watson & K. Beswick (Eds.), *Mathematics: Essential research, essential practice* (Proceedings of the 30th annual conference of the Mathematics Education Research Group of Australasia, Hobart, Vol 2, pp. 775-783). Adelaide: MERGA.

Watego, L. (2005). An Indigenous perspective on mathematics contextualisation in a pre-school: From safety to empowerment. In P. Clarkson, A. Downton, D. Gronn, M. Horne, A. McDonough, R. Pierce, & A. Roche (Eds.), *Building connections: Theory, research and practice.* (Proceedings of the 28th annual conference of the Mathematics Education Research Group of Australasia, Melbourne, Vol. 2, pp. 767-774). Sydney: MERGA.

Young-Loveridge, J. (2005). *Seeing the mathematics of Maori boys in a new light: The benefits of diagnostic interviews.* Paper presented at the NZARE National Conference, Dunedin, New Zealand.

Zevenbergen, R., Mousley, J., & Sullivan, P. (2004). Making the pedagogic relay inclusive for Indigenous Australian students in mathematics classrooms. *International Journal of Inclusive Education, 8*(4), 391-405.

AFFILIATIONS

Tamsin Meaney
University of Otago, New Zealand

Colleen McMurchy-Pilkington
University of Auckland, New Zealand

Tony Trinick
University of Auckland, New Zealand

GLORIA STILLMAN, JILL BROWN, & PETER GALBRAITH

RESEARCH INTO THE TEACHING AND LEARNING OF APPLICATIONS AND MODELLING IN AUSTRALASIA

This chapter commences by considering perspectives on teaching through modelling and applications, followed by a review of research on specific aspects within the field. Studies exploring connections between modelling and complex situations, including topics such as teachers' models of learning and student sense making in modelling situations are then examined. In a section on methodological tools in modelling research alternative developments, including a framework for identifying student blockages and reflection tool design, are reviewed. A wider view on curriculum change involving mathematical modelling precedes the final section in which the current situation in modelling and applications research is reviewed.

Key words: applications, context, mathematical modelling, real world, task design, technology

INTRODUCTION

Applications and modelling research has been part of the agenda of the international mathematics education community for a substantial period, where it has been the focus of Working or Topic Study Groups at ICME Congresses for more than 20 years. During that time a variety of curriculum initiatives at all levels of education have been undertaken, that vary from complete courses, to the more common infusion of aspects of mathematical modelling within a wider curriculum. Although serious efforts to include applications and mathematical modelling have been spasmodic in the mathematics curricula of many education systems in the Australasian region, research into teaching and learning through applications and mathematical modelling has been ongoing because of its potential to add another dimension to the mathematical experience and skill of learners. It is currently being pursued with renewed vigour in many parts of the world (Kaiser, Blomhoj, & Sriraman, 2006), boosted by the 14th ICMI study on applications and modelling in mathematics education held in Dortmund, Germany, in 2004. The pre and post study volumes (Henn & Blum, 2004; Blum, Galbraith, Henn, & Niss, 2007) provide a significant documentation of the state of affairs in this field of research and practice where a number of the Australasian research community have been active for some time (e.g., English, Fox, & Watters, 2005; Galbraith & Stillman, 2006; Ikeda, Stephens, & Matsuzaki, 2007; Izard, 2007). Another major stimulus for researchers in this field has been the biennial International Conferences on the

H. Forgasz et al. (eds.), Research in Mathematics Education in Australasia 2004-2007, 141-164.

Teaching of Mathematical Modelling and Applications (ICTMA). At ICME 10 in Copenhagen, the ICTMA community, which had sponsored the conferences for over twenty years, participated as an Affiliated Study Group of ICMI for the first time, and research on applications and modelling from Australasia (e.g., Galbraith, 2004; Stillman, 2004b) was showcased in several forums. Against such a continually emerging field of interest, this chapter will present and critique research in this region or by researchers from Australasia from a number of perspectives.

PERSPECTIVES ON TEACHING THROUGH MODELLING AND APPLICATIONS

The term mathematical modelling, as it is used in curricular discussions and implementations does not have a unique meaning. We acknowledge that the term has been given a variety of idiosyncratic meanings for localised purposes, but in this chapter the intention is to focus on work associated with interpretations that are accepted and pursued within the international community of workers in the field (see e.g., Blum et al., 2007).

Five (at least) approaches to the use of mathematics with connections to the real world can be identified within mathematics education:

1. Using contextualised examples to motivate the study of mathematics;
2. Modelling as curve fitting – particularly using CAS;
3. Using modelling as a vehicle for teaching mathematics;
4. Modelling as course content, in which preserving the integrity of modelling as a mathematical activity is central within its educational purposes; and
5. Emergent modelling as illustrated by activity in the Netherlands. [While this approach has some links to 1 and 3 its emphasis is slightly different as indicated by the title of a recent paper: *Emergent modelling as a precursor to mathematical modelling* (Gravemeijer, 2007).]

In this review we shall be discussing Australasian work, which in practice relates to each of the above (except 5). Meanings and distinctions will be elaborated within the discussion of specific programs and papers. Within overall alternative approaches to applications and modelling, specific activities pertaining to the performance of students and teachers occur, some of which have a common although varied presence across approaches. This review will also address some such specifics that have been a focus of activity within the Australasian context.

Using contextualised examples to motivate the study of mathematics

Examples of an interpretation that sees mathematical modelling primarily as a means of motivating, developing, and illustrating the relevance of particular mathematical content is found in Chinnappan and Thomas (2003) and Pierce (2005). When the appropriate mathematics has been motivated or illustrated, frequently the process stops, whether or not important real-world aspects of a problem situation remain incomplete. A computer simulation of the Flying Gonzo (http:://www.funny-games.biz/flying-gonzo.html) showing the trajectory of Gonzo as he is shot out of a cannon, for example, might be used to show that a quadratic model is appropriate for projectile motion (Goos & Bennison, 2007).

Pierce and Stacey (2006) explored why teachers chose problems set in real world contexts, and the teaching priorities that influenced the way such problems were treated. They were interested in the extent to which the 'halo effect' – interpreted as the achievement of positive feelings towards mathematics through the use of problem contexts that appeal to students' interests – influenced pedagogical choices. It emerged that the goals were overwhelmingly affective rather than cognitive, so that mathematical learning was frequently subservient, with more weight attached to how students were viewing a task, and their associated feelings about mathematics, than to exploiting its potential for enhancing mathematical understanding. In many cases the treatment of problems rendered terms like mathematical modelling and applications inappropriate to describe the intention and practice of the classroom activity. These outcomes illustrate that the mere use of a real-world context in presenting mathematics does not necessarily mean that mathematical modelling or applications work is being conducted in some significant way, a misunderstanding not uncommon among those with only a passing acquaintance with the field.

Modelling as curve fitting

This approach has become increasingly common with the availability of regression alternatives in software and graphical calculators. Galbraith (2007) gives an example involving the derivation of an equation to fit the seasonal variations in daylight time for two Australian cities. Using the trigonometric regression option on a TI-83 calculator gives a curve of best fit for the real data for which the maximum occurs on December 15, and the year has 377 days. The model generated by this means is a purely technical artefact whose parameters vary with the particular data set, and which can be generated in complete ignorance of the principles underlying the real situation – indeed undertaken in complete ignorance of where a table of data comes from. It raises a profound theoretical issue – the relative authority of empirical data versus theoretical structure. Curve fitting remains an important ability to foster within the wider modelling enterprise. However, using curve fitting as a synonym for mathematical modelling creates a dangerous aberration of the modelling concept. In particular, the subversion of the requirement of reality, by choices available on the menus of calculators or computers, represents a substantial distortion of the purpose of modelling, and leads to bad modelling habits as well as to inappropriate outcomes.

Modelling as vehicle

The terms modelling as vehicle and modelling as content have been used by Julie (e.g., 2002) to distinguish between two important but distinct emphases with respect to how mathematical modelling is introduced and used within education. When used as a vehicle, modelling contexts are chosen so that the mathematics of interest is embedded in the associated examples. The principal driving force in defining the boundary of the activity is curriculum content, and the genuineness of applications is made subservient to this perceived need. Note however that useful modelling skills may be utilised within this approach as indicated below.

Within this interpretation of modelling the next two papers describe studies into the teaching of linear functions using problems based in real-world contexts. Bardini and Stacey (2006) focus on year 9 student conceptions of the parameters '*m*' and '*c*' within the function rule: $y = mx + c$. Symbolic, numerical, and graphical representations of this relationship were considered. It was found that some problem specific features may have affected the way in which meanings of '*m*' and '*c*' were conceptualised. Bardini, Pierce, and Stacey (2004) also involved an approach to linear functions using contextualised problem settings, this time with year 8 above average ability students, but involved as an additional goal the use of graphical calculators to facilitate learning. Intercepts, slopes, points of intersection, and intervals required interpretation in context, across a variety of problem settings. Of primary interest was the level of understanding and facility with algebra that students developed during the five week study. Findings were: they learned to write algebraic rules in conventional formats, were comfortable selecting symbols that made sense in terms of the problem settings, and showed understanding of the function property of expressing one variable quantity in terms of another. The technology generally provided the students with support for the exploration of problems in real-world settings, and positive attitudes were recorded. A reader might observe that five weeks is a substantial investment of time for the intended purposes.

In both cases the effectiveness of learning was assessed by means of pre and post testing. The reported instruction featured direct teaching aimed at individual learning – students were expected to make decisions about contextualised problems on their own, whereas in reality, for example, a household decision about which plumber to hire would usually be taken by a couple or family group, after examining, and perhaps debating, the merits of respective quotes.

Confusion was exhibited by students, as in giving competing interpretations to the parameters '*m*' and '*c*' in a practical context, and this is precisely the kind of situation where individual understanding can be clarified by peer discussion. Such an approach is supported both by the process of modelling as a team activity, and by Vygotskian theory whereby learning in a group ZPD (Vygotsky, 1962) is scaffolded by members sharing incomplete but complementary expertise. In summary, useful learning outcomes were achieved in both studies – more might have been possible if authentic rather than curtailed modelling practices had been engaged.

Modelling as content

This perspective is motivated by the desire for students to develop and apply skills appropriate to obtaining a mathematically productive outcome for a problem with genuine real-world connections. Here the solution to a problem must take seriously the context outside the mathematics classroom, within which the problem is located, and its evaluation involves giving serious attention to that context. This approach (e.g., English, 2007; Ikeda et al., 2007; White, 2005, 2006) does not view applications and modelling merely as a means for achieving some other mathematical learning end, but for developing abilities that are significant in themselves. The goal is to equip students with skills that enable them to apply and communicate mathematics in relation to the solving of problems in their world.

While the content and vehicle approaches differ in some key emphases, their goals are not necessarily always mutually exclusive, since in seeking to solve genuine problems the need for new mathematical content may emerge, while real-world contexts can provide legitimate vehicles for the structured introduction of desired mathematics. Galbraith and Stillman (2006), for example, see the first interpretation as encompassing the second, and these researchers claim their aim is to make their work applicable both to those who value the goal of teaching students to apply their mathematics to life-related situations, and to those who, restricted by choice or circumstances to more limited opportunities, are nonetheless keen to provide motivating applications and genuine illustrations of the mathematics they teach.

It should be pointed out that neither the content nor vehicle approach argues in some abstract sense that all mathematical curricular content must be justified in terms of relevance – mathematical modelling has a role to play in meeting certain important goals, but other significant mathematical skills and purposes are important as well.

Within this genre, Galbraith, Stillman, Brown, and Edwards (2007) in taking the approach that the real-world problem must be taken seriously, acknowledge two important factors that impact on their research. Firstly, real data are usually messy, and secondly, access to technology can change the types of problems students, at various levels of schooling, can access and also impacts on the sophistication of solution processes available. Consequently, their research seeks to identify and document characteristic levels of performance in modelling and applications tasks (Stillman, 2004a; Stillman & Brown, 2007); the occurrence or removal of blockages during solution (Galbraith & Stillman, 2006; Galbraith, Stillman, & Brown, 2006); the use of numerical, graphical, and algebraic approaches; quality of argumentation; and the interactions between modelling, mathematical content, and technology (Galbraith, Stillman, Brown, & Edwards, 2007).

Within their approach to mathematical modelling as content, Galbraith, Stillman, Brown, and Edwards (2007) and other researchers (e.g., McNab, Moss, Woodruff, & Nason, 2006; White, 2006) include the appropriate use of technology as central, and its integration with mathematics within the modelling process as creating essential challenges requiring more research. Hence, they are particularly interested in exploring aspects represented by intersections in Figure 1.

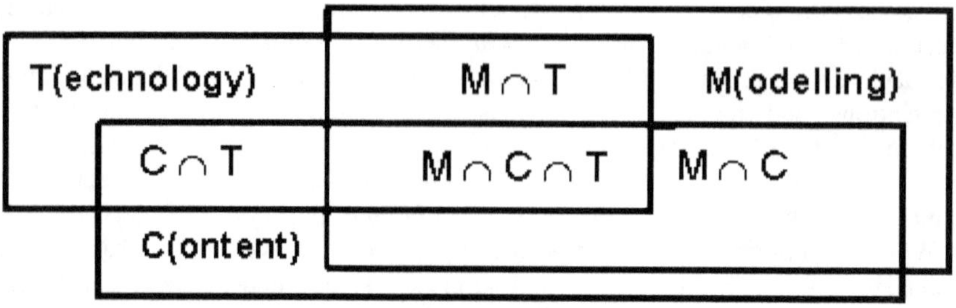

Figure 1. Interactions between modelling (M), mathematics content (C), and technology (T) (Galbraith, Stillman, Brown, & Edwards, 2007).

Galbraith Stillman, Brown, and Edwards (2007) argue that for beginning modellers the mathematics required for solution needs to be within the range of known and practised knowledge and techniques. However, it may not be clear exactly which mathematics is appropriate for a situation under investigation – these decisions are part of the modelling process. Additionally, the presence of a technology-rich teaching and learning environment (TRTLE) impacts on the modelling process, changing the mathematics that is accessible to students.

TRTLE's afford new ways of engaging students in learning mathematics. The formulation phase of an extended modelling task can present many challenges for beginning modellers especially at Year 9. However, teacher support to meet these challenges allows meaningful engagement with such tasks and first hand experience of how mathematical modelling is an essential form of reasoning about real situations. Stillman and Brown (2007) present findings from the implementation of an extended modelling task in three Year 9 classes, each from a different school, where teacher and student moves in successfully negotiating the formulation phase are examined. In the classes where the majority of students had less difficulty with formulation, the teachers used a combination of physical demonstrations and dynamic computer simulations thus linking enactive and iconic worlds. In effect, these teachers used these demonstrations to help students "put themselves mentally in the situation" (Leiß, 2006, p. 87) in order to develop and refine a suitable model for the situation.

A TRTLE is a place where the intention is that students and teachers enact myriad affordances of the environment (i.e., offerings of the environment for both facilitating and impeding learning through potential relationships between the teacher and/or student and the environment) for supporting the learning of mathematics, as an integrated part of the learning and modelling process. Affordances include the multiple representation-ability offered by function graphers, for example, allowing a multiple representation approach, which offers additional opportunities for interpretation across representations as well as from each representation back to the situation being modelled. The construct, affordances, is being used by researchers (e.g., Brown, 2005, 2006) to highlight the important differences between what is available, and what is actually taken up by teachers and students.

In an example located within the tertiary sector, Gruenwald, Sauerbier, Zverkova, and Klymchuk (2007) describe a joint project focussing on first year university engineering mathematics students. Students at two universities, Wismar, Germany and Auckland University of Technology were presented with "real ecology models developed by professional mathematicians" (p. 315). Students in both countries were given models related to local environmental issues. As models were almost complete, the students were not required to engage in the full modelling cycle. The German students had to complete formulation and set up the (differential) equation describing a pollution concentration model. The New Zealand students were required to write an extensive interpretation giving recommendations about permitted fishing quotas. Both student groups had to find the solution analytically. Data collection included task scripts from 147 students and an anonymous questionnaire completed by 63 students. The results from the two groups were similar and hence combined. The questionnaire showed only 48% of responding students perceived the project as practical and 35% as relevant to their career in mechanical or electrical engineering. The researchers acknowledged that their "assumption that the engineering students [would] relate to important ecology problems from their region in personal and emotional ways was too optimistic" (p. 321). Analysis of student questions and scripts identified difficulties in the formulation, solving and interpretive stages of the modelling cycle. Many students could not select appropriate techniques, formulae or information, nor apply their knowledge in an unfamiliar context or with unfamiliar notation. The authors see implications for their own teaching including the need for their students to appreciate the importance of enhancing their modelling competencies.

RESEARCH ON SPECIFIC ASPECTS WITHIN THE FIELDS OF APPLICATIONS AND MATHEMATICAL MODELLING

Thus far we have considered broad interpretations within which various teaching and learning programs involving applications and mathematical modelling have been developed and employed. We now turn attention to specific teaching and learning foci, which are central to the field and which have been the subject of recent activity by Australasian researchers.

Senior secondary students and strategy selection in applications tasks

Little modelling and applications research appears to have been reported in the final years of secondary schooling during the review period. Stillman (2004a) has reported research involving "the study of mathematical applications in life-related situations" (p. 50) by Year 11 and 12 students in a large provincial Australian city. Data collection included the video recording of 41 students as they attempted to solve up to four applications tasks, supplemented by 64 semi-structured simulated recall interviews undertaken immediately following task completion. Task access conditions were identified by applying qualitative data analysis techniques using a cognitive/metacognitive framework developed by the author. Major conditions facilitating task access were recall of a similar or parallel task; prior knowledge of task context; recognition of mathematical cues in the form of trigger words or visual features; being able to visualise the situation; and engagement with the task context. Major conditions impeding task accessibility included language problems (technical language, comprehension difficulties, and unusual wording of the task); mathematical or contextual cuing words not being salient; difficulties with the integration of given or derived contextual information or in recalling a concept, formula, procedure or relevant prior knowledge of the task context. Strategies used by students to facilitate task access included retrieval strategies to recall information, recognition strategies, mental imaging strategies, perception, and the use of elaboration.

Stillman's (2004a) framework proved useful in identifying and examining these conditions and the strategies that the students employed in taking advantage of facilitating conditions or successfully overcoming initial impediments to progress. In general a well-developed repertoire of cognitive and metacognitive strategies and a rich store of mathematical knowledge, real-world knowledge and experiences, and comprehension skills facilitated access. Once moderate skill was achieved in accessing these applications, coordination and integration of multiple representations, further cues, and mathematical processes and procedures became critical.

This research, consistent with an earlier study (Williams, 2002), suggests that whilst task difficulty varies from student to student, task complexity is fixed as it is determined by task attributes; however, "these attributes may be related to particular solution methods rather than the task per se" (Stillman, 2004a, p. 63). This is an important finding with implications for teachers of senior secondary students in their task selection, and also for researchers as this is clearly an avenue for further work.

Task design

Galbraith (2007) discusses three of the models of modelling introduced earlier in this chapter, namely, modelling as vehicle and modelling as content and the restricted definition of modelling as "fitting curves to sets of data points" (p. 48). In response to concerns about the integrity of mathematical modelling, Galbraith proposes a set of principles to be used when searching for mathematical modelling contexts "that provide legitimate experience for students within a program driven by a philosophy that values real-world authenticity" (p. 54). They were derived from examining outcomes and attributes of successful modelling problems used in research studies. The five principles together with a final pedagogical design principle are:

- There is some genuine link(s) with the real world of the students.
- There is opportunity to identify and specify mathematically tractable questions from a general problem statement.

- Formulation of a solution process is feasible, involving the use of mathematics accessible to students, the making of necessary assumptions, and the assembly of necessary data.
- Solution of the mathematics for the basic problem is possible, together with interpretation.
- An evaluation procedure is available that enables checking for mathematical accuracy, and for the appropriateness of the solution with respect to the contextual setting.
- The problem may be structured into sequential questions that retain the integrity of the real solution. (p. 55)

Designing extended real-world investigative tasks that present manageable and engaging challenges for lower secondary students is a substantial challenge. Stillman (2006) reports on a teaching experiment related to implementation of two extended real world tasks with a Year 9 class that focussed on what students perceive as challenges during the solution of such tasks. In both designing the tasks and in providing timely task scaffolding during task implementation, there were anomalies between the expected student moves and challenges, and what actually transpired. Some students took up the challenges carefully crafted into the task as expected, but for others these same challenges did not eventuate as the significance of particular task requirements was missed, or the mathematical implications of interim results which should generate challenge were not realised. Unforeseen challenges arose for other students as they discovered different complexities in their unanticipated interpretation of tasks.

Mathematical modelling and multiple-choice questions

While it might initially appear that multiple-choice items and mathematical modelling are incompatible, such items have been used in two distinct ways – as assessment (Izard, 2007) and as an introduction to modelling (Ikeda et al., 2007). Izard argues that multiple choice items can be used to assess aspects of students' understanding of mathematical modelling, whereas Ikeda et al. (2007) suggest they have a role in the often crowded curriculum, in order to provide students with at least some insight into mathematical modelling.

Izard (2007) employed the modelling cycle as he sought to develop appropriate modes for assessing progress in mathematical modelling. Arguing that, "performance on real-life tasks provides evidence of achievement" (p. 160), Izard suggests an alternative to the usual assessment practices of comparing students with students is to "compare each student's achievement with the curriculum intentions" (p. 166). Seven hundred students undertook 21 modelling test items, where each partial credit multiple-choice item focussed on "student skills at key developmental stages without the students being required to carry out a complete modelling task" (p. 159). From the original item bank of questions, Izard details how various subsets of these can be employed and shows how to select items that contribute to understanding of what each student can do.

In contrast, Ikeda et al. (2007) use multiple-choice items "as a basis for classroom teaching about mathematical modelling" (p. 101). Their pilot study, involving eight high school students in Japan, builds on earlier work (Ikeda & Stephens, 2001) which suggested well-structured questions discussed in both small group and whole class situations supported understanding of modelling. Ikeda et al. (2007) found that "when combined with group discussion and careful teacher direction, the use of multi-choice modelling tasks... was quite effective in helping to shape students' thinking about key features and stages of mathematical modelling" (p. 108). They acknowledge that, whilst this learning experience was no substitute for actually engaging in extended modelling tasks, such scenarios have a place in the classroom. Firstly, they give students some modelling experiences where time does not permit extended tasks. Secondly, implementation of extended modelling tasks presents a challenge for teachers but the use of the multiple-choice items allows teachers, who otherwise may not bring any mathematical modelling into their classrooms, to engage students in some aspects.

Turner (2007), using results from PISA 2003 (OECD, 2004) on both short response and multiple choice items where elements of modelling were needed for solution, points out that the level of complexity of mathematical modelling activities that 15 year old students can cope with is low despite curriculum documents in many countries indicating that the mathematical thinking underpinning such tasks is highly valued. Tasks not involving modelling were easier for students. However, the kind of thinking demanded by different phases of the modelling cycle impacted on difficulty. Items requiring only interpretation and reflection phases were of intermediate difficulty whereas items requiring formulation of a model or manipulation of a given model were most difficult. Turner suggests breaking the modelling process into components would make it more accessible to more students.

STUDIES EXPLORING CONNECTIONS BETWEEN MODELLING AND COMPLEX SITUATIONS

The research agenda of the models and modelling group based at QUT has been to develop and demonstrate the efficacy of this perspective in the primary school (Years 1 to 7 in Queensland) and lower secondary school through implementation studies firstly in Years 5, 6, 7 (English, 2004, 2006b; English & Fox, 2005), then Years 3, 4, 5 (English, Fox, & Watters, 2005; English & Watters, 2004, 2005a, 2005b) and finally Years 4 to 9 (English, 2006a, 2007). The first of these studies, involved semester-long sequences of model-eliciting activities aimed at developing students' higher-order mathematical understandings and abilities. In particular it was hoped that these modelling activities would enable students to develop conceptual tools for success beyond school.

> A perspective on task characteristics leads to an instructional sequence of activities that begins by engaging students with non-routine problem situations that elicit the development of significant mathematical constructs and then extending, exploring and refining those constructs in other problem

> situations leading to a generalizable system (or model) that can be used in a range of contexts... The students' descriptions, explanations, and justifications form an integral component of the models they produce. In contrast to many of the problem-solving situations students meet in school, modelling activities are inherently social experiences, where students work in small teams to develop a product that is explicitly sharable. Numerous questions, issues, conflicts, resolutions, and revisions arise as students develop, assess, and prepare to communicate their products. (English & Doerr, 2004, p. 217)

The boundary between the kind of modelling described here (involving model – eliciting activities) and emergent modelling (e.g., Gravmeijer, 2007) is blurred and broad. As the latter points out, 'emergent modelling' represents an attempt to improve on 'didactical modelling' in endeavouring to make abstract mathematics more accessible for students. It typically starts with a real situation that is used to stimulate progressively deepening conceptual understanding of concepts made relevant by the choice of the situation. Model-eliciting activities maintain a stronger situational link by revisiting the original context in terms of a problem it has posed. And a sequence of model-eliciting activities aims to enhance the abilities of learners as mathematical modellers, which is not necessarily the goal of the former.

Research perspectives into how teachers might develop an understanding of the utility of this approach to models and modelling as both a teaching and learning approach is also necessary (Watters, English, & Mahoney, 2004). Class teachers' models of their students' mathematical learning was a major focus of the first implementation study and a smaller concurrent study (Doerr & English, 2006; English & Doerr, 2004). The second major implementation study investigated the enhancement of primary school children's (and their teachers') capacities to engage in basic mathematical modelling as a foundation for working with more complex situations. A third implementation study, located in a regional P-12 coeducational college, aims to trace students' development through involvement in modelling activities, develop teachers' means of interpreting and supporting their students' ways of thinking mathematically, and produce a practice based collaborative professional development model focussing on designing and implementing complex problem-solving activities. Additional studies (e.g., Doyle, 2006a; McNab et al., 2006) have undertaken related or independent investigations in the primary school context, while Fox (2006) advocated future studies explore the development of a modelling program for even younger students beginning in the preparatory classroom.

Teachers' models of learning

English and Doerr (2004) posit that "effective teachers need: (1) an understanding of the multiple ways that students' thinking might develop; (2) ways of listening to that development; and (3) ways of responding with pedagogical strategies that will support that development" (p. 216). Two papers (English & Doerr, 2004; Doerr & English, 2006) describe how two middle years teachers working with modelling tasks, developed their understanding of how students worked, and changed their own perspectives and practices. Both developed their pedagogical content knowledge in ways that were partly determined by their respective mathematical content knowledge, but were strongly influenced by a developing appreciation of what their students were doing and saying. It was observed (English & Doerr, 2004) that one teacher shifted to "an interpretive and hermeneutic orientation" but associated evidence was not provided, and the claim was not repeated in the later paper. The classification in these terms derives from a characterisation by Davis (1996), whereby: an evaluative teacher is primarily concerned with diagnosing and correcting misunderstandings; an interpretive teacher with accessing students' understandings through listening; whilst a teacher with a hermeneutic orientation continually interacts with students, listening to ideas and engaging them in negotiation of meaning and understanding.

Doerr and English (2006) also examined how three design principles for modelling tasks impacted on teachers' learning. Teacher approaches included (for one teacher) the use of various perspectives (e.g., viewpoints of a mathematician, a manufacturer) to engage students in realistic sense making for the task situation, while both teachers shifted from "listening for the purpose of evaluating students work to listening for the purpose of engaging students in self-evaluation of their work" (p. 28). The teachers' use of representational forms reflected differing views of the nature of mathematics (number based versus wider conceptual views), with the latter providing more open and challenging opportunities for both teacher and students to work increasingly with interpretive rather than evaluative orientations.

Students' models for making sense in modelling situations

English (2004, 2006b) reports the findings from the students' perspective of the second year of the first implementation study. One model eliciting activity, one model exploration activity and two model-application/model-adaptation activities were used with Year 6 students over a six month period. The main findings (English, 2004) were that students applied mathematical ideas in different ways to generate models independently of instruction and the modelling activities provided rich experiences for students where they engaged in interpreting and re-interpreting the problem statement, clarifying and revisiting the goal, making decisions subjectively, applying and refining mathematical procedures, posing hypotheses and problems, making assumptions, posing arguments and counter arguments, considering all options, asking for justification, thinking metacognitively, and applying previous models. The tasks used, where students worked collaboratively and were required to present and justify their model, provided many opportunities for students to use mathematics to solve real problems, and to make generalisations and utilise this learning in later tasks. English (2006b) deals with the first of the model adaptation activities, the creation of a consumer guide for snack chips. In this task students had opportunities to transfer and refine mathematical constructs developed in previous tasks and to refine any representational systems they might have developed. A detailed analysis of the iterative cycles of development for one group is presented. The mathematisation processes displayed by all groups is then examined. Children's development included creating systems for operationally defining constructs; selecting, categorising, and ranking factors; quantifying quantitative and qualitative data and transforming quantities.

English and Fox (2005) describe the third and final year of this first implementation project when the students were in Year 7. Again, one model eliciting activity, one model exploration activity and two model-application/model-adaptation activities were used with the final problem being the focus of the paper. Being more experienced modellers, students focused more on building their models and not "getting bogged down initially in debating irrelevant contextual issues, [and so]... made more use of examples and counter-examples in justifying and questioning the ideas they proposed" (p. 327) as they clarified, justified, and argued about the validity of their models. Clearly, this first longitudinal study gave strong support for the inclusion of mathematical modelling in the primary years. Furthermore, "the interdisciplinary nature of mathematical modelling means that we can create problems that can help unify some of the myriad core ideas within the primary curriculum" (English, 2007, p. 282) across discipline areas.

Use of literacy strategies in modelling tasks

Modelling tasks used in these studies require students to elicit the main ideas about an authentic real-life situation, such as a lawn mowing business, from a variety of narrative, expository and graphic texts (usually involving tables). The information is then to be organised meaningfully so they can share, explain and describe ideas as they work in groups and communicate their models to other groups through written and oral presentation. Literacy processes are thus an integral part of this type of modelling experience. "To be mathematically literate necessitates relevant selection, understanding and organization of this information so that one can acquire relevant knowledge" (Doyle, 2005, p. 40). Doyle proposed that top-level structuring (TLS) be used as a strategy in text based modelling problems for identifying and recalling the main ideas in the background information and problem statements and when students had to "select and justify a main idea, or to discuss, argue and communicate their position in a planful, strategic way" (p. 43).

Doyle (2006a, 2006b) reports the findings from an intervention study involving two Year 4 classes with moderate literacy and numeracy levels, who had not been exposed to modelling previously. The introduction of TLS involving the four basic organisational text structures of comparison, cause/effect, problem/solution, and list/description together with the signalling words associated with these, differentially within the two classes resulted in an increase in the amount of peer questioning, by (and of) student groups, increased attention to task, and greater levels of explanation, questioning, and justification during presentations. Additional evidence of the positive impact of TLS was identified in the subsequent structuring of written work.

In a further report on the same study (Doyle, 2007), the critical role of the teacher when using modelling tasks in both constructing tasks appropriate for learning and responding appropriately to classroom interactions to enhance rather than detract from students' mathematising is emphasised. Doyle sees mathematical modelling with TLS as a suitable means for teachers to provide a mathematical sense-making environment for students.

Transfer of collective argumentation from the classroom to a modelling competition

Brown and Redmond (2007) explore the effects of a novel situational context – a mathematical modelling challenge for mixed school teams–on student use of Collective Argumentation (Brown & Renshaw, 2000). The three Year 5/6 students who are the focus of the paper had been using Collective Argumentation in class but were allocated to the same group with another student who had not. Brown and Redmond infer from their interpretation of student-student interactions in transcripts of videotapes taken during the modelling challenge, that the focus students viewed doing mathematics in the challenge "as providing a forum where personal understandings can be expressed, reconstructed, shared and co-authored" (pp. 169-170) despite being in a competition. Thus, their modelling activity displayed the hallmarks of "a culture of inquiry", namely "cooperation and collaborative relationships" (p. 170) stimulated by their regular classroom experience.

METHODOLOGICAL TOOLS IN MODELLING RESEARCH

In this section we consider recent research that has focused on the development of research tools for facilitating the investigation of aspects of teaching and learning in the field of applications and modelling.

Framework for identifying blockages in transitions

Galbraith, Stillman, Brown, and Edwards (2007) have developed and refined a new research tool (Galbraith & Stillman, 2006; Galbraith, Stillman, & Brown, 2006) in attempting to address their goals of (a) identifying and classifying critical aspects of modelling activity within transitions between stages in the modelling process, and (b) identifying pedagogical insights for implementation through such activities as task design and organisation of learning.

Typically a genuine modelling task begins with a messy real world problem which is then transformed and hopefully solved mathematically as the modeller carries out various processes in the modelling cycle. In Figure 2 the modelling cycle is depicted as a series of products, connected by arrows representing transitions that involve cognitive activity essential to the modelling process. This diagram includes arrows in the reverse direction, emphasising the reality that the modelling cycle is far from unidirectional, and needs to account for metacognitive aspects as modellers engage in reflective activity.

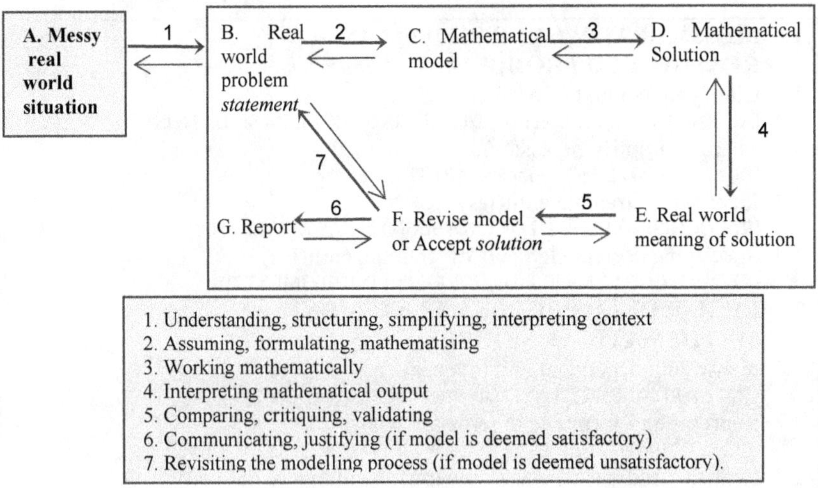

Figure 2. Modelling process (Stillman, Galbraith, Brown, & Edwards, 2007).

Initially, the tool consisted of an empty frame of the transitions between phases in the solution process as described in the modelling cycle. Using preliminary data analysis an emergent framework was developed (Galbraith, Stillman, & Brown, 2006; Galbraith, Stillman, Brown, & Edwards, 2007) for identifying potential places where student blockages could occur in transitions between phases of the modelling cycle. Details for the first two transitions illustrated with data from various tasks used at year 9 level are shown in Figure 3. Generic elements are in regular type, instantiations for a particular task in small capitals. Continuing analysis of data from tasks (Stillman, Galbraith, Brown, & Edwards, 2007) has led to further refinements including the need for a further construct, level of intensity of a blockage (Stillman, Brown, & Galbraith, 2007).

The Framework has direct application arising from the way the research has been conducted – to identify specifically, activities and content with which modellers need competence in order to successfully apply mathematics at their level. This is what the Framework systematically documents. Given that the elements in the framework were identified by observing students working (and, in particular, wrestling with blockages to progress), there are two immediate potential applications. First are the insights obtained into student learning, and how these can inform our understanding of the ways that students act when approaching modelling problems. Second are associated pedagogical insights. By identifying difficulties with generic properties, the possibility arises to anticipate where, in given problems, blockages of different types might be expected. This understanding can then contribute to planning teaching and task design, in particular the identification of prerequisite knowledge and skills, preparation for intervention at key points if required, and scaffolding of significant learning episodes.

1. MESSY REAL WORLD SITUATION → REAL WORLD PROBLEM STATEMENT
1.1 Clarifying context of problem
[ACTING OUT, SIMULATING, DISCUSSING PROBLEM SITUATION]
1.2 Making simplifying assumptions
[RUNNERS WILL MOVE IN STRAIGHT LINES]
1.3 Identifying strategic entit(ies)
[RECOGNISING LENGTH OF LINE SEGMENT AS KEY ENTITY]
1.4 Specifying correct elements of strategic entit(ies)
[IDENTIFYING SUM OF TWO CORRECT LINE SEGMENTS]

2. REAL WORLD PROBLEM STATEMENT → MATHEMATICAL MODEL
2.1 Identifying dependent and independent variables
[TOTAL RUN LENGTH AND DISTANCE FROM CORNER]
2.2 Representing formulae in terms of 'knowns'
[LENGTH EXPRESSED IN TERMS OF FIELD EDGE DISTANCES]
2.3 Realising independent variable must be uniquely defined
[X-CANNOT BE DISTANCE FROM BOTH A AND B]
2.4 Choosing technology to enable calculation
[RECOGNISING HAND METHODS ALONE ARE IMPRACTICAL]
2.5 Choosing technology to automate formulae for multiple cases
[LISTS HANDLE MULTIPLE X-VALUES]
2.6 Choosing technology to produce graphical output
[SPREADSHEET OR GRAPHING CALCULATOR]

Figure 3. Initial transitions in Galbraith, Stillman, Brown, & Edwards (2007) Framework.

Reflection tool design

English (2006b) noted that: "reflection tools [are] being recognised increasingly as an essential component of the curriculum" (p. 306). Such tools not only serve to help students focus on important aspects of mathematical modelling, they also act as a tool for research. In this study students used a reflection tool referred to as "a written critique" where they were required to "reflect on the adequacy of their own and their peers' models for the problem at hand, provide constructive feedback on how the model could be improved, and identify other problem situations in which the model could be applied" (p. 307). In addition to the other data analysis reported above, students' "written critiques were reviewed for their comments on their peers' models and for how they would apply the models to new problem situations" (p. 309). This analysis provided insight into students' thinking and developing understanding of the modelling process. Reported student reflections were classified according to three criteria: queries, suggestions for improvement, and application of the model to another situation. In the first category, seven of the eleven comments "addressed the generalisability of the model" (p. 317). The importance of generalisability also was apparent in students' suggested improvements. Other suggestions included "the need for a more detailed explanation of model development and for more effective tables" (p. 318). In the final category, students' responses showed some evidence of "understanding of how a model could be applied to other problem situations" (p. 319). This tool has potential for providing additional insight into student thinking and developing understanding of mathematical modelling as students' engage in the critical phase of reflection on their model and the models of others.

CURRICULUM CHANGE RELATED TO APPLICATIONS AND MODELLING

In this section we address research that investigates the impact of initiatives introducing aspects of applications and mathematical modelling into curricula.

Several Australian states have attempted to include applications and modelling topics more centrally within their curricula. An on-going study by Stillman (2004b, 2007) is being conducted into the implementation of applications and mathematical modelling curricula at the upper secondary level in two states (Victoria and Queensland) where the path of curriculum change has been different with markedly different outcomes. Data collected included curriculum documents, interviews with key curriculum people and practising teachers, classroom artefacts and video recordings of classroom practice. From the preliminary results of document analysis, Stillman (2004b) raised speculative possibilities for further investigation. Possible influences on whether the change would be sustained included: how teachers are positioned as professionals, conservative political groups, whether curriculum change is discipline oriented versus whole curriculum, and changes in society. Suggested conditions supporting curriculum change included: freedom for teachers to act and accept responsibility, support of the statutory authority, high levels of professional expertise of teachers, atmosphere supporting change, provision of resources, and effective evaluation mechanisms (p. 265).

Stillman (2007) reported some of the findings resulting from the Victorian case study. Following examination of curriculum documents and materials from the 1980s onwards (see Stillman, 2004b), purposeful samples of six key curriculum figures (e.g., members of expert advisory committees or curriculum managers for the statutory board), six teachers in key implementation roles (e.g., as seconded project officers for the statutory board, and state or regional chairs of verification panels), and six classroom teachers were selected. These participants were interviewed using tailored semi-structured interview schedules, all were asked about their experiences during the change and subsequently and about their beliefs regarding conditions that promoted or hindered the introduction and ongoing use of mathematical applications and modelling in upper secondary classrooms in the state. Practising teachers were also asked about the impact of the changing role of applications and modelling in the various mathematics Study Designs on their teaching and assessment practices. Classroom artefacts (e.g., tasks) typifying their current practice in the area of mathematical applications, and modelling at the upper secondary level were also collected. A grounded theory approach to data analysis was taken with the identification of influences on change and conditions impeding and sustaining curriculum change in this area the goal of this analysis. Two of the major conditions that threatened the ability of the Victorian system to sustain the change were (a) the extent of the change and (b) the pace of the change.

The topic of curriculum change is also an appropriate location for discussion of the 'relevance paradox'. Clarke (2006) recalls the 'paradox' identified by Niss (1994), in which the "objective relevance of mathematics in society was contrasted with its subjective irrelevance as perceived by many students" (p. 380). Clarke challenged whether this was a western perspective, in part by citing research in which students from a Chinese school valued mathematics as applicable, even though their own mathematics was taught in a purely mathematical context. In this respect it should be noted that a substantial number of mathematical modelling programs and approaches have been ongoing in China (Ye, Blum, Houston, & Jiang, 2003), so that inferences from a particular school classroom in terms of national characteristics are risky. The South African initiative to contextualise the curriculum around themes of genuine societal significance (e.g., substance abuse and HIV-AIDS) is used as an example of the "blurring of the boundary between the mathematics and the everyday" (Sethole, Adler, & Vithal, 2002, p. 11, cited in Clarke, 2006, p. 381), seen as dissolving the need for explicitly connecting mathematics with its applications. On the one hand this argument appears to suggest that the mere presence of ideas about relevance, or of contextualised mathematical situations embedded in curricula will dissolve the paradox. While of assistance, we know that the mere presence of such material in curricula does not mean that the essentials of mathematical modelling will be engaged in, in any serious way – by no means does it follow that students will learn strategies and skills they can use to solve problems in their lives. This latter outcome involves far more than recognition that mathematics has applications, where indeed the associated mathematics can be trivialised.

A second point follows from the first. What happens in the classroom is the enacted curriculum, which is often not the intended curriculum, and the ultimate burden of successful implementation devolves to individual teacher activity and thence ultimately to teacher education. Unless mathematical modelling is treated with integrity within teacher education programs there should be no expectation that the best of curricular intentions will be realised. In summary, those in the field of mathematical modelling continue to draw sharp distinctions between their purposes and the more limited goals frequently associated with the 'relevance' debate.

CONCLUSION: CURRENT STATE OF PLAY IN MODELLING AND APPLICATIONS RESEARCH

In their review of the current situation regarding applications and mathematical modelling in education worldwide, Niss, Blum, and Galbraith (2007) identified ten issues presenting current challenges, and articulated specific questions within each that require ongoing research and development. These ten issues are:

- epistemology of applications and modelling;
- views of the modelling process;
- goals of applications and modelling and their curricular embedding;
- modelling competencies and beliefs;
- mathematical competencies and their relation to applications and modelling; modelling pedagogy;
- teaching/learning practice of applications and modelling;
- implementation of applications and modelling in practice;
- assessment and evaluation of applications and modelling;
- and technology with and for applications and modelling.

The latter nine issues are explicitly represented, to different extents, in the papers contributing to this review of Australasian research, and the first is inevitably engaged when reflecting on the outcomes and challenges presented by the research studies. The vast majority of this research has been conducted in primary and lower secondary schooling contexts. No research was identified as reported in the period in the contexts of modelling in vocational/technical education and teacher education.

In summarising the contents of this chapter, we note below representative aspects of the above issues that are being researched in the Australasian community. The most concentrated areas of Australasian research activity has been in modelling competencies, teaching/learning practice of modelling in primary and lower secondary classrooms, and implementation of applications and modelling in practice. Specific research efforts are pursuing answers to questions such as the following:

- How are different versions of the modelling process being applied within particular contexts and levels of education, and what are their achievements and ongoing challenges?
- What pressures do teachers indicate are significant in deciding when and how they will incorporate applications and modelling tasks in their teaching? What purposes do they have in using such tasks?
- How have efforts at incorporating modelling at a curriculum level unfolded, and what impediments have impacted on the outcomes achieved?
- What blockages inhibit successful modelling practice, and how can these be addressed through identification and development of necessary sub-competencies?
- In what ways do mathematical, technological, and modelling knowledge and competencies interact in the solution of contextualised problems?
- Which particular mathematical competencies are mathematical modelling approaches most suited to enhance?
- What pedagogies are most productive for the successful implementation of modelling approaches?
- To what extent have teaching approaches and teachers been influenced through the inclusion of applications and modelling material in courses?

Although there has been a clear upsurge in research in Australasia during this time period by both established researchers in the field and the entry of others, some areas such as beliefs about the purpose of modelling and applications in the curriculum, connections between mathematical and modelling competencies and assessment need further attention. There is also a need for further research into the competencies pre-service and in-service teachers need to acquire to effectively teach through a modelling and applications approach and how these competencies might be acquired.

REFERENCES

Bardini, C., & Stacey, K. (2006). Students' conceptions of *m* and *c*: How to tune a linear function. In J. Novotná, H. Moraová, M. Krátká, & N. Stenhliková (Eds.), *Proceedings of the 30th annual conference for the Psychology of Mathematics Education* (Vol. 2, pp. 113-120). Prague, Czech Republic: Charles University.

Bardini, C., Pierce, R., & Stacey, K. (2004). Teaching linear functions in context with graphics calculators: Students' responses and the impact of the approach on their use of algebraic symbols. *International Journal of Science & Mathematics Education, 2*(3), 353-376.

Blum, W., Galbraith, P. L., Henn, H-W., & Niss, M. (Eds.). (2007). *Modelling and applications in mathematics education: New ICMI studies.* (Series No. 10). New York: Springer.

Brown, J. (2005). Affordances of a technology-rich teaching and learning environment. In P. Clarkson, A. Downton, D. Gronn, M. Horne, A. McDonough, R. Pierce, & A. Roche (Eds.), *Building connections: Theory, research and practice* (Proceedings of the 28th annual conference of the Mathematics Education Research Group of Australasia, Melbourne, pp. 177-184). Sydney: MERGA.

Brown, J. (2006). Manifestations of affordances of a technology-rich teaching and learning environment (TRTLE). In J. Novotná, H. Moraová, M. Krátká, & N. Stehlíková (Eds.), *Proceedings of the 30th annual conference for the Psychology of Mathematics Education* (Vol. 2, pp. 241-248). Prague, Czech Republic: Charles University.

Brown, R., & Redmond, T. (2007). Collective argumentation and modelling mathematics practices outside the classroom. In J. Watson & K. Beswick (Eds.), *Mathematics: Essential research, essential practice.* (Proceedings of the 30th annual conference of the Mathematics Research Group of Australasia, Hobart, Vol. 1, pp. 163-171). Adelaide: MERGA.

Brown, R., & Renshaw, P. (2000). Collective argumentation: A sociocultural approach to reframing classroom teaching and learning. In H. Cowie & G. van der Aalsvoord (Eds.), *Social interaction in learning and instruction: The meaning of discourse for the construction of knowledge* (pp. 52-66). Amsterdam, The Netherlands: Pergamon.

Chinnappan, M., & Thomas, M. (2003). Teachers' function schemas and their role in modelling. *Mathematics Education Research Journal, 15*(2), 151-170.

Clarke, D. (2006). Using international research to contest prevalent oppositional dichotomies. *Zentralblatt für Didaktik der Mathematik, 38*(5), 376-387.

Davis, B. (1996). *Teaching mathematics: Toward a sound alternative.* New York: Garland.

Doerr, H., & English, L. D. (2006). Middle-grade teachers' learning through students' engagement with modelling tasks. *Journal of Mathematics Teacher Education, 9*(1), 5-32.

Doyle, K. (2005). Mathematical problem solving: A need for literacy. In B. Bartlett, F. Bryer, & D. Roebuck (Eds.), *Stimulating the "action" as participants in participatory research* (Proceedings of the 3rd annual international conference on cognition, language, and special education research. (Vol. 2, pp. 39-45). Gold Coast, QLD: School of Cognition, Language, and Special Education, Griffith University.

Doyle, K. (2006a). Creating mathematical models with structure. In J. Novotná, H. Moraová, M. Krátká, & N. Stenhliková (Eds.), *Proceedings of the 30th annual conference for the Psychology of Mathematics Education* (Vol. 2, pp. 457-464). Prague, Czech Republic: Charles University.

Doyle, K. (2006b). Organisational structure for mathematical modelling. In P. Grootenboer, R. Zevenbergen, & M. Chinnappan (Eds.), *Identities, cultures and learning spaces* (Proceedings of the 29th annual conference of the Mathematics Education Research Group of Australasia, Canberra, pp. 187-194). Adelaide: MERGA.

Doyle, K. (2007). The teacher, the tasks: Their role in students' mathematical literacy. In J. Watson & K. Beswick (Eds.), *Mathematics: Essential research, essential practice*. (Proceedings of the 30th annual conference of the Mathematics Research Group of Australasia, Hobart, pp. 246-254). Adelaide: MERGA.

English, L. D. (2004). Mathematical modelling in the primary school. In I. Putt, R. Faragher, & M. McLean (Eds.), *Mathematics education for the third millennium: Towards 2010*. (Proceedings of the 27th annual conference of the Mathematics Education Research Group of Australasia, Townsville, pp. 207-214). Sydney: MERGA.

English, L. D. (2006a). Introducing young children to complex systems through modelling. In P. Grootenboer, R. Zevenbergen, & M. Chinnappan (Eds.), *Identities, cultures and learning spaces* (Proceedings of the 29th annual conference of the Mathematics Education Research Group of Australasia, Canberra, pp. 195-202). Adelaide: MERGA.

English, L. D. (2006b). Mathematical modelling in the primary school: Children's construction of a consumer guide. *Educational Studies in Mathematics, 63*, 303-323.

English, L. D. (2007). Interdisciplinary modelling in the primary mathematics curriculum. In J. Watson & K. Beswick (Eds.), *Mathematics: Essential research, essential practice*. (Proceedings of the 30th annual conference of the Mathematics Research Group of Australasia, Hobart, pp. 275-284). Adelaide: MERGA.

English, L. D., & Doerr, H. M. (2004). Listening and responding to students' ways of thinking. In I. Putt, R. Faragher, & M. McLean (Eds.), *Mathematics education for the third millennium: Towards 2010* (Proceedings of the 27th annual conference of the Mathematics Education Research Group of Australasia, Townsville, pp. 215-222). Sydney: MERGA.

English, L. D., & Fox, J. L. (2005). Seventh-graders' mathematical modelling on completion of a three-year program. In P. Clarkson, A. Downton, D. Gronn, M. Horne, A. McDonough, R. Pierce, & A. Roche (Eds.), *Building connections: Theory, research and practice* (Proceedings of the 28th annual conference of the Mathematics Education Research Group of Australasia, Melbourne, pp. 322-329). Sydney: MERGA.

English, L. D., Fox, J. L., & Watters, J. J. (2005). Problem posing and solving with mathematical modelling. *Teaching Children Mathematics, 12*(3), 156-163.

English, L. D., & Watters, J. J. (2004). Mathematical modelling with young children. In M. J. Hoines & A. B Fuglestad (Eds.), *Proceedings of the 28th annual conference for the Psychology of Mathematics Education* (Vol. 2, pp. 335-342). Bergen, Norway: Bergen College of Education.

English, L. D., & Watters, J. J. (2005a). Mathematical modelling in the early school years. *Mathematics Education Research Journal, 16*(3), 58-79.

English, L. D., & Watters, J. J. (2005b). Mathematical modelling with 9-year-olds. In H. L. Chick & J. L Vincent (Eds.), *Proceedings of the 29th Conference for the Psychology of Mathematics Education,* (Vol. 2, pp. 297-304). Melbourne: University of Melbourne.

Fox, J. (2006). A justification for mathematical modelling experiences in the preparatory classroom. In P. Grootenboer, R. Zevenbergen, & M. Chinnappan (Eds.), *Identities, cultures and learning spaces* (Proceedings of the 29th annual conference of the Mathematics Education Research Group of Australasia, Canberra, pp. 221-228). Adelaide: MERGA.

Galbraith, P. (2004). Applications and modelling in mathematics education: Progress to celebrate – so much more to do. *Abstracts: Plenary and regular lectures, 10th International Congress on Mathematics Education* (ICME-10) (p. 49). Copenhagen, Denmark: ICME-10.

Galbraith, P. (2007). Dreaming a 'possible dream': More windmills to conquer. In C. Haines, P. Galbraith, W. Blum, & S. Khan (Eds.), *Mathematical modelling (ICTMA12): Education, engineering and economics* (pp. 44-62). Chichester, UK: Horwood.

Galbraith, P., & Stillman, G. (2006). A framework for identifying student blockages during transitions in the modelling process. *Zentralblatt für Didaktik der Mathematik, 38*(2), 143–162.

Galbraith, P., Stillman, G., & Brown, J. (2006). Identifying key transition activities for enhanced engagement in mathematical modelling. In P. Grootenboer, R. Zevenbergen, & M. Chinnappan (Eds.), *Identities, cultures and learning spaces* (Proceedings of the 29th annual conference of the Mathematics Education Research Group of Australasia, Canberra, pp. 237-245). Adelaide: MERGA.

Galbraith, P., Stillman, G., Brown, J., & Edwards, I. (2007). Facilitating middle secondary modelling competencies. In C. Haines, P. Galbraith, W., Blum, & S. Khan (Eds.), *Mathematical modelling (ICTMA12): Education, engineering and economics* (pp. 130-140). Chichester, UK: Horwood.

Goos, M., & Bennison, A. (2007). Technology-enriched teaching of secondary mathematics: Factors influencing innovative practice. In J. Watson & K. Beswick (Eds.), *Mathematics: Essential research, essential practice.* (Proceedings of the 30th annual conference of the Mathematics Research Group of Australasia, Hobart, pp. 275-284). Adelaide: MERGA.

Gravmeijer, K. (2007). Emergent modelling as a precursor to mathematical modelling. In W. Blum, P. Galbraith, M. Niss, & H-W. Henn (Eds.), *Modelling and applications in mathematics education: The 14th ICMI Study* (New ICMI Studies Series, Vol. 10, pp. 137-144). New York: Springer.

Gruenwald, N., Sauerbier, G., Zverkova, T., & Klymchuk, S. (2007). Models of ecology in teaching engineering mathematics. In C. Haines, P. Galbraith, W. Blum, & S. Khan (Eds.), *Mathematical modelling (ICTMA12): Education, engineering and economics* (pp. 314-322). Chichester, UK: Horwood.

Henn, H-W., & Blum, W. (Eds.). (2004). *International Commission on Mathematics Instruction* (ICMI) *Study 14: Applications and modelling in mathematics education* (Pre-conference volume). Dortmund, Germany: ICMI.

Ikeda, T., & Stephens, M. (2001). The effects of students' discussion in mathematical modelling. In J. P. Matos, W. Blum, K. Houston, & S. P. Carriera (Eds.), *Modelling and mathematics education: (ICTMA 9) Applications in science and technology* (pp. 381-390). Chichester, UK: Horwood.

Ikeda, T., Stephens, M., & Matsuzaki, A. (2007). A teaching experiment in mathematical modelling. In C. Haines, P. Galbraith, W. Blum, & S. Khan (Eds.), *Mathematical modelling (ICTMA12): Education, engineering and economics* (pp. 101-109). Chichester, UK: Horwood.

Izard, J. (2007). Assessing progress in mathematical modelling. In C. Haines, P. Galbraith, W. Blum, & S. Khan (Eds.), *Mathematical modelling (ICTMA12): Education, engineering and economics* (pp. 158-166). Chichester, UK: Horwood.

Julie, C. (2002). Making relevance relevant in mathematics teacher education. *Proceedings of the second international conference on the teaching of mathematics at the undergraduate level,* (CD). Hoboken, NJ: Wiley.

Kaiser, G., Blomhoj, M., & Sriraman, B. (2006). Towards a didactical theory for mathematical modelling. *Zentralblatt für Didaktik der Mathematik, 38*(2), 82-85.

Leiß, D. (2006). Teacher intervention versus self-regulated learning? *Teaching Mathematics and its Applications, 24*(2-3), 75-89.

McNab, S. L, Moss, J., Woodruff, E., & Nason, R. A. (2006). "We were nicer, but we weren't fairer!" Mathematical modelling exploring "fairness" in data management with 5th and 6th grade students. In G. Burrill (Ed.), *2006 NCTM yearbook: Thinking and reasoning with data and chance* (pp. 171-184). Reston, VA: National Council for Teachers of Mathematics.

Niss, M. (1994). Mathematics in society. In Biehler, R., Scholz, R. W., Straesser, R., Winkelmann, B. (Eds.), *The didactics of mathematics as a scientific discipline* (pp. 367-378). Dordrecht, The Netherlands: Kluwer.

Niss, M., Blum, W., & Galbraith, P. (2007). Introduction. In W. Blum, P. Galbraith, M. Niss, & H-W. Henn (Eds.), *Modelling and applications in mathematics education: The 14th ICMI Study (New ICMI Studies Series)*, (Vol. 10, pp. 3-33). New York: Springer.

Pierce, R. (2005). Algebraic insight underpins the case of CAS for modelling. *The Montana Mathematics Enthusiast, 2*(2), 107-117.

Pierce, R., & Stacey, K. (2006). Enhancing the image of mathematics by association with simple pleasures from real world contexts. *Zentralblatt für Didaktik der Mathematik, 38*(3), 214-225.

OECD (2004). *Learning for tomorrow's world – First results from PISA 2003*. Paris, France: OECD.

Stillman, G. (2004a). Strategies employed by upper secondary students for overcoming or exploiting conditions affecting accessibility of applications tasks. *Mathematics Education Research Journal, 16*(1), 41-71.

Stillman, G. (2004b). Sustained curriculum change: The example of the implementation of applications and modelling curricula in two Australian states. In H-W. Henn & W. Blum, (Eds.), *International Commission on Mathematics Instruction Study 14: Applications and modelling in mathematics education* (Pre-conference volume, pp. 261-266). Dortmund, Germany: ICMI.

Stillman, G. (2006). The role of challenge in engaging lower secondary students in investigating real world tasks. In E. Barbeau & P. Taylor (Eds.), *Proceedings of the ICMI Study 16: Mathematical challenges, Trondheim, Norway,* (Pre-conference volume). Retrieved March 22, 2007, from: http://www.amt.edu.au/icmis16participantpapers.html

Stillman, G. (2007). Upper secondary perspectives on the issues. In W. Blum, P. Galbraith, M. Niss, & H-W. Henn (Eds.), *Modelling and applications in mathematics education: The 14th ICMI Study*. (New ICMI Studies Series) (Vol. 10, pp. 463-468). New York: Springer.

Stillman, G., & Brown, J. (2007). Challenges in formulating an extended modelling task at year 9. In H. Reeves, K. Milton & T. Spencer (Eds.), *Mathematics: Essential for learning, essential for life*. (Proceedings of the 21st biennial conference of the Australian Association of Mathematics Teachers, pp. 224-231). Adelaide: AAMT.

Stillman, G., Brown, J., & Galbraith, P. (2007). *Identifying challenges within transition phases of mathematical modelling activity at year 9*. Paper presented at the 13th International Conference on the Teaching of Mathematical Modelling and Applications held at Bloomington IN, USA. Retrieved January 18, 2008 from: http://site.educ.indiana.edu/Papers/tabid/5320/Default.aspx

Stillman, G., Galbraith, P., Brown, J., & Edwards, I. (2007). A framework for success in implementing mathematical modelling in the secondary classroom. In J. Watson & K. Beswick (Eds.), *Mathematics: Essential research, essential practice*. (Proceedings of the 30th annual conference of the Mathematics Research Group of Australasia, Hobart, pp. 688-707). Adelaide: MERGA.

Turner, R. (2007). Modelling and applications in PISA. In W. Blum, P. Galbraith, M. Niss, & H-W. Henn (Eds.), *Modelling and applications in mathematics education: The 14th ICMI Study*. (New ICMI Studies Series) (Vol. 10, pp. 434-440). New York: Springer.

Vygotsky, L. S. (1962). *Thought and language*. Cambridge, MA: MIT Press. [Edited and translated by Eugenia Hanfmann & Gertrude Vahar.]

Watters, J. J., English, L. D., & Mahoney, S. (2004, April). *Mathematical modelling in the elementary school*. Paper presented at the annual conference of the American Educational Research Association. San Diego.

White, A. L. (2005). Enriching secondary school mathematics through modelling with graphics calculators. In M. Coupland, J. Anderson, & T. Spencer (Eds.), *Making mathematics vital* (Proceedings of the 20th biennial conference of the Australian Association of Mathematics Teachers, pp. 402-408). Adelaide: AAMT.

White, A. L. (2006). Shaping the future of secondary school mathematics through modelling with the graphics calculator. In H. S. Dhindsa, I. J. Kyeleve, O. Chukwu, & J. S. H. Q. Perera (Eds.), *Shaping the future of science, mathematics and technical education* (pp. 139-146) (Proceedings of the eleventh international conference, Sultan Hassanal Bolkiah Institute of Education, Universiti Brunei Darussalam). Brunei: University Brunei Darussalam.

Williams, G. (2002). Developing a shared understanding of task complexity. In L. Bazzini & C. Whybrow Inchley (Eds.), *Mathematical literacy in the digital era: Research and classroom practice aiming at mathematics education for all*. (Proceedings of the 53rd conference of the International

Commission for the Study and Improvement of Mathematics Teaching, pp. 263-268). Milano: Ghisetti e Corve Edirori.

Ye, Q-X., Blum, W., Houston, S. K., & Jiang, Q-Y. (Eds.). (2003). *Mathematical modelling in education and culture: ICTMA 10.* Chichester, UK: Horwood.

AFFILIATIONS

Gloria Stillman
The University of Melbourne, Australia

Jill Brown
ACU National, St Patrick's Campus, Australia

Peter Galbraith
The University of Queensland, Australia

MIKE THOMAS & MOHAN CHINNAPPAN

TEACHING AND LEARNING WITH TECHNOLOGY: REALISING THE POTENTIAL

In this chapter we have examined research that focused on two major themes that involve technology and mathematics learning: teacher knowledge and practice, and students' conceptual development. Our review indicates that there is a high level of enthusiasm from both students and teachers to embrace a variety of technologies, a trend consistent with findings of the previous MERGA review. The increase in the number of studies on technology and mathematics is healthy and the findings of these studies highlight the importance of the teacher and suggest alternative ways of adopting such tools. We would also like to see more studies that provide specific information about a) how students' learning can be enhanced and b) what role teacher knowledge plays in producing robust mathematical understandings that are required of numerate students.

Key words: affordances and constraints, beliefs and attitudes, digital media, pedagogical practices, pedagogical technology knowledge

INTRODUCTION

For the last two or three decades, debates and reform statements on mathematics and mathematics teaching have emphasized the need to improve access to mathematics for all students. The use of technology to support active participation of learners has featured prominently in the arguments about the democratizing of mathematics at all levels of schooling. Hence, in recent years there has been a great deal of research on this topic internationally, and Australasia has been no exception. In turn, this approach has struck a positive chord among practitioners and there has been a high level of adoption of different forms of technology in order to mediate thinking and, generally, establishing meaningful learning experiences. The focus in the present review is on the crucial role of the teacher when employing technological tools (specifically computers, calculators and the Internet), *how* such teaching scaffolds learning, and the outcomes in terms of *what* is learnt by students. While the predominant thrust of the literature is clearly aimed at the secondary school, a number of papers that address primary and tertiary learning are included. In fact one issue that is expected to have greater visibility in the future, and which will be the topic of a special issue of the Mathematics

H. Forgasz et al. (eds.), Research in Mathematics Education in Australasia 2004-2007, 165-193.

Education Research Journal (MERJ) in 2008, is that of the transition from school to university, with the role of technology as one variable.

While many in the teaching community have embraced the use of technology, the literature reviewed in this chapter indicates that the research community is concerned with the quality of learning that eventuates from its use, and the teacher characteristics that produce the flexibility to integrate a variety of technologies in their pedagogy. There is still a need to realise the potential of technology in the learning of mathematics.

TEACHING WITH TECHNOLOGY

While, as argued by Lynch (2006), the tools of technology impact on what and how children learn, it is equally important that research brings into focus teachers' knowledge, understanding of the context, and concept that is being enacted in the Information and Communication Technologies (ICT) environment. One theoretical perspective that researchers have found to be useful is Gibson's (1977) presentation of relationships between agents, applied to teachers and students in this context, and the environment they seek to interact with. A key feature of this interaction is the presence of *affordances* (attributes of the environment that contribute to the potential for an interaction to occur) and *constraints* (characteristics of the affordance that provide structure and guidance for the interaction, see Kennewell, 2001). In this mix, according to Greeno (1994, p. 338), "An affordance relates attributes of something in the environment to an interactive activity by an agent who has some ability". In short, affordances speak about the potential for action, while constraints impose the structure for that action. Thus, an example of an affordance, given by Gibson (1977) is the provision of mailboxes for posting letters. In a technology classroom setting, the presence of technology is an affordance, with student or teacher instrumentation, time available to use the technology, and the content of curriculum as examples of constraints. The role of the teacher in this theoretical model, according to Kennewell (2001), is to orchestrate the affordances and constraints so that learning takes place. Other researchers have used different words to describe the nature of constraints with regard to technology use in mathematics learning. For example, Forgasz (2006a) talks about encouraging and inhibiting factors, while Thomas (2006c) uses the terminology of obstacles to technology use. However, there seems to be a difference between a constraint, which implies the presence of an affordance, and a factor that inhibits the presence of an entity with its potential for affordance. We suggest reserving the use of the term *obstacle* for this latter idea, namely something that prevents the presence of an affordance-producing entity in the classroom situation. More details on the notion of affordances in a technology-rich mathematics learning environment are available in the discussion by Brown, Stillman, and Herbert (2004).

There are many factors influencing a teacher's use (or otherwise) of technology in the classroom, including their affective beliefs and attitudes, perceptions of the nature of mathematical knowledge and how it should be learned, mathematical content knowledge, and pedagogical content knowledge (PCK) (Shulman, 1986). One may assume that if a teacher possesses a limited knowledge of a concept and its related sub-concepts (Chinnappan & Thomas, 2003) then they will find it more difficult to provide the kind of environment and experiences that will assist students in the construction of rich conceptual thinking. Pedagogical content knowledge refers to understanding the mathematical ideas that are involved in a particular topic and how these relate to the principles and techniques required to teach and learn it, including appropriate structuring of content and relevant classroom discourse and activities (Chinnappan & Thomas, 2003; Cooney, 1999; Shulman, 1986; Simon, 1995).

PEDAGOGICAL TECHNOLOGY KNOWLEDGE (PTK)

Shulman's (1986) notion of PCK provides insight into transformations that teachers need to make with their mathematics subject-matter knowledge so that that knowledge is germane to learning. However, PCK does not fully capture the integration of knowledge of technology with subject-matter knowledge by teachers of mathematics in contexts where learning is supported with technological aids. The quality of this knowledge base plays a significant role in teachers' confidence and competence in using technological tools to teach. Hong and Thomas (2006) and Thomas and Hong (2005a) have referred to this as *pedagogical technology knowledge* (PTK). A teacher's PTK applied to mathematics includes the principles, conventions, and techniques required to teach mathematics through the technology. In this review, PTK is presented as an emerging framework that can be used by researchers to generate explanations about teachers' use of technology to scaffold mathematical learning. The notion of PTK necessitates a new mindset on the part of teachers, a *shift of mathematical focus* to a broader perspective of the implications of the technology for the learning of the mathematics. This PTK partly comprises the teacher's perspective on the mathematics and the technology, and partly their familiarity with it. They need to appreciate that technological tools may be employed in teaching in qualitatively different ways, and that students may form a variety of different relationships with them depending on the teacher's directional emphasis, or *teacher privileging* (Kendal & Stacey, 1999, 2001), which has been shown to shape student preferences so that they follow the teacher's lead. We argue that teachers who have built up a sound store of PTK are more likely to feel comfortable in accessing tools of technology in designing mathematical learning experiences. This privileging can have a positive impact on students' uptake of technology in exploring mathematics.

A second key aspect of PTK involves the transforming of a technological tool into an instrument (Rabardel, 1995). This engages the teacher in actions and decisions to adapt the tool to a particular mathematical task by considering what it can do and how. This concerns more than just the mechanics of the getting the syntax and semantics of the input/output, the algebraic expectation, and coping with the difficulties of navigating between screens and between menu operations.

Most importantly, it means seeing how the technology can mediate between the user and the mathematics; an epistemic mediation. Hence, Guin and Trouche (1999) stress the need for instrumentation and conceptualisation to occur concurrently in the classroom, meaning that teachers need to have sufficiently developed PTK to focus technological activity on specific conceptions. In work somewhat in correspondence with the concept of PTK, Pierce and Stacey (2004b) outline four key aspects of knowledge and skills for teaching effectively with Computer Algebra System (CAS) technology: mathematical, machine, technical and personal, and maintain that the technical and personal need the most attention by new teachers. In order to assist development of the above range of skills, they proposed a framework for teaching to promote effective use of CAS by monitoring progress.

TECHNOLOGY AND PEDAGOGY

Mathematics teachers have received much encouragement from professional organizations such as the Australian and New Zealand Associations of Mathematics Teachers to use technological tools in supporting students' engagement with mathematics, and although they often claim to be supportive of the use of technology in their teaching (Forgasz, 2006a; Thomas, 2006b) the degree and type of use in the classroom often does not always correlate with this response (Becker, 2000a, 2000b; Ruthven & Hennessy, 2002) and some are not totally convinced of its value (Forgasz, 2006b). There is a range of possible reasons for this, and some factors influencing the uptake and implementation of technology in mathematics teaching have been identified. For example, Goos (2005) lists some of these as: skill and previous experience in using technology; time and opportunities to learn (pre-service education, guidance during practicum and beginning teaching, professional development); access to hardware (computers and calculators), software, and computer laboratories; availability of appropriate teaching materials; technical support; support from colleagues and school administration; curriculum and assessment requirements and how teachers interpret these for students perceived to have different mathematical abilities; knowledge of how to integrate technology into mathematics teaching; and beliefs about mathematics and how it is learned. Forgasz (2006a) agrees, with her computer survey listing access to computers and/or computer laboratories as the most prevalent inhibiting factor, with lack of professional development and technical problems, including lack of technical support next. Thomas' (2006c) survey of computer use in all New Zealand secondary schools found that teachers make similar statements, citing availability of computers as the major issue, followed by a lack of software, training, and confidence. The first two may be described as obstacles, while the last two are constraints. While research has highlighted many of these factors, further investigation is needed to describe the relationships and interactions between them.

Teacher variables

Some research also points out the influence of teacher beliefs and attitudes on their teaching practices that incorporates technology. In her study, Forgasz (2006a) found that teacher confidence, experience, skills, or enjoyment of computers comprised the third highest factor encouraging computer use. In their small-scale study of teachers using graphic calculators (GCs), Hong and Thomas (2006) noted that even teachers who have fewer affordances, such as school resources, are not well supported by their heads of department, and those who do not have strong personal GC skills can do quite well in implementing GC use. They suggest that a teacher's personal attitudes and beliefs, if strong enough, can override other negatives, constraints and obstacles. In particular, they felt that a strong belief in the value of technology in learning mathematics coupled with confidence and commitment on the part of teachers to be open to learning could be crucial factors (Thomas & Hong, 2005a). However, it is not clear what other personal qualities and attitudes are involved, and whether all teachers can attain the levels required to surmount a negative environment. Goos' (2003) report on the experience of one teacher supports the importance of these attitudes, citing a willingness to take risks with the calculator, feeling comfortable admitting mistakes, and accepting help from students. That is, allowing herself to be a co-learner with students appeared to assist the teacher to use technology in the learning process.

When teachers do use technology in their teaching there are various options for employing it, with the range only limited by their PTK. Kissane (2004), for example, has described four different pedagogical styles for e-activities, namely exposition, controlled practice, open exploration, and discovery. Goos, Galbraith, Geiger, and others (Galbraith, 2006; Galbraith & Goos, 2003; Geiger, 2005, 2006; Goos, 2005; Goos & Bennison, 2007; Goos, Galbraith, Renshaw, & Geiger, 2000), have continued to apply Vygotsky's (1978) and Valsiner's (1997) concepts of Zone of Proximal Development (ZPD), Zone of Free Movement (ZFM) and Zone of Promoted Action (ZPA) in their research to focus on teachers' thinking and modes of working with technology, using the metaphors of technology as master, servant, partner, and extension of self. Within ZPD, the teacher, as the more experienced and knowledgeable person, could deploy technology as the master to direct learners while simultaneously guiding learners as a servant. Both ZFM and ZPA capture teacher-learner actions that involve applications of technology to experiment with and work mathematically in ways that would result in the formation of new understandings. The last of these, which is not often achieved, occurs when technology is "seamlessly incorporated into a teacher's pedagogical and mathematical repertoire such as through the integration of a variety of technology resources into course planning and the everyday practices of the mathematics classroom" (Goos, 2005, p. 40). Under this model it is the ZFM that gives rise to affordances and constraints, both personal to the teachers, and related to their environments. Geiger (2006) presents examples of how *mathematically enabled technologies and applications* (META) can extend the notion of ZPD to include the technology as a contributing member of a student group. While technology can be used as learning tools in both zones, within ZFM the learner has greater freedom in the manner that the tool can be manipulated. Thus, technology as used within ZFM acts as a contributing member for the learning outcome.

Pierce (2005b) has shown that to progress from technology as master takes effort on the part of teachers, and they need to develop the habit of discerning strategic use of CAS as well as modelling its effective use to students. In their latest research study, Goos and Bennison (2007) set out to explore how and why secondary school mathematics teachers use digital technologies to help students learn. Using a case-study methodology, the uptake of digital technologies by two effective users of technology was investigated, and their beliefs about mathematics and use of technology to promote learning was analysed within the framework of the ZPD. One of the teachers extended students' understanding of quadratic functions within a learning environment that was generated by computer simulation of projectile motions. This learning object, which allowed students to investigate the relationship between angle of projections versus distance, was accessed from a website. Students also used graphic calculators to tabulate and plot data. The second teacher launched a stand-alone piece of software, *Autograph*, to investigate the relationship between symbolic and graphical representations of a range of functions. Both teachers reported that the affordance of a high level of support from the school administration was important for them to integrate technology into their practice. The teachers also felt it was their belief that technology was a better tool than traditional teaching aids in improving students' access to mathematics, which led them to embrace computer-based learning environments in practice. Both *Autograph* and the projectile simulation proved to be effective learning objects in helping students visualise complex algebraic functions. This research is consistent with other work (Forgasz, 2003; Goos, 2003; Hong & Thomas, 2006) that focused on teachers' beliefs about the pedagogical role of technology in affording or constraining students' rich learning experiences. The data here also show that, from a methodological perspective, in-depth studies of practices that use technology in the classroom have the potential to throw light on issues concerning how and why teachers adopt them, and the kind of teacher's PTK that could drive these activities. Brown (2005a) has also employed the ideas from Valsiner's zone theory, but coupled with the notion of affordances. She investigated how a teacher in a *technology-rich teaching and learning environment* (TRTLE) perceived its affordances. While the teacher took up affordances in his teaching and promoted them to students, he constructed a tight ZPA and limited the ZFM to optimise student experiences and constrain students to his ZPA. However, one factor that consistently needs attention is whether the success reported in studies can translate to teachers in general, or whether the research participants are exceptional in some ways.

An internet-based environment

A few researchers have looked at the use of the Internet by teachers (Chinnappan, 2006; Loong, 2003; Loong & White, 2003). In their research Loong and White

(2003) analysed Internet use by six regular users and concluded that the Internet is under-utilised but has the potential to play a significant role in learning in mathematics. Loong (2003) describes some of the ways that 63 teachers reported using the Internet in their teaching. The dominant use was for accessing and analyzing data of different real-life contexts. This was somewhat consistent with findings from a study reported by Thomas (2006c) who found that while teachers in general used the Internet for gathering information, 74% did not use it for remediation or drill and practice. Chinnappan (2006) considered web-based instruction and analysed the understandings developed by a cohort of pre-service teachers as they exchanged ideas about teaching the concept of multiplication to primary children. The discourse among the participants in the study was carried out over the Internet using the courseware, *WebCT*^{TM}. Qualitative analyses of the talk among these prospective teachers showed an emerging consensus that teachers should use a blend of on-line and stand-alone software in order to generate multiple representations of the multiplication concept for students. In a related study, Rider and Hunting (2006) discussed the role of the multimedia software package, *VideoPaper* (a multimedia document combining video, text and images), in supporting teacher professional growth through self-dialogue and peer communication. This is a multimedia tool that can be displayed on any web browser that permits teachers to share their *VideoPaper*. In implementating this on-line tool, it was found that teachers tended to develop deeper insights into their own practice and shared issues with colleagues.

Among the newer technologies now being researched is the use of Interactive White Boards (IWBs) (Tanner, Jones, Kennewell, & Beauchamp, 2005; Zevenbergen & Lerman, 2007). Zevenbergen and Lerman (2007) took the view that IWBs are artefacts, or tools, that can mediate mathematics learning, just like other technologies. However, with the aid of IWBs teachers are able to access a wide-range of dynamic digital resources to engage students in the learning process. In the study reported, the authors examined a principal dimension of Productive Pedagogies called Intellectual Quality (Gore, Griffiths, & Ladwig, 2002). Although, teachers could have, and did, use the range of resources in the IWB environment, the results of this study showed that "the overall pedagogy remained similar to most lessons we have observed in the more traditional modes of teaching" (Zevenbergen & Lerman, 2007, p. 860). The authors comment that such an approach is not consistent with the quality of learning expected in Intellectual Quality. The results of this research, and others (e.g., Thomas, Hong, Bosley, & delos Santos, 2007) highlight the need for practitioners to develop as part of their PTK better strategies for the use of digital technologies in their classrooms if the potential of powerful emerging tools for effective pedagogy is to be realised. The lesson here is having access to and using technology may not bring about the improved quality of learning that one might expect with sophisticated tools. The monitoring and management of the technology is equally important for students to develop deep and substantive learning.

Professional development

One of the key issues raised by various researchers with regard to teaching is the need for targeted pre-service training and in-service professional development of mathematics teachers to address the challenge of using technology in the classroom for developing conceptual thinking (Fitzallen, 2005; Forgasz, 2006b). A number of studies have focused on this, with Forster (2005) developing a questionnaire to measure teacher preparedness for teaching with Information and Communication Technology (ICT). Two cohorts of 38 and 42 pre-service secondary teachers were given the questionnaire before and after an eight week course. The results showed that the trainees felt more prepared to teach with ICT after the unit of work than before it, validating the instruction provided. The potential interactions between classroom teachers' content, pedagogical and technological knowledge were investigated in a study by Chinnappan (2005b). A cohort of prospective teachers was trained in the use of the Web-based software called $B_{10}B$, freeware software that allows students and teachers to construct bars to represent groupings of numbers and operations involving whole numbers. This environment allowed teachers and students to focus on the notion of place value. In this particular study, Chinnappan (2005b) invited the teachers to examine the tool for the purpose of teaching multiplication of two-digit numbers. The participants in this study provided alternative representations of multiplications as well the commutative property. The results of the study showed that future primary teachers could gain, and reveal, important insights into multiplication concepts through the use of an IT-based environment. In spite of the above, it appears that more research into mathematical concept development through technology use would pay dividends.

In a study of in-service professional development, Fitzallen (2005) used teacher profiling and portfolios to examine the ability of four teachers to apply ICT to learning. Her conclusions help in seeing that personal ICT skills alone are not sufficient to develop good PTK, and that professional development needs to address specifically how technology can promote conceptual mathematical thinking. Interestingly, Vale (2007) used teacher self-reports as data to show that there was a major difference in the way practising and pre-service teachers used digital technology in practice. The practising teachers had 1-25 years of experience while the pre-service teachers were enrolled in the Graduate Diploma (Secondary) program. The former group tended to use technology more flexibly and innovatively to engage students, while the latter group was more concerned with demonstration of concepts. In her three-year study of over 70 practising secondary teachers, Forgasz (2006b) found that they believed their computer skills (one feature of PTK) were adequate, and they were not nervous about using computers in the classroom (a measure of confidence). They generally felt that competence with computers was a more important factor in successful use of computers than strong mathematics, and that this favoured boys over girls (see also Forgasz, 2004b). This raises the question of whether professional development should address possible gender inequities, and if so, how. The results here suggest that experience and confidence are important factors in determining how effectively technology can be utilised in scaffolding learning, and this latter idea agrees with the findings of Thomas et al. (2007) who believed that teacher confidence in teaching with technology is one of the key factors in realising its potential. In each case, the experience required includes development of rich PCK and PTK. However, the interactions between these two areas of knowledge, and how one impacts on the other as teachers enter into practice, is an important area for research, and an issue that this review indicates has received limited attention so far.

In order for professional development in technology to be successful, Garner and Leigh-Lancaster (2003) proposed that it needs to address the congruence of pedagogy, curriculum, and assessment. They reported the positive effect of this on the teaching and learning perspectives of one teacher of Victorian Certificate of Education (VCE) mathematics. We suggest that professional development and pre-service training that meet this congruence are relatively rare, and hence there is a continuing need for research into the nature of the technology provision for both potential and existing teachers (Thomas et al., 2007). One aspect that could be considered is the method of delivery of the PD. Goos and Benison (2005) trialed the use of an online discussion to build a community of practice for teachers, and this is an avenue that may be worth further exploration. The literature reviewed above also indicates that the link between teacher confidence, quality of PTK, and use of technology by teachers has not been investigated with sufficient rigour. For example, a question that could be investigated is whether there are particular perspectives on mathematics and technology use that predispose one to personal constructions of schemes and techniques for a given mathematical task, and if so, how this process develops, and what the role of confidence is in it.

LEARNING WITH TECHNOLOGY

The over-riding question of technology use in the learning of mathematics is whether, and to what extent, it can influence conceptualization of the subject matter; to do so, does technology need to be employed in a particular manner, and how does its use impinge on individual abstraction and concept construction? Further, we want to know what relationships should be encouraged between teacher, student, and technology in a technology-rich classroom. In their major chapter summarizing the use of CAS in algebra learning, Thomas, Monaghan and Pierce (2004) suggested that there were three crucial questions to ask when considering implementation of CAS:

- How does CAS use influence student conceptualisation?
- What is the nature of students' development and ability to work with CAS as an instrument?
- How does the way students work on tasks by hand inform their work in a CAS environment and vice versa, and how does each of these interact with students' mathematical thinking and understanding?

These CAS-specific questions clearly have application to all types of technology. The second question directs attention to the work of Artigue (1997) and Guin and Trouche (1999) in using the ideas of Rabardel (1995) and Vérillon and Rabardel (1995) to describe the distinction between *tool* and *instrument* use of technology in mathematics education. Trouche (2000, 2003) promoted the importance of instrumentation action schemes, and maintained that instrumentation and conceptualisation are dependent on each other for the production of these schemes. Hence, for instrumentation to occur, classroom activity should be clearly directed towards particular conceptions. Lagrange (1999, 2000, 2001) stresses the important role of mental schemes and *techniques,* or mathematical activity between tasks and theories, in instrumental genesis. Commenting on the symbiotic nature of personal schemes and general techniques Lagrange (1999, p. 197) noted that "techniques without schemes are ineffective because they are not likely to evolve and cannot produce knowledge", and hence classroom discussion on techniques is essential to help students develop suitable schemes. These issues produce a level of technical difficulty in beginning to use CAS (e.g., Pierce & Stacey, 2004a) that can be overcome by students who have the right attitude, but predicting who these are based on notions of conventional mathematical ability is difficult.

Recently, some research has begun to focus more on the interrelationship of instrumentation and other factors. For example, Kieran and Drijvers (2006) investigated the synergy between *task, techniques, and theory* (TTT) in the emergence of mathematical understanding for students using CAS calculators. Following Chevallard (1999) and Artigue (2002) they suggested that the task must "play an equally fundamental role" (Kieran & Drijvers, 2006, p. 209) as technique and theory, so that there is a co-emergence of by-hand and CAS techniques and theory, and that the design of activities is crucial in this process. This relationship between the three T's, Task, Technique and Theory, is such that they suggest that "… the epistemic value of CAS techniques by themselves may depend both on the nature of the task and the limits of students' existing understanding" (Kieran & Drijvers, 2006, p.258). One important aspect of framing these learning activities, or tasks, encouraged by FitzSimons (2005) and arising from activity theory, is to consider their mediating role in supporting meaning making for learners as part of a community of practice. This may help provide a bridge between the techniques of the community and the schemes of the individual.

Learner expectations and attitudes

Two of the key variables in terms of predicting students' ability to succeed with technology, particularly CAS, have been algebraic insight and expectation (Pierce 2005a; Pierce & Stacey, 2004c), as employed in the RITEMATHS and CAS-CAT projects. Algebraic insight (AI) has two aspects, algebraic expectation, comprising the recognition of basic conventions and processes, and the identification of structure and key features (Pierce, 2005c), along with the ability to link representations. It has been suggested that AI provides a useful framework for monitoring students' learning and progress when working with technology (Pierce & Stacey, 2004c).

Another area of emphasis in research into student learning has been that of student attitudes to technology use. Generally, in the workplace, younger people are more familiar with technology and better disposed to its use than older workers (Zevenbergen, 2004). Zevenbergen believed that these positive attitudes can be used in education through a change in focus that employs the student perspective of "technology as a tool through which the mundane can be reduced and more global aspects of the problems can be addressed" (p. 115). Of course, to be able to measure different student attitudes a reliable scale is necessary. Pierce, Stacey, and Barkatsas (2006) and Barkatsas (2005) aimed to produce a scale suitable for use with students as young as 14, and with a wide range of abilities, to monitor five affective variables: mathematics confidence; attitude to learning mathematics with technology; confidence with technology; and behavioural and affective engagement in learning mathematics. The resulting Mathematics Technology Attitudes Scale (MTAS) comprises just 20 questions and can be administered in only ten minutes. Following a trial with 350 students from grades 8-10 the researchers reported reasonable Cronbach alpha reliability scores of 0.87, 0.89, 0.79, 0.72 and 0.65 for each sub-scale respectively, making this a scale that may be useful for measuring student attitudes when using technology.

A number of researchers have considered student attitudes to be important and report on their influence in learning. Vale and Leder (2004) and Vale (2003b), for example, used a 15-point scale sourced from other researchers, to measure gender differences in student attitudes to computer-based mathematics and perceptions of success. Results from the relatively small sample of 49 students showed that they were positive about the use of computers in mathematics, although girls tended to view computer-based learning less favourably than boys; these results proved to be consistent with previous findings, including those of Vale (2003a). Forgasz (2003, 2004a) carried out a large-scale survey of over 1600 grade 7-10 students about their attitudes to mathematics, computers, and computer use for learning mathematics. She found that students were less convinced than their teachers of the benefits of computer use in mathematical understanding, with female students less convinced than males. Further analysis of the results showed that there were some equity issues associated with computer ownership and the measured attitudes, with socio-economic status and geographic location implicated in both. Forgasz (2003) found that males, students with higher self-rating of mathematics achievement, and, interestingly, students from non-English speaking backgrounds (NESB) were all advantaged with regard to computer ownership, and had more positive attitudes on all three scales. The agreement here on girls having less positive attitudes to computers leads one to ask what effects this might have in respect of their mathematical futures, and how this situation might be addressed. The investigation of why students from NESB have positive disposition towards mathematics and the utility of technology in the mathematics classroom is an important area for future research.

Computer Algebra System (CAS) use

Use of CAS, particularly with reference to its entry into the external high stakes examination arena, with all the possibilities for positive and negative outcomes that this affords (see Hong, Thomas, & Kiernan, 2000), has led some researchers to consider the kinds of work students do with CAS. In their study, Thomas and Hong (2004) primarily found a low level of student activity with CAS, mostly for checking work completed by-hand. The specific categories of CAS use they identified were:

- performing a direct, straightforward procedure,
- checking of procedural by-hand work,
- performing a direct complex procedure, for ease of use, or because the procedure is too difficult by hand,
- performing a procedure within a more complex process, possibly to reduce cognitive load, and
- investigating a conceptual idea.

Ball and Stacey (2005) interviewed five students on their types of CAS use in examinations and found that the students decided question by question whether or not to use it. This may be seen as a positive approach, in agreement with the framework of Pierce and Stacey (2004b). However, the students still seemed to place an emphasis on by-hand skills, using CAS to supplement and check them. While there is evidence that many teachers see the benefits of calculators primarily in terms of improved speed, efficiency, and accuracy (Forster, 2004a; Thomas et al., 2007), Forster (2004b) further showed a need for algebraic knowledge and skills to underpin students' efficient use of the GC. The difficult question of how to encourage students to see CAS not just as a procedural crutch but as a conceptual tool is one that requires additional research resources.

In earlier work, Ball (2003) observed that the use of CAS, with its own peculiar syntax, may impact on students' use of conventional syntax. This led her to suggest that what it is that must be recorded in order both to achieve a solution and to communicate how the problem has been solved should be reconsidered. Building on this idea, Ball and Stacey (2004, 2005) proposed the rubric RIPA (*Reasons–Inputs–Plan–(some)Answers*) as a guide for teaching students how to record their solutions, arguing that when cognition is shared with CAS it is no longer sufficient to get students to write down *their* working, since it is difficult to extricate it from the CAS contribution. The claim of Ball and Stacey (2004) – although statistical evidence is lacking – is that the records of CAS students contain more use of function notation, such as $f(x)$, than those not using CAS. They concluded that a new mathematical practice is emerging with CAS, and this process of learning to communicate mathematics requires active guidance by teachers and a different curriculum emphasis. Evans, Norton, and Leigh-Lancaster (2005) conducted an analysis of student performance on the Victorian pilot CAS examinations in the high stakes 12th grade subject Mathematical Methods, where some questions on the paper were common with the examination taken by students not using CAS thus enabling comparisons to be made. The researchers concluded that CAS was scaffolding the students, assisting them to engage in extended response analysis questions with relatively good success. In support of this stance, more recent research findings from Norton, Leigh-Lancaster, Jones, and Evans (2007) support the claim that students who use CAS develop at least the same level of skills as those who use GCs.

Primary mathematics

Within the primary mathematics curriculum, researchers have been interested in the use of computer-based technology in scaffolding the growth of high levels of numeracy. Children's understanding of numbers and the use of these understandings in making sense of, and solving, real-life problems provide evidence of children's developing numeracy. Chinnappan (2005a) tackled this larger issue by focusing on young children's grasp of fractions. Transition to fractions from whole numbers presents considerable difficulties for many learners as, in the case of fractions, they are required to visualize numbers that have parts and wholes. The visual representations of fractions constructed by a group of grade 5 children in a *Javabars* environment were examined by Chinnappan (2005a). *Javabars* is Internet-based freeware developed at the University of Georgia. This dynamic learning environment helped children construct, partition and manipulate bars of different perimeters and areas (area analogues). The partitioning of a bar or a number of bars revealed children's developing understanding of fractions. The children who participated in the study could utter words for commonly taught fractions such as ¼ and ½. However, they had developed limited understanding of what these numbers actually mean as revealed by their inability to map symbolic representations of these fractions with the corresponding *Javabars* area analogues.

Spreadsheets and ICT environments

In one stream of research, the instructional effect of the use of a particular technology such as spreadsheets (Calder, 2005; Williams, 2006) and *VideoPaper* (Rider & Hunting, 2006) has been examined. However, such investigations have spawned further questions about the effect of these tools on the content and context of student learning. Lynch's (2006) analysis focused on these issues in developing alternative frameworks for researchers in mathematics education. The central point in this debate is that the use of technology impacts on *what* is learnt and *how* it is learnt. Both these themes have been addressed by other researchers (Calder, Brown, Hanley, & Darby, 2006; Galbraith, 2006; Zevenbergen & Lerman, 2006). We concur with Lynch's (2006, p. 41) view that "technology use necessarily transforms learning" and that there is a need to consider the design of studies that brings more focus to this question.

The nature of the engagements of students and pre-service teachers as they worked within a collaborative learning environment embedded in spreadsheets was analysed in studies reported by Calder (2004a, 2004b, 2005) and Calder, Brown, Hanley, and Darby (2006). The results showed that it is important to understand that even though spreadsheets have the potential to open up investigations, they might also constrain them (Calder, 2004a). One way that they may shape the investigation is by influencing student understanding of the problem (Calder, 2005), evoking a generalisation response. Calder et al. (2006) adopted an interpretive framework as they focussed on patterns of social interactions among the participants that provided evidence that students constructed and tested informal conjectures as they worked through open-ended mathematical activities. The results led the researchers to suggest that the pedagogical medium had the effect of opening up unique avenues of exploration, which provided the kind of cognitive and social support needed for conjecturing. This study provided evidence consistent with the arguments advanced by Lynch (2006) that teachers and researchers need to focus on outcomes of technology use and not lose sight of how students learn and the resulting changes in patterns of social interactions at all levels of schooling.

The use of ICT to foster mathematics learning has been embraced by a number of researchers. A common strand in their studies is the role of ICT-based learning environments in supporting active knowledge construction by students and the increase in levels of participation in the learning process. The pedagogical value of stand-alone mathematical software in classroom practice has prompted researchers to examine strategies to support teachers for more effective use of such tools, as well as identifying reasons why teachers do not use computer-based lessons.

Williams (2006) worked with grade 5 teachers on the use of spreadsheets in the teaching of chance and data – a major strand in Australian K-6 mathematics curricula. The study identified four factors that could account for the reluctance of teachers to incorporate technology in the classroom. These factors, in decreasing order of importance, were: confidence level, time, reliability, and logistics. In a similar study, but that focused on the teaching of numeracy, Zevenbergen and Lerman (2006) reported less than satisfactory uptake of ICT tools despite the availability of facility. This study covered a broad range of schools (high-tech vs low-tech; high socio-economic status [SES] vs low SES). The authors suggested that teachers may lack the confidence and skill levels required to implement ICT-based numeracy lessons, supporting the claim made in William's (2006) study.

Technology-rich teaching and learning environments

Whether the notion of affordances is helpful in analysing learning situations in *technology-rich teaching and learning environments* (TRTLEs) in the RITEMATHS program has been the subject of considerable work by Brown and others (Brown, 2004, 2005a, 2005b, 2006a, 2006b; Brown, Stillman, & Herbert, 2004). One task was to explore the value of affordances in constructing a framework for the RITEMATHS program (Brown et al., 2004) for which mathematics learning was approached through technology and real world problems. This program included an experiment in two schools where linear functions were used with grade 9 students alongside the computer program, *GridPic*, and graphic calculators (Brown, 2004). Among the 11 GC affordances described were: the use of the home screen for calculations and storing values; generating numerical values in a table; entering generated data in lists and using *LinReg* to determine the algebraic function; and using a graphing representation to identify a function as linear and to find the sign of the gradient or the *y*-intercept (Brown, 2004, 2005a, 2005b).

How technology can support and elicit conceptual understanding among students is a major concern of this review. Towards this end, there are three recent studies in which this issue was addressed by investigating students' actions and thinking in a range of contexts. First, using a case-study approach, Highfield and Mulligan (2007) investigated pre-schoolers' pattern-generating activities within virtual manipulative and dynamic interactive software such as *Pattern Blocks*. It was envisaged that the digital technologies would provide greater freedom for children to explore and create new patterns in comparison to conventional concrete manipulatives. The results of the study showed that children who worked within the technology-mediated environment tended to create more patterns and transformations than those children who interacted with concrete aids. In what might be deemed to be routine tasks such as reproduction of patterns, the two groups did not show significant differences. The results led the authors to conclude that technology does in fact promote engagement by children in qualitatively superior learning activities that could contribute to the development of pre-algebra concepts. In the second study, set within a spreadsheet environment, students were encouraged to test and seek evidence for their conjectures (Calder, 2007). The apparent tension between the visual output of the spreadsheet and what was expected by the students led to *visual perturbances*. This showed how a commonly-available technological tool can be used to introduce cognitive conflicts and encourage learners to be reflective. In the third study, Brown (2007) examined one grade 9 student's understanding of relations embedded in linear functions. This student was in a class where students had their own laptops and TI-83/83Plus GCs. The lesson involved students completing two extension tasks: *Cunning Running* and *Shot on Goal*. Analyses of the data that were based on teacher interviews, classroom observations, assessment scripts and student task scripts showed that TRTLE assisted students to elicit deep understandings. These understandings manifested in the students being able to link multiple representations of functions that captured a real-life phenomenon. The author also pointed out that students should learn when to use technology and when to use by-hand methods. This emphasis on reflectivity was evident in another study reported by Tanner and Jones (2007) on learning behaviours exhibited by students in the age group 5-14 in video-stimulated reflective dialogue. The dialogue also showed that students were adept at using mathematical terms such as variables.

In related research targeting specific mathematical conceptions, Vincent (2003, 2005) used the dynamic geometry software, Cabri Géomètre, with 29 grade 8 students to examine their geometric reasoning in this context. She records that while some students did abstract invariance by observation in the *dragging* mode of operation, there was also a failure by others to do so. Even those who did *see* the geometric invariance did not necessarily relate it to mathematical properties

(Vincent, 2003). It seems that the interpretation necessary to move from a drawing to a figure by imposing mathematical properties on the drawing (Laborde, 1993; Thomas, 2006a) requires a level of sophistication in what we pay attention to (Mason, in press). Ways to encourage and promote the discipline of noticing in a technology environment could form a useful research topic. Vincent, Chick, and McRae (2005) used argumentation profile charts to analyse students' reasoning and found that these provided valuable insight into how students approach problem-solving tasks. This may provide a way to investigate what students are paying attention to and how they are doing so. These three studies provide support for the argument that to realise their potential teachers need to utilise tools in such a manner that learners go beyond routine activities and engage in inquiries that could lead to the discovery of new relations. While the authors are careful in recommending the use of technology with very young children, it would seem that pedagogy needs to focus on deeper thinking and knowledge integration.

A consideration of the potential of the *JavaMathWorlds* technology to enhance grade 12 students' conceptual understanding of rate of change was the aim of research by Herbert and Pierce (2005). Results from 12 students indicated that they had an improved understanding of the concept of rate of change in both motion and non-motion contexts, providing some evidence for benefits from the dynamic, multi-representational nature of the software. There does not appear to be a great deal of Australasian research into the learning of statistics with technology, which is strange given that it is consistently the area of greatest use in schools. One notable exception is the work of Fitzallen (2006) who has developed a theoretical framework for statistical thinking and reasoning (Pfannkuch & Wild, 2004) using the *TinkerPlots*[TM] *Dynamic Data Exploration* software in order to construct an assessment instrument that is currently being validated. The framework provides specificity about the type of thinking that teachers need to focus on as students work in this environment. The emphasis on general and specific knowledge that learners need to develop is a strength of the model which is consistent with the constructivist approach. A solid basis of pedagogical practice, as part of teachers' PTK, is a prerequisite for understanding the learning/teaching interface provided by technological tools. In this context, the need to develop theoretical frameworks such as Fitzallen's (2006) that provide guidance and a better rationale for practice is a healthy development. In this regard, Graham and Thomas (2005) considered some theoretical principles underlying use of GCs in statistics teaching and learning, and the need for constructing representational versatility of student thinking, and also gave examples of how this might be achieved. In a similar vein, Forster (2004a) reported on findings regarding the use of GCs by a group of year 12 students while they were exploring vector calculus. She reported that GCs support the construction of dynamic graphs, that in turn allow students to coordinate 'the phenomenal and conceptual' aspects of a reality. In a related study that focused on the usefulness of GCs in examinations, Mueller and Forster (2004) investigated changes in questions posed to students in examinations. Interestingly, it was reported that examiners were compelled to reformulate the examination questions involving linear equations in order to assess a set of key learning outcomes. The authors suggest that there is need to monitor both the positive and unintended consequences of introducing new technologies in examinations so that students are motivated to use GCs in appropriate manner. Forster (2006, 2007) also considered grade 12 students' learning of descriptive statistics using dynamic Java applets, and their learning of trends in bivariate data using spreadsheets. She concluded that the dynamic transformations the applets provided, along with linked representations and visual clues, demonstrated the superiority of the computer over the GC for the graphical work. Similarly the spreadsheet seemed superior to the GC for the work on trends, due to poor GC contrast and resolution, and an absence of scales. As Forster (2006) noted, GCs have not had the facility for direct manipulation of graphs. It will be interesting to see if future research shows whether the potential of a new generation of CAS, such as the TI-NSpire™, that does have the capability to manipulate graphs in a manner dynamically linked to algebraic representations, can be harnessed in a similar way to computers.

TERTIARY STUDIES

It is interesting to record that, in the period under consideration, the learning of mathematics with technology at the tertiary level was the focus of various studies. Although some concern has been expressed about the level of integration of technology at this level, particularly in universities (Oates, 2004; Thomas & Holton, 2003), most still espouse its incorporation into creative teaching (Holton, 2005). Oates (2004) gathered data on six variables to show a wide variation in the use of technology in tertiary mathematics courses. His research also pointed out the need for a refined taxonomy to describe what is really meant by a technology-integrated curriculum at this level (or indeed at school). Stewart (2005), for example, showed that most of her sample of 502 first year mathematics undergraduates did not take up the option of CAS calculator use. This was due to the cost, doubt over use in the examination, and satisfaction with by-hand methods. This reluctance to use CAS was in agreement with Pierce, Herbert and Giri's (2004) findings with 45 undergraduates using CAS calculators in examinations. In addition, they reported on the negative effects of by-hand methods on student satisfaction, failing to appreciate CAS affordances, and teacher privileging (defined in Kendal & Stacey, 2001). This means that simply making CAS available is not enough to ensure its adoption (Stewart & Thomas, 2004, 2005). In another study of 340 first and second year science and engineering majors using computer-based CAS (Maple), Stewart, Thomas, and Hannah (2005) were able to describe four categories of students based on their level of instrumentation of CAS in their university mathematics. These were students who:

- are openly opposed to computer work, believing strongly in the superiority of by-hand working for doing and understanding mathematics.
- are happy to use the computer, but have limited instrumentation, seeing its value only as a tool for checking answers already found, to avoid careless errors, or to take the tedium out of by-hand work.

- rely too much on the computer and would find the mathematics harder without it as a support.
- are positive towards computer use and are trying different methods to integrate it into their by-hand working.

The characteristics of the second and third groups are similar to groups of students in secondary schools, including those who use CAS primarily for checking by-hand answers (Thomas & Hong, 2004, 2005b; Stewart & Thomas, 2005), or those, described by Hong, Thomas, and Kiernan (2000), whose work is likely to suffer if CAS support is removed. Cretchley (2006) considered the attitudes of 87 undergraduate mathematics students to computer-based CAS (Matlab) use. She found that there was a strong correlation (0.76) between technology and by-hand performance, but little correlation between technology performance and confidence or attitude to technology in learning mathematics. She also reported no evidence that previous use of computers advantaged students, confirming prior results (Berger & Cretchley, 2005) in which the attitudes of 57 South African mathematics undergraduate students to computer-based CAS (Mathematica) were examined. In that study no significant difference was found between attitude to technology use of those with previous experience and those without. In contrast, Stewart, Thomas, and Hannah (2005) found that compared to than inexperienced users, second year engineering students, with experience of CAS, were making greater movement towards CAS-based instrumented techniques, were less inclined to work through them by hand, preferring to use the CAS to do long or difficult calculations, were more interested in using CAS to save time, and were less likely to display anti-computer sentiments. One possible way to engage undergraduates in CAS use was suggested by Coupland (2004) and Coupland and Crawford (2006), They maintained that engagement occurs when students have a purpose for the engagement. This is consistent with activity theory (Engeström, 1999; Vygotsky, 1978) which predicts engagement under these conditions. However, having a purpose alone seems to be insufficient, since the quality of the purpose that students perceive is still open to question. It appears that there is a need for further investigation of the attitudes and practices of undergraduates with regard to CAS use throughout their tertiary careers. It may be, for example, that there are differential attitudes and practices among undergraduates with differing backgrounds and in different subject majors; this might prove to be a fruitful research area.

CONCLUSIONS AND FUTURE DIRECTIONS FOR RESEARCH

This review has shown that researchers and practitioners alike access a range of technological tools to scaffold student learning and enhance participation in mathematics. The range of tools used indicates a willingness to experiment with, and explore, more effective teaching approaches. Both the theoretical and empirical studies examined in this chapter raise several areas for future investigation that have their genesis in the way teachers and learners wish to use technology. First there is a consistent stream of findings to support the claim that students, at all levels, would like to use technology tools as aids to *unpack* the deeper and more difficult aspects of learning mathematics and using mathematics to solve problems. A majority of studies reported here tended to skirt this important issue about deep mathematical understandings, and how technology use would help learners develop such understandings. There is a need to develop effective methodologies for the investigation of how technology is and can be used during instruction. Our review showed a lack of innovation on methodological issues that can be useful for technology-related studies. Additionally, future research has to draw on existing work on mathematical conceptual understanding with a view to assessing the potential pedagogical impact of using particular technologies. In this regard, Calder's (2007) study on students' ability to conjecture and develop justifications in a technology-driven environment is an important positive development.

Furthermore, this review has indicated that the research and teaching community at large are enthused in their uptake of technology in designing and implementing mathematics lessons. Technology-mediated mathematics learning is very much in vogue among teachers and learners, but teachers need support and guidance in classroom implementation to realise its potential. If we accept that a pedagogical use of technology requires a shift of focus to support for the learning of substantive mathematics, then action is needed. Teachers need support and assistance to acquire and develop knowledge, especially PTK that will assist them and their students to engage in technology-mediated mathematical thinking, and hence interact more fully with the beautiful structure of mathematics. One of the primary means to develop this PTK is through pre- and in-service training with technology that targets mathematics learning, increases teacher confidence, and hence impinges positively on learning outcomes. Further, as Forgasz (2006b) noted, the implications of research for this training are far reaching, and with challenges: "Identifying the needs, adapting courses appropriately, and aptly skilling all pre-service teacher educators are among challenges to be addressed" (p. 464). Thus, the nature of PTK and its development through both pre- and in-service courses should constitute important themes for future research endeavours.

Studies about the use of CAS technology provided conflicting results about students' willingness to embrace it as a learning tool. Students seem to have different reasons for their decisions to use CAS calculators and how to use them. It would seem that students' prior knowledge of mathematics and their general ability levels may influence the way CAS technology is used. It is suggested that an examination of the reasons underlying teachers' and students' adoption of CAS technology in fostering mathematical development is another area that is ripe for further research.

This literature reviewed here also suggests caution in the desire to embrace technology in fostering high levels of learning outcomes. In an assessment of technologically-enriched teaching approaches to mathematics, Galbraith (2006, p. 277) pointed out that we should not be "transfixed by technological blindness" in our haste to embrace it. Similarly, FitzSimons (2005, p. 774) concluded that "it is essential that the use of new learning technologies does not come to dominate any learning activity as a whole." It should be remembered that it is not the technology per se that illuminates the mathematics, but rather the illumination is in its role in epistemic mediation, in the mathematical thinking that a trained teacher can assist students to access through technology.

REFERENCES

Andrews, M., Herbert S., & Pierce, R. (2005). Real contexts, graphics calculators and linear functions. In J. Mousley, L. Bragg, & C. Campbell (Eds.), *Mathematics – celebrating achievement.* (Proceedings of the 42nd annual conference of the Mathematical Association of Victoria, CD-ROM). Melbourne: MAV.

Artigue, M. (1997). Le logiciel 'DERIVE' comme revelateur de phenomenes didactiques lies a l'utilisation d'environments informatiques pour l'apprentissage [The symbolic manipulator Derive as a revealer of didactic phenomena in the use of calculating environments for learning]. *Educational Studies in Mathematics, 33,* 133-169.

Artigue, M. (2002). Learning mathematics in a CAS environment: The genesis of a reflection about instrumentation and the dialectics between technical and conceptual work. *International Journal of Computers for Mathematical Learning, 7,* 245-274.

Ball, L. (2003). *Communication of mathematical thinking in examinations: A comparison of CAS and non-CAS student written responses.* Paper presented at the 2003 Symposium on Computer Algebra in Mathematics Education. Retrieved February 17, 2008 from: http://www.lkl.ac.uk/research/came/events/reims/1-Reaction-Ball.doc

Ball, L., & Stacey, K. (2004). A new practice evolving in learning mathematics: Differences in students' written records with CAS. In P. Clarkson, A. Downton, D. Gronn, M. Horne, A. McDonough, R. Pierce, & A. Roche (Eds.), *Building connections: Theory, research, and practice* (Proceedings of the 28th annual conference of the Mathematics Education Research Group of Australasia, Melbourne, Vol. 1, pp. 177-184). Sydney: MERGA.

Ball, L., & Stacey, K. (2005). Students' views on using CAS in senior mathematics. *Building connections: Theory, research, and practice* In P. Clarkson, A. Downton, D. Gronn, M. Horne, A. McDonough, R. Pierce, & A. Roche (Eds.), *Building connections: Theory, research, and practice* (Proceedings of the 28th annual conference of the Mathematics Education Research Group of Australasia, Melbourne, Vol. 1, pp. 121-128). Sydney: MERGA.

Barkatsas, A. (2005). A new scale for monitoring students' attitudes to learning mathematics with technology (MTAS). In P. Clarkson, A. Downton, D. Gronn, M. Horne, A. McDonough, R. Pierce, & A. Roche (Eds.), *Building connections: Theory, research, and practice* (Proceedings of the 28th annual conference of the Mathematics Education Research Group of Australasia, Melbourne, pp. 129-136). Sydney: MERGA.

Becker, H. J. (2000a). How exemplary computer-using teachers differ from other teachers: Implications for realizing the potential of computers in schools. *Contemporary Issues in Technology and Teacher Education [Online serial], 1*(2), 274-293. Retrieved February 17, 1008 from: http://www.citejournal.org/vol1/iss2/seminal/article1.htm (Originally published in *Journal of Research on Computing in Education, 26*(3), 291-321)

Becker, H. J. (2000b). *Findings from the teaching, learning and computing survey: Is Larry Cuban right?* Paper presented at the 2000 School Technology Leadership Conference of the Council of Chief State Officers, Washington, DC.

Berger, M., & Cretchley, P. (2005). Technology and mathematics learning at university at university level: A South African perspective. *African Journal of Research in Mathematics, Science, and Technology Education, 9*(2), 97-108.

Brown, J. (2004). Enabling teachers to perceive the affordances of a technologically rich learning environment for linear functions in order to design units of work incorporating best practice. *Proceedings of the 9th Asian Technology Conference in Mathematics.* Singapore: ATCM (CD).

Brown, J. (2005a). Affordances of a technology-rich teaching and learning environment. In P. Clarkson, A. Downton, D. Gronn, M. Horne, A. McDonough, R. Pierce, & A. Roche (Eds.), *Building connections: Theory, research, and practice* (Proceedings of the 28th annual conference of the Mathematics Education Research Group of Australasia, Melbourne, Vol. 1, pp. 177-184). Sydney: MERGA.

Brown, J. (2005b). Identification of affordances of a technology-rich teaching and learning environment. In H. Chick, & J. Vincent (Eds.), *Proceedings of the 29th annual conference for the Psychology of Mathematics Education* (Vol. 2, pp.185-192). Melbourne: University of Melbourne.

Brown, J. (2006a). Looking through zones at the teacher's role in technology-rich teaching and learning environments (TRTLE's). In C. Hoyles, J-B. Lagrange, L. H. Son, & N. Sinclair (Eds.), *Proceedings of the 17th ICMI Study: "Technology revisited"*, Hanoi, December 3-8, 2006) (pp. 56 – 65) (CD). Hanoi, Vietnam: Hanoi University of Technology & ICMI. Retrieved February 28, 2008 from: http://icmistudy17.didirem.math.jussieu.fr/lib/exe/fetch.php?id=start&cache=cache&media=icmi17proceedingspart2.pdf

Brown, J. (2006b). Manifestations of affordances of a technology-rich teaching and learning environment (TRTLE). In J. Novotná, H. Moraová, M. Krátká, & N. Stehlíková (Eds.), *Proceedings of the 30th annual conference for the Psychology of Mathematics Education* (Vol. 2, pp. 241-248). Prague: Charles University.

Brown, J. (2007). Early notions of functions in a technology-rich teaching and learning environment (TRTLE). In J. Watson & K. Beswick (Eds.), *Mathematics: Essential research, essential practice* (Proceedings of the 30th annual conference of the Mathematics Education Research Group of Australasia, Hobart, Vol. 1, pp. 153-162). Adelaide: MERGA.

Brown, J., Stillman, G., & Herbert, S. (2004). Can the notion of affordances be of use in the design of a technology enriched mathematics curriculum? In I. Putt, R. Faragher, & M. McLean (Eds.), *Mathematics education for the third millennium: Towards 2010* (Proceedings of the 27th annual conference of the Mathematics Education Research Group of Australasia, Townsville, Vol. 1, pp. 119-126). Sydney: MERGA.

Calder, N. (2004a). Shaping understanding: How does investigating in a spreadsheet environment affect the conversations of initial training students doing mathematics. In I. Putt, R. Faragher, & M. McLean (Eds.), *Mathematics education for the third millennium: Towards 2010* (Proceedings of the 27th annual conference of the Mathematics Education Group of Australasia, Townsville, Vol. 1, pp. 143-150). Sydney: MERGA.

Calder, N. (2004b). Using spreadsheets with 8 year olds: Easy to visualise? *Teachers and Curriculum, 7*, 39-44.

Calder, N. (2005). "I type what I think and try it": Children's initial approaches to investigation through spreadsheets. In P. Clarkson, A. Downton, D. Gronn, M. Horne, A. McDonough, R. Pierce, & A. Roche (Eds.), *Building connections: Theory, research, and practice* (Proceedings of the 28th annual conference of the Mathematics Education Group of Australasia, Melbourne, Vol. 1, pp. 185-192). Sydney: MERGA.

Calder, N. (2007). Visual perturbances in digital pedagogical media. In J. Watson & K. Beswick (Eds.), *Mathematics: Essential research, essential practice* (Proceedings of the 30th annual conference of the Mathematics Education Research Group of Australasia, Hobart, Vol. 1, pp. 172-181). Adelaide: MERGA.

Calder, N., Brown, T., Hanley, U., & Darby, S. (2006). Forming conjectures within a spreadsheet environment. *Mathematics Education Research Journal, 18*(3), 110-116.

Chevallard, Y. (1999). L'analyse des pratiques enseignantes en theorie anthropologique du didactique. *Recherches en Didactique des Mathematiques, 19,* 221-266.

Chinnappan, M. (2005a). Children's mappings of part-whole construct of fractions. In P. Clarkson, A. Downton, D. Gronn, M. Horne, A. McDonough, R. Pierce and A. Roche (Eds.), *Building connections: Theory, research and practice* (Proceedings of the 28th annual conference of the Mathematics Education Research Group of Australasia, Melbourne, Vol. 1, pp 241-248). Sydney: MERGA.

Chinnappan, M. (2005b). Prospective teacher's representation of multiplication. In P. Clarkson, A. Downton, D. Gronn, M. Horne, A. McDonough, R. Pierce and A. Roche (Eds.), *Building connections: Theory, research and practice* (Proceedings of the 28th annual conference of the Mathematics Education Research Group of Australasia, Melbourne, Vol. 1, pp 249-256). Sydney: MERGA.

Chinnappan, M. (2006). Using productive pedagogies framework to build a community of learners online in mathematics education. *Distance Education, 27*(3), 355-370.

Chinnappan, M., & Thomas, M. O. J. (2003). Teachers' function schemas and their role in modelling. *Mathematics Education Research Journal, 15*(2), 151-170.

Chinnappan, M., & Thomas, M. O. J. (2004). Use of graphic calculator to promote the construction of algebraic meaning. In R. M Ali, A. A. Kamil, A. Baharum, A. Mustafa, A. I. M. Ismail, & V. Ravichandran (Eds.), *Proceedings of the 2nd National Conference on Graphing Calculators* (pp. 21-28). Penang: Science University of Malaysia.

Cooney, T. J. (1999). Conceptualizing teachers' way of knowing. *Educational Studies in Mathematics, 38*, 163-187.

Coupland, M. (2004). *Learning with new tools*. Unpublished PhD thesis, University of Wollongong, Australia.

Coupland, M., & Crawford, K. (2006). Many dimensions: The complex picture of student encounters with a computer algebra system. In P. Grootenboer, R. Zevenbergen, & M. Chinnappan (Eds.), *Identities, cultures and learning spaces* (Proceedings of the 29th annual conference of the Mathematics Education Research Group of Australasia, Canberra, Vol. 1, pp. 155-162). Adelaide: MERGA.

Cretchley, P. (2006). Students using computer software for mathematics assessment: A comparison of technology and pen-and-paper performance. In *Proceedings of the 3rd International Conference on the Teaching of Mathematics at the Undergraduate Level.* (CD). Istanbul, Turkey: Turkish Mathematical Society.

Delos Santos, A. G., & Thomas, M. O. J. (2003). Perspectives on the teaching of derivative with graphic calculators. *Australian Senior Mathematics Journal, 17*(1), 40-58.

Engeström, Y. (1999). Activity theory and individual and social transformation. In Y. Engeström, R. Mietinnen, & R. Punamaki (Eds.), *Perspectives on activity theory* (pp. 19-38). Cambridge, UK: Cambridge University Press.

Evans, M., Norton, P., & Leigh-Lancaster, D. (2005). Mathematical methods Computer Algebra System: 2004 pilot examinations and links to a broader research agenda. In P. Clarkson, A. Downton, D. Gronn, M. Horne, A. McDonough, R. Pierce, & A. Roche (Eds.), *Building connections: Theory, research, and practice* (Proceedings of the 28th annual conference of the Mathematics Education Research Group of Australasia, Melbourne, Vol. 1, pp. 329-336). Sydney: MERGA.

Fitzallen, N. (2005). Integrating ICT into professional practice: A case study of four mathematics teachers. In P. Clarkson, A. Downton, D. Gronn, M. Horne, A. McDonough, R. Pierce, & A. Roche (Eds.), *Building connections: Theory, research, and practice* (Proceedings of the 28th annual conference of the Mathematics Education Research Group of Australasia, Melbourne, Vol. 1, pp. 353-360). Sydney: MERGA.

Fitzallen, N. (2006). A model of students' statistical thinking and reasoning about graphs in an ICT environment. In P. Grootenboer, R. Zevenbergen, & M. Chinnappan (Eds.), *Identities, cultures and learning spaces* (Proceedings of the 29th annual conference of the Mathematics Education Research Group of Australasia, Canberra, Vol. 1, pp. 203-210). Sydney: MERGA.

FitzSimons, G. E. (2005). Technology mediated post-compulsory mathematics: An activity theory approach. *International Journal of Mathematical Education in Science and Technology, 36*(7), 769-777.

Forgasz, H. (2003). Equity and beliefs about the efficacy of computers for mathematics learning. In N. A. Pateman, B. J. Dougherty, & J. Zilliox (Eds.), *Proceedings of the 27th conference for the Psychology of Mathematics Education* (Vol. 2, pp. 381-388). Honolulu, HI: University of Hawai'i.

Forgasz, H. (2004a). Equity and computers for mathematics learning: Access and attitudes. In M. J. Høines & A. B. Fuglestad (Eds.), *Proceedings of the 28th conference for the Psychology of Mathematics Education* (Vol. 2, pp. 399-406). Bergen, Norway: Bergen University College.

Forgasz, H. (2004b). *Computers for mathematics learning and gender stereotypes.* Paper presented at Topic Study Group 26 [Gender and mathematics education], ICME-10, Copenhagen (Denmark), July. Retrieved February 17, 2008 from: http://www.icme-organisers.dk/tsg26/1Forgasz.doc

Forgasz, H. J. (2005). Why study grade 11 mathematics: What have computers got to do with it? In M. Goos, C. Kanes, & R. Brown (Eds.), *Mathematics education and society.* (Proceedings of the 4th International Mathematics Education and Society Conference, pp. 166-175). Gold Coast, Qld: Centre for Learning Research, Griffith University.

Forgasz, H. (2006a). Factors that encourage and inhibit computer use for secondary mathematics teaching. *Journal of Computers in Mathematics and Science Teaching, 25*(1), 77-93.

Forgasz, H. J. (2006b). Teachers, equity, and computers for secondary mathematics learning. *Journal for Mathematics Teacher Education, 9*(5), 437-469.

Forster, P. A. (2004a). Conceptualising motion through dynamic graphing. In W-C. Yang, S-C. Chu, T. de Alwis, & K-C. Ang (Eds.), *Proceedings of the 9th Asian Technology Conference in Mathematics* (pp. 196-205). Blacksburg, VA: ATCM.

Forster, P. A. (2004b). Efficient use of graphics calculators in high school calculus. *International Journal of Technology in Mathematics Education, 11*(1), 13-23.

Forster, P. A. (2005). Technologies for bringing about a statistical stance towards data analysis. In F. Olivera & R. Sutherland (Eds.), *Visions of mathematics education: Embedding technology in learning.* (Proceedings of the 7th international conference on technology in mathematics teaching, pp. 3-10). Bristol: University of Bristol.

Forster, P. A. (2006). Assessing technology-based approaches for teaching and learning mathematics. *International Journal of Mathematical Education in Science and Technology, 37*(2), 145-164.

Forster, P. A. (2007). Technologies for teaching and learning trend. *International Journal for Mathematical Education in Science and Technology, 38*(2), 143-161.

Galbraith, P. (2006). Students, mathematics, and technology: Assessing the present – challenging the future. *International Journal of Mathematical Education in Science and Technology, 37*(3), 277-290.

Galbraith, P., & Goos, M. (2003). From description to analysis in technology aided learning: A contribution from zone theory. In L. Bragg, C. Campbell, G. Herbert, & J. Mousley (Eds.), *Mathematics education research: Innovation, networking, opportunity.* (Proceedings of the 26th annual conference of the Mathematics Education Research Group of Australasia, Geelong, pp. 364-371). Sydney: MERGA.

Garner, S., & Leigh-Lancaster, D. (2003). A teacher-researcher perspective on CAS in senior secondary mathematics. In L. Bragg, C. Campbell, G. Herbert, & J. Mousley (Eds.), *Mathematics education research: Innovation, networking, opportunity.* (Proceedings of the 26th annual conference of the Mathematics Education Research Group of Australasia, Geelong, pp. 372-379). Sydney: MERGA.

Geiger, V. (2005). Master, servant, partner and extension of self: A finer grained view of this taxonomy. In P. Clarkson, A. Downton, D. Gronn, M. Horne, A. McDonough, R. Pierce, & A. Roche (Eds.), *Building connections, theory, research and practice* (Proceedings of the 28th annual conference of the Mathematics Education Research Group of Australasia, Melbourne, pp. 369-376). Sydney, MERGA.

Geiger, V. (2006). Standing on the outside: A tale of how technology can engage those working on the margins of a community of inquiry. In P. Grootenboer, R. Zevenbergen, & M. Chinnappan (Eds.), *Identities, cultures and learning spaces* (Proceedings of the 29th annual conference of the Mathematics Education Research Group of Australasia, Canberra, pp. 246-253). Sydney: MERGA.

Gibson, J. J. (1977). The theory of affordances. In R. Shaw, & J. Bransford (Eds.), *Perceiving, acting and knowing: Towards an ecological psychology* (pp. 67-82). Hillsdale, NJ: Erlbaum.

Goos, M. (2003). Learning to teach mathematics with technology: A study of beliefs-in-context. In L. Bragg, C. Campbell, G. Herbert, & J. Mousley (Eds.), *Mathematics education research: Innovation, networking, opportunity.* (Proceedings of the 26th annual conference of the Mathematics Education Research Group of Australasia, Geelong, pp. 404-411). Sydney: MERGA.

Goos, M. (2005). A sociocultural analysis of the development of pre-service and beginning teachers' pedagogical identities as users of technology, *Journal of Mathematics Teacher Education, 8*, 35-59.

Goos, M., & Bennison, A. (2005). The role of online discussion in building a community practice for beginning teachers of secondary mathematics. In P. Clarkson, A. Downton, D. Gronn, M. Horne, A. McDonough, R. Pierce, & A. Roche (Eds.), *Building connections: Theory, research, and practice* (Proceedings of the 28th annual conference of the Mathematics Education Research Group of Australasia, Melbourne, Vol. 1, pp. 385-392). Sydney: MERGA.

Goos, M., & Bennison, A. (2007). Technology-enriched teaching of secondary mathematics: Factors influencing innovative practice. In J. Watson and K. Beswick (Eds.), *Mathematics: Essential research, essential practice* (Proceedings of the 30th annual conference of the Mathematics Education Research Group of Australasia, Hobart, Vol. 1, pp. 315-324). Adelaide: MERGA.

Goos, M., Galbraith, P., Renshaw, P., & Geiger, V. (2000). Reshaping teacher and student roles in technology rich classrooms. *Mathematics Education Research Journal, 12*(3), 303-308.

Gore, J. G., Griffiths, T., & Ladwig, J. G. (2002). *Productive pedagogy as a framework for teacher education: Towards better teaching.* Paper presented at the annual conference of the Australian Association for Research in Education, Melbourne, December. Retrieved February 17, 2008 from: http://www.aare.edu.au/01pap/gor01501.htm

Graham, A., & Thomas, M. O. J. (2005). Representational versatility in learning statistics. *International Journal of Technology in Mathematical Education, 12*(1), 3-14.

Greeno, J. G. (1994). Gibson's affordances. *Psychological Review, 101*(2), 336-342.

Guin, D., & Trouche, L. (1999). The complex process of converting tools into mathematical instruments: The case of calculators. *International Journal of Computers for Mathematical Learning, 3,* 195-227.

Herbert, S., & Pierce, R. (2005). Potential of technology and a familiar context to enhance students' concept of rate of change. In P. Clarkson, A. Downton, D. Gronn, M.Horne, A. McDonough, R. Pierce and A. Roche (Eds.), *Building connections: Theory, research and practice.* (Proceedings of the 28th annual conference of the Mathematics Education Research Group of Australasia, Melbourne, Vol. 2, pp. 435-442. Sydney: MERGA.

Highfield, K., & Mulligan, J. (2007). The role of dynamic interactive technological tools in pre-schoolers' mathematical thinking. In J. Watson and K. Beswick (Eds.), *Mathematics: Essential research, essential practice* (Proceedings of the 30th annual conference of the Mathematics Education Research Group of Australasia, Hobart, Vol. 1, pp. 372-381). Adelaide: MERGA.

Holton, D. (2005). Tertiary mathematics education for 2024. *International Journal of Mathematical Education in Science and Technology, 36*(2-3), 303-313.

Hong, Y. Y., & Thomas M. O. J. (2004). Integrating CAS calculators in teaching: An inexperienced teacher's approach. *Proceedings of the 9th Asian Technology Conference in Mathematics* (pp. 262-271). Blacksburg, VA: ATCM.

Hong, Y. Y., & Thomas, M. O. J. (2006). Factors influencing teacher integration of graphic calculators in teaching. *Proceedings of the 11th Asian Technology Conference in Mathematics* (pp. 234-243). Hong Kong.: ATCM.

Hong, Y. Y., Thomas, M. O. J., & Kiernan, C. (2000). Super-calculators and university entrance calculus examinations. In M. Chinnappan & M. O. J. Thomas (Eds.), *Mathematics Education Research Journal, 12*(3), 321-336.

Kendal, M., & Stacey, K. (1999). Varieties of teacher privileging for teaching calculus with computer algebra systems. *The International Journal of Computer Algebra in Mathematics Education, 6*(4), 233-247.

Kendal, M., & Stacey, K. (2001). The impact of teacher privileging on learning differentiation with technology. *International Journal of Computers for Mathematical Learning, 6,* 143-165.

Kennewell, S. (2001). Using affordances and constraints to evaluate the use of informal and communications technology in teaching and learning. *Journal of Information Technology for Teacher Education, 10*, 101-114.

Kieran, C., & Drijvers, P. (2006). The co-emergence of machine techniques, paper-and-pencil techniques, and theoretical reflection: A study of CAS use in secondary school algebra. *International Journal of Computers for Mathematical Learning, 11*, 205-263.

Kissane, B. (2004). The design and use of eActivities for learning mathematics. In W-C. Yang, S-C. Chu, T. de Alwis, & K-C. Ang (Eds.), *Proceedings of the 9th Asian Technology Conference in Mathematics: Technology in mathematics* (pp. 367-376). Blacksburg, VA: ATCM

Laborde, C. (1993). The computer as part of the learning environment: The case of geometry. In C. Keitel, & K. Ruthven (Eds.), *Learning from computers: Mathematics education and technology* (pp. 48-67). Berlin: Springer-Verlag.

Lagrange, J-B. (1999). Learning pre-calculus with complex calculators: Mediation and instrumental genesis. In O. Zaslavsky (Ed.), *Proceedings of the 23rd annual conference for the Psychology of Mathematics Education* (Vol. 1, pp. 193-200). Haifa, Israel: Technion – Israel Institute of Technology..

Lagrange, J-B. (2000). L'intégration d'instruments informatiques dans l'enseignement: Une approche par les techniques [The integration of calculators into teaching: An approach using techniques]. *Educational Studies in Mathematics, 43*(1), 1-30.

Lagrange, J-B. (2001). A multi-dimensional study of the use of IC technologies: The case of computer algebra systems. In J. Novotná (Ed.), *Proceedings of the second conference of the European Society for Research in Mathematics Education* (pp. 170-182). Mariánské Lázne, Czech Republic: Charles University.

Loong, E. (2003). Australian secondary school teachers' use of the Internet for mathematics. In L. Bragg, C. Campbell, G. Herbert, & J. Mousley (Eds.), *Mathematics education research: Innovation, networking, opportunity* (Proceedings of the 26th annual conference of the Mathematics Education Research Group of Australasia, Geelong, pp. 484-491). Sydney: MERGA.

Loong, E., & White, B. (2003). Teaching mathematics using the Internet. In L. Bragg, C. Campbell, G. Herbert, & J. Mousley (Eds.), *Mathematics education research: Innovation, networking, opportunity*. (Proceedings of the 26th annual conference of the Mathematics Education Research Group of Australasia, Geelong, pp. 492-498). Sydney: MERGA.

Lowrie, T. (2003). Posing problems in ICT-based contexts. In L. Bragg, C. Campbell, G. Herbert, & J. Mousley (Eds.), *Mathematics education research: Innovation, networking, opportunity*. (Proceedings of the 26th annual conference of the Mathematics Education Research Group of Australasia, Geelong, pp. 499-506). Sydney: MERGA.

Lynch, J. (2006). Assessing effects of technology usage on mathematics learning. *Mathematics Education Research Journal, 18*(3), 29-43.

Mason, J. (in press). Doing ≠ construing and doing + discussing ≠ learning: The importance of the structure of attention (*Proceedings of ICME-10), Copenhagen.*

McCann, B., & Pierce, R. (2005). BMX bicycles, dirt jumps, movies and mathematics. In J. Mousley, L. Bragg, & C. Campbell (Eds.), *Mathematics – celebrating achievement* (Proceedings of the 42nd annual conference of the Mathematical Association of Victoria, CD-ROM). Melbourne: MAV.

Mueller, U. A., & Forster, P. A. (2004). Applicable mathematics examination questions: The impact of graphics calculators. In W-C. Yang, S–C. Chu, T. de Alwis, & K–C. Ang (Eds.), *Proceedings of the 9th Asian Technology Conference in Mathematics*, (pp. 133-142). Blacksburg, VA: ATCM.

Norton, P., Leigh-Lancaster, D., Jones, P., & Evans, M. (2007). Mathematical methods and mathematical methods computer algebra systems (CAS) 2006 – Concurrent implementation with a common technology free examination. In J. Watson & K. Beswick (Eds.), *Mathematics: Essential research, essential practice* (Proceedings of the 30th annual conference of the Mathematics Education Research Group of Australasia, Hobart, Vol. 2, pp. 543-550). Adelaide: MERGA.

Oates, G. (2004). Measuring the degree of technology use in tertiary mathematics courses. In W-C. Yang, S-C. Chu, T. de Alwis, & K-C. Ang (Eds.), *Proceedings of the 9th Asian Technology Conference in Mathematics* (pp. 282-291). Blacksburg, VA: ATCM.

Pfannkuch, M., & Wild, C. (2004). Towards an understanding of statistical thinking. In D. Ben-Zvi, & J. Garfield (Eds.), *The challenge of developing statistical literacy, reasoning and thinking* (pp. 17-46). Dordrecht, The Netherlands: Kluwer Academic Publishers.

Pierce, R. (2005a). Algebraic insight underpins the use of CAS for modelling. *The Montana Mathematics Enthusiast, 2*(2), 107-117.

Pierce, R. (2005b). Using CAS to enrich the teaching and learning of mathematics. In S-C. Chu, H-C. Lew, & W-C. Yang (Eds.), Enriching technology and enhancing mathematics for all. *Proceedings of the 10th Asian Conference on Technology in Mathematics* (pp. 47-58). Blacksburg VA: ATCM.

Pierce, R. (2005c). Linear functions and a triple influence of teaching on the development of students' algebraic expectation. In H. L. Chick, & J. L. Vincent (Eds.), *Proceedings of the 29th annual conference for the Psychology of Mathematics Education* (Vol. 4, pp. 81-88). Melbourne: PME.

Pierce, R. Herbert, S., & Giri, J. (2004). CAS: Student engagement requires unambiguous advantages. In I. Putt, R. Faragher, and M. I. McLean (Eds.), *Mathematics education for the third millennium: Towards 2010* (Proceedings of the 27th annual conference of the Mathematics Education Group of Australasia, Townsville, pp. 462-469). Sydney: MERGA.

Pierce, R., & Stacey, K. (2004a). Learning to use CAS: Voices from a classroom. In M. J. Høines, & A. B. Fuglestad (Eds.), *Proceedings of the 28th conference of the International Group for the Psychology of Mathematics Education* (Vol. 4, pp. 25-33). Bergen, Norway: Bergen University College.

Pierce, R., & Stacey, K. (2004b). A framework for monitoring progress and planning teaching towards effective use of computer algebra systems. *International Journal of Computers for Mathematical Learning, 9*(1), 59-93.

Pierce, R., & Stacey, K. (2004c). Monitoring progress in algebra in a CAS Active context: Symbol sense, algebraic insight and algebraic expectation. *International Journal for Technology in Mathematics Education, 11*(1) 3-11.

Pierce, R., Stacey, K., & Ball, L. (2004). Mathematics from still and moving images. In B. Tadich, P. Sullivan, S. Tobias, C. Breen (Eds.), Towards excellence in mathematics. *Proceedings of the 41st annual conference of the Mathematical Association of Victoria* (pp. 386-395). Brunswick, Victoria.

Pierce, R., Stacey, K., & Barkatsas. A. (2006). A scale for monitoring students' attitudes to learning mathematics with technology. *Computers & Education.*

Rabardel, P. (1995). *Les hommes et les technologies, approche cognitive des instruments contemporains.* Paris: Armand Colin.

Rider, R., & Hunting, R. (2006). The VidepPaper: Issues in implementation of a multimedia tool for professional self-dialogue and communication in mathematics education. In P. Grootenboer, R. Zevenbergen, & M. Chinnappan (Eds.), *Identities, cultures and learning spaces* (Proceedings of the 29th annual conference of the Mathematics Education Research Group of Australasia, Canberra, Vol. 2, pp. 440-446). Adelaide: MERGA.

Ruthven, K., & Hennessy, S. (2002). A practitioner model of the use of computer-based tools and resources to support mathematics learning and teaching, *Educational Studies in Mathematics, 49*, 47-88.

Shulman, L. C. (1986). Those who understand: Knowledge growth in teaching. *Educational Researcher, 15,* 4-41.

Simon, M. (1995). Reconstructing mathematics pedagogy from a constructivist perspective. *Journal for Research in Mathematics Education, 26*(2), 114-145.

Stacey, K. (2003). What should students record when solving problems with CAS? Reasons, information, the plan and some answers. In J. T. Fey, A. Cuoco, C. Kieran, L. Mullin, & R. M. Zbiek (Eds.), *Computer algebra systems in secondary school mathematics education* (pp. 289-303). Reston, VA: The National Council of Teachers of Mathematics.

Stewart, S. (2005). Concerns relating to the use of CAS at university level. In P. Clarkson, A. Downton, D. Gronn, M. Horne, A. McDonough, R. Pierce, & A. Roche (Eds.), *Building connections: Research, theory and practice* (Proceedings of the 28th annual conference of the Mathematics Education Research Group of Australasia, Melbourne, Vol. 2, pp. 704-711). Sydney: MERGA.

Stewart, S., & Thomas, M. O. J. (2004). The learning of linear algebra concepts: Instrumentation of CAS calculators. *Proceedings of the 9th Asian Technology Conference in Mathematics* (pp. 377-386). Blacksburg, VA: ATCM.

Stewart, S., & Thomas, M. O. J. (2005). University student perceptions of CAS use in mathematics learning. In H. L. Chick, & J. L. Vincent (Eds.), *Proceedings of the 29th annual conference for the Psychology of Mathematics Education* (Vol. 4, pp. 233-240). Melbourne: University of Melbourne.

Stewart, S., Thomas, M. O. J., & Hannah, J. (2005). Towards student instrumentation of computer-based algebra systems in university courses. *International Journal of Mathematical Education in Science and Technology, 36*(7), 741-750.

Tanner, H., & Jones, S. (2007). Learning from children about their learning with and without ICT using video-stimulated dialogue. In J. Watson & K. Beswick (Eds.), *Mathematics: Essential research, essential practice* (Proceedings of the 30th annual conference of the Mathematics Education Research Group of Australasia, Hobart, Vol. 2, pp. 708-716). Adelaide: MERGA.

Tanner, H., Jones, S., Kennewell, S., & Beauchamp, G. (2005). Interactive whole class teaching and interactive white boards. In P. Clarkson, A. Downton, D. Gronn, M. Horne, A. McDonough, R. Pierce, & A. Roche (Eds.), *Building connections: Theory, research, and practice* (Proceedings of the 28th annual conference of the Mathematics Education Research Group of Australasia, Melbourne, Vol. 2, pp. 720-727). Sydney: MERGA.

Thomas, M. O. J. (2006a). Developing versatility in mathematical thinking, In A. Simpson (Ed.), *Proceedings of retirement as process and concept: A festshcrift for Eddie Gray and David Tall* (pp. 223-241). Prague, Czech Republic: Charles University.

Thomas, M. O. J. (2006b). Teachers using computers in mathematics: A longitudinal study. In C. Hoyles, J-B. Lagrange, L. H. Son, & N. Sinclair (Eds.), *Technology revisited* (Proceedings of the Seventeenth ICMI Study Conference, CD version, pp. 545-554). Hanoi, Vietnam: Hanoi University of Technology.

Thomas, M. O. J. (2006c). Teachers using computers in the mathematics classroom: A longitudinal study. In J. Novotna, H. Moraova, M. Kratka, & N. Stehlikova (Eds.) *Proceedings of the 30th annual conference for the Psychology of Mathematics Education* (Vol. 5, pp. 265-272). Prague, Czech Republic: Charles University.

Thomas, M. O. J., & Holton, D. (2003). Technology as a tool for teaching undergraduate mathematics. In A. J. Bishop, M. A. Clements, C. Keitel, J. Kilpatrick, & F. K. S. Leung (Eds.), *Second international handbook of mathematics education* (Vol. 1, pp. 347-390). Dordrecht: Kluwer.

Thomas, M. O. J., & Hong, Y. Y. (2004). Integrating CAS calculators into mathematics learning: Issues of partnership. In M. J. Høines, & A. B. Fuglestad (Eds.), *Proceedings of the 28th annual conference for the Psychology of Mathematics Education* (Vol. 4, pp. 297-304). Bergen, Norway: Bergen University College.

Thomas, M. O. J., & Hong, Y. Y. (2005a). Teacher factors in integration of graphic calculators into mathematics learning. In H. L. Chick, & J. L. Vincent (Eds.), *Proceedings of the 29th annual conference for the Psychology of Mathematics Education* (Vol. 4, pp. 257-264). Melbourne: University of Melbourne.

Thomas, M. O. J., & Hong, Y. Y. (2005b). Learning mathematics with CAS calculators: Integration and partnership issues. *The Journal of Educational Research in Mathematics, 15*(2), 215-232.

Thomas, M. O. J., Hong, Y. Y., Bosley, J., & delos Santos, A. (2007). Calculator use in the mathematics classroom: A longitudinal study, In W-C Yang, T. de Alwis, & J-C Chuan (Eds.), *Proceedings of the 12th Asian Technology Conference in Mathematics* (C, pp. 37-47). Taiwan: ATCM.

Thomas, M. O. J., Monaghan, J., & Pierce, R. (2004). Computer algebra systems and algebra: Curriculum, assessment, teaching, and learning. In K. Stacey, H. Chick, & M. Kendal (Eds.), *The teaching and learning of algebra: The 12th ICMI study* (pp. 155-186). Norwood, MA: Kluwer Academic Publishers.

Trouche. L. (2000). La parabole du gaucher et de la casserole à bec verseur: Ètude des processus d'apprentissage dans un environnement de calculatrices symboliques [The left-handed parabola and the saucepan lip: A study of a learning module using a symbolic calculator environment]. *Educational Studies in Mathematics, 41*(3), 239-264.

Trouche, L. (2003). *Managing the complexity of human/machine interaction in a computer based learning environment: Guiding students' process command through instrumental orchestrations.* Retrieved February 17, 2008 from: http://www.lkl.ac.uk/research/came/events/reims/2-Presentation-Trouche.doc

Vale, C. (2003a). Computers in mathematics: A super-highway to social justice. In L. Burton (Ed.), *Which way social justice in mathematics education?* (pp. 277-301). Westport, CT: Praeger Publishers.

Vale, C. (2003b). Gender and attitudes to computer use in secondary mathematics. In L. Bragg, C. Campbell, G. Herbert, & J. Mousley (Eds.), *Mathematics education research: Innovation, networking, opportunity.* (Proceedings of the 26th annual conference of the Mathematics Education Research Group of Australasia, Geelong, pp. 680-687). Sydney: MERGA.

Vale, C. (2007). Pedagogical practices with digital technologies: Pre-service and practicing teachers. In J. Watson & K. Beswick (Eds.), *Mathematics: Essential research, essential practice* (Proceedings of the 30th annual conference of the Mathematics Education Research Group of Australasia, Hobart, Vol. 2, pp. 727-734). Adelaide: MERGA.

Vale, C. M., & Leder, G. C. (2004). Student views of computer based mathematics in the middle years: Does gender make a difference? *Educational Studies in Mathematics, 56*, 287-312.

Valsiner, J. (1997). *Culture and the development of children's action: A theory of human development.* (2nd ed.) New York: John Wiley and Sons.

Vérillon, P., & Rabardel, P. (1995). Cognition and artifacts: A contribution to the study of thought in relation to instrumented activity. *European Journal of Psychology of Education, 10*(1), 77-101.

Vincent, J. (2003). Year 8 students' reasoning in a Cabri environment. In L. Bragg, C. Campbell, G. Herbert, & J. Mousley (Eds.), *Mathematics education research: Innovation, networking, opportunity.* (Proceedings of the 26th annual conference of the Mathematics Education Research Group of Australasia, Geelong, pp. 696-703). Sydney: MERGA.

Vincent, J. (2005). Interactive geometry software and mechanical linkages: Scaffolding students' deductive reasoning. In W. Masalski (Ed.), *Technology-supported mathematics learning environments* (2005 yearbook of the National Council of Teachers of Mathematics, pp. 93-112). Reston, VA: National Council of Teachers of Mathematics.

Vincent, J., Chick, H. L., & McRae, B. (2005). Argumentation profile charts as tools for analysing students' argumentations. In H. L. Chick, & J. L. Vincent, (Eds.), *Proceedings of the 29th annual conference for the Psychology of Mathematics Education* (Vol. 4, pp. 281-288). Melbourne: University of Melbourne.

Vygotsky, L. S. (1978). *Thought and language.* Cambridge, MA : MIT Press.

Williams, A. (2006). Spreadsheets in Year 5 chance and data: A professional development experience. In P. Grootenboer, R. Zevenbergen, & M. Chinnappan (Eds.), *Identities, cultures and learning spaces* (Proceedings of the 29th annual conference of the Mathematics Education Research Group of Australasia, Canberra, Vol. 2, pp. 559-566). Adelaide: MERGA.

Zevenbergen, R. (2004). Technologizing numeracy: Intergenerational differences in working mathematically in new times. *Educational Studies in Mathematics, 56*, 97-117.

Zevenbergen, R., & Lerman, S. (2006). Using ICTs to support numeracy learning across diverse settings. In P. Grootenboer, R. Zevenbergen, & M. Chinnappan (Eds.), *Identities, cultures and learning spaces* (Proceedings of the 29th annual conference of the Mathematics Education Research Group of Australasia, Canberra, Vol. 2, pp. 591-598). Sydney: MERGA.

Zevenbergen, R., & Lerman, S. (2007). Pedagogy and interactive whiteboards: Using activity theory to understand tensions in practice. In J. Watson & K. Beswick (Eds.), *Mathematics: Essential research, essential practice* (Proceedings of the 30th annual conference of the Mathematics Education Research Group of Australasia, Hobart, Vol. 2, pp. 853-862). Adelaide: MERGA.

AFFILIATIONS

Mike Thomas
The University of Auckland, New Zealand

Mohan Chinnappan
The University of Wollongong, Australia

GLENDA ANTHONY & MARGARET WALSHAW

CHARACTERISTICS OF EFFECTIVE PEDAGOGY FOR MATHEMATICS EDUCATION

Characterising effective teaching is an important objective for mathematics education. In this chapter we offer a review of a range of Australasian studies about how teachers effectively arrange for learning within their classrooms. In synthesising the literature we provide evidence of the sort of pedagogical practices that contribute to desirable outcomes for diverse students. Structured around two key organisers, the review describes elements of practice in relation to the community of the classroom, and the kinds of tasks and tools that enhance students' thinking. The review shows that sound teacher knowledge underwrites the structural, material, and intellectual choices teachers make in their classrooms. It also reveals that informed teacher decisions contribute to the development of students' mathematical competencies and identities.

Key words: classroom communities, classroom discourse, designing effective tasks, language, pedagogy, small groups, tools and representations

INTRODUCTION

This chapter is about the way effective teachers arrange for learning within their classrooms. We draw on Australasian research studies to provide evidence of the sort of pedagogical practices that contribute to desirable outcomes for diverse students. In our review we describe elements of practice that take into account the social and cultural community of practice in which the teaching is embedded. Our starting point is in the understanding that teachers who produce positive student outcomes have done so through their belief in the rights of all students to have access to education in a broad sense – understanding of the big ideas of curriculum and an appreciation of its value and application in everyday life.

This investigation into effective mathematics pedagogies comes at an important moment in light of the current trend towards evidence-informed practice and policy (see government funded reports by, for example, Anthony & Walshaw, 2007; Doig, McCrae, & Rowe, 2003; Ingvarson, Beavis, Bishop, Peck, & Elsworth, 2004). Within this climate, effective or quality pedagogical practices are those that help address pressing pedagogical challenges by producing desirable outcomes for students. The review is timely, not only as a response to calls for research accountability and transparency, but in the sense of addressing a concerning trend of systematic underachievement amongst particular groups of students. Identifying

H. Forgasz et al. (eds.), Research in Mathematics Education in Australasia 2004-2007, 195-222.

and explaining effective practice that meets the needs of all learners is more urgent than at any previous time.

While our lens is focused on the teacher within the classroom, this is by no means meant to imply that the school professional community as well as the relationships that community establishes within the home, does not have a marked effect on teacher effectiveness, and hence on learner outcomes. We make this point, drawing attention to the work of a number of researchers (e.g., Bicknell, 2006; Bobis & Anderson, 2006; Callingham, 2005; Clarke & Robbins, 2004; Goos & Bennison, 2006; Goos & Jolly, 2004; Lowrie, 2005; Mulligan & Busatto, 2004; Pegg, Graham, & Bellert, 2005; Pritchard & McDiarmid, 2006; Rogers, 2006; Trinick, 2005; Walshaw & Anthony, 2006). To that end, we have conceptualised teaching as nested within an evolving network of systems, in which the activities of the students and the teacher, as well as the school community, the home, the processes involving the mandated curriculum and education-at-large, are mutually constituted through their interactions with each other. Within this network, it is the work of the classroom teacher that assumes significant importance and it is this aspect of the system that remains the focus of this review.

Our review of the Australasian literature for this period includes both the small-scale as well as the larger studies undertaken by Australasian researchers. It is guided by two key organisers: community and task. Through this organisation we explore how teachers work effectively within their classrooms and how they develop students' mathematical competencies and identities.

CLASSROOM COMMUNITY

In this first section we explore how teachers arrange for learning within their classroom communities. We begin by investigating how teachers work at establishing a web of relationships within the classroom community with a view to enhancing desirable outcomes in students. It is in the classroom community that students develop the sense of belonging that is essential if they are to engage with mathematics. Within this community the effective teacher creates a space for individual thinking and for collaborative mathematical explorations. Our review shows how language plays a key part in shaping students' mathematical experiences. We also explore how teachers use student explanation and justification as a basis for decision making about the best ways to advance student understanding.

Relationship building and participation

The classroom community is the cornerstone for developing a sense of belonging in mathematics classrooms. Researchers (e.g., Averill & Clark, 2006; Doyle, 2007: Goos, 2004; Hyett, 2005; Pegg, Panizzon, & Lynch, 2007) have shown that teachers who produce effective classroom communities care about student engagement. They do more than comply with "the politeness norm that dominates most current teacher discourse" (Ball & Cohen, 1997, p. 27). Teaching practice that is founded in an ethic of care takes pains to avoid a permanent dependency from students towards teachers; rather it pays attention to developing interrelationships that are oriented towards enhancing students' capacity to think, reason, communicate, reflect upon, and critique what they do and say in class.

A substantial number of research studies have demonstrated that teachers who care, are those who identify, recognise, respect and value the mathematics of diverse cultural groups (e.g., Goos, 2004; Howard & Perry, 2005; Trinick, 2005). Caring for and building the confidence of students from diverse cultural backgrounds involves moving closer to each other; it implies mutual responsibility and a recognition that students learn from teachers and that teachers also have something to learn from their students (Walls, 2006). From their research with Australian indigenous students, Cooper, Baturo, and Warren (2005) argue for inclusive pedagogical arrangements that incorporate learning contexts that are authentic for the targeted student groups and that encourage student participation.

Teachers do not always create classroom norms of participation consistently across groups. For example, in a survey of New Zealand schools, Bicknell and Riley (2005) found that gifted and talented Māori students are typically under-served in terms of culturally appropriate programmes. In another study Zevenbergen (2005) interviewed students from six diverse schools and found that whereas higher stream/set students, in general, tended to thrive in their classes, lower stream students believed their mathematical experiences were more restrictive and less enriching. The lower streamed classes followed a protracted curriculum and the lesser cognitive demands placed on lower stream students had the effect of constraining their development of a positive mathematical disposition. Bartholomew (2005) also found that some students, more than others, appear to participate more enthusiastically within the classroom community. In her two year study of a top stream/set secondary school class, Bartholomew reported that many boys tended to dominate discussions and many girls appeared reluctant to contribute. Whilst the girls were performing academically just as well as the boys, many perceived themselves as struggling with the work.

Reporting on the *The Mathematics Enhancement Project,* Barton, Paterson, Kensington-Miller, and Bartholomew (2005) offer preliminary support for enhanced participation and achievement in mathematics from predominantly Māori, Pasifika and new immigrant senior secondary school students. Over 80% of students in the project are bilingual, and more than half of these understand at least three languages. Enhancing students' participation in mathematics is also a focus of the *Overcoming Barriers* project undertaken by Sullivan, Mousley, and Zevenbergen (2006a). The project operates through teacher development and classroom-based research to support students who are disadvantaged because of socio-economic factors or because of missed prior learning opportunities.

The daily practices and rituals of the classroom play an important part in how students perceive and learn mathematics. Students create 'insider' knowledge of mathematical behaviour and discourse from the established norms. Seah (2005) found that for immigrant teachers, establishing norms of participation involved a negotiation of personal values with those of the host culture. By scaffolding the development of the negotiated patterns of participation, teachers regulated the mathematical opportunities available in the classroom. Based on research within eighth-grade classrooms, Clarke (2004) investigated participation patterns through the practice of *kikan-shido* (instruction between desks). For teachers in the study across Asian and Western classrooms (O'Keefe, Hua, & Clarke, 2006), *kikan-shido* involved monitoring student activity, guiding student activity, organisational activity as well as social talk.

Teachers who provide inclusive participation rites within the classroom are themselves active searchers and inquirers of mathematics, committed to lifelong learning and reflection in their practice, as well as to positive relationships (Carroll, 2005). Smith (2006) reported on active inquiry into her own teaching practice, with a view towards teaching for understanding. It meant making decisions about the balance between explicit instruction and student-centred inquiry-based learning. As a result of the shift, moments of explicit instruction became more confined to the end rather than the beginning of the lesson. Listening to students' group discussions and their whole-class 'think-aloud' sessions allowed her time to record insights from each individual student for acting upon in the future, resulting in a more inclusive and integrated approach to teaching and learning.

In summary, research has shown that in building productive relationships and arranging for active student participation, effective teachers truly care about the development of students' mathematical identities and competencies. Through an inclusive and integrated approach to teaching and learning they work hard to develop productive relationships and create norms of participation consistently across groups.

Working in small groups

Many researchers have shown that student participation in small groups can provide the context for social and cognitive engagement. Reporting on the in-class grouping practices of 206 teachers involved in the *Early Numeracy Research Project*, Horne (2004) noted that teachers tended to use multiple organisational approaches and apportioned a similar time for individual, pairs, or larger groups work. Groups were established with a view towards their potential in developing students' mathematical competencies and identities. In a study undertaken by Higgins (2005) a teacher set up the learning environment as a collective of groups working at varying levels of mathematical sophistication, each contributing to the overall class discussion and debate.

Peers serve as an important resource for developing mathematical thinking and for finding out about the nature of task demands. Holton and Clarke (2006) maintain that peers, as well as the teacher, scaffold the learning process by providing a "metaphorical bridge between the social and the personal" (p. 128). Studies undertaken by Goos (2004) and Walshaw (2005), both situated in secondary classrooms, reported on students' belief in the advantages of interacting with a friend on mathematical problems. Rawlins (2007), likewise, found that although students in secondary mathematics classes felt comfortable asking the teachers for help, their first preference was to discuss mathematical problems with their classmates. It was through their interaction with their friends, in particular, that students learned to make conjectures as well as learned to engage in argumentation and validation.

Students' willingness to contribute in the public arena of the classroom is influenced by the nature of the community established; however, it is also affected by a students' ability to function in social situations. Barnes (2005) found that within small groups both poor social relationships and poor communication limited student mathematical engagement. Her investigation into classroom group processes at the senior secondary school level provides evidence that some students are frequently interrupted and that their efforts are ignored by others during group work. Barnes reports that these students, marginalised within the small group, learned less and tended to lose confidence in their mathematical ability.

A number of studies have provided evidence that students can benefit from independent learning opportunities – including time alone to think and work quietly away from the demands of a group. Williams (2006), for example, reported on one Year 6 student, Eden, who was expected to work with a peer using separate computers for an activity on straight line graphs. Williams showed how, during an extended period of silence, as he watched 'from the sideline' his peer's involvement in an interactive computer activity, Eden developed his creative insights about straight line geometry. Williams concluded that: "Had Eden worked in a pair as expected, he may not have had the cognitive autonomy to pursue his ideas" (p. 572).

In summary, research has shown that effective teachers use both individual and group processes to enhance students' cognitive thinking. By constantly monitoring the effectiveness of their structural and participation arrangements within the classroom, teachers ensure that a space is provided for all students to engage more fully in the creation of mathematical knowledge.

Shaping students' mathematical language

Quality teaching bridges students' intuitive understandings and the mathematical understandings sanctioned by the world-at-large. Language plays a central role in building those bridges. The teacher who makes a difference for diverse learners helps them retrieve information from written and spoken language so that they might access mathematical concepts (Meaney & Flett, 2006). Such a teacher is focused on shaping the development of novice mathematicians who speak the precise and generalisable language of mathematics (Warren, Young, & de Vries, 2007). Woodward and Pfannkuch (2007) found that the specialised language in mathematics and in statistics can be problematic for learners. Particular words, grammar and vocabulary used in school mathematics can hinder access to the meaning sought and the objective for a given lesson.

Bower (2005) compared the language used by one secondary school teacher in her mathematics class with the language she used in her physics class. The teacher's methods in the two classes differed in terms of style, context, grammatical structure and teacher-student participation and, in particular, the mathematics talk highlighted procedures rather than conceptual understanding. Based on her investigation with a Year 10 class, Meaney (2006) concluded that developing conceptual understanding demands that a teacher move "from prompting students to recognise the new language to facilitating experiences in which the new language is needed for efficient communication of mathematical ideas" (p. 43). A fruitful approach aimed at clarifying descriptions and explanations is for the teacher to purposefully provide information or ask questions. Lobato, Clarke, and Ellis (2005) refer to this as an aspect of teacher telling.

McChesney (2005) noted that effective teachers establish classroom environments in which conventional mathematical language migrates from the teacher to the students. Over time, students' contributions begin to appropriate the language and the understandings of the wider mathematical community. Norton and Irvin (2007) explored this process, focusing on algebraic language in a teacher/researcher intervention at the Year 9 level. Activities over the 6 week period were chosen with the expressed purpose of developing productive algebraic discourse and enhancing student collaboration that would lead to improved algebraic understanding. The study reported higher levels of student engagement with algebra.

Responsibility for distinguishing between terms and phrases, and sensitising students to their particular nuances, weighs heavily with the teacher, who profoundly influences the mathematical meanings made by students in the class (e.g., Norton & Irvin, 2007). Christensen (2004) reported on the language difficulties encountered by bilingual students. Specifically, *Te Poutama Tau* students face problems in using *mua* and *muri* for number sequencing, and in using *whakamua* and *whakamuri* for forward and backward directional counting.

A number of researchers have investigated the challenges of teaching mathematics in multilingual contexts (e.g., Clarkson, 2006; Latu, 2005). Many, like Barwell, Barton, and Setati (2006) ask: "How different is it learning mathematics in English if it is not your home language?" (p. 2). Using both English and Mandarin versions of a test, Neville-Barton and Barton (2005) found that, on average, Chinese and Mandarin-speaking students were disadvantaged in the English test by 15 percent. The syntax of mathematical discourse, and especially word order and interpreting context, created more difficulties than did vocabulary. Importantly, the teachers of the students in their study had not been aware of these problems. Clarkson (2006) reported that language switching (code switching), involving words, phrases and sentences, was a useful pedagogical strategy that enhanced understanding for Vietnamese bilingual students, particularly when they could not understand a mathematical concept or when the task level increased.

In summary, research has shown that effective teachers shape the development of novice mathematicians by sharing with their students the meanings and technical use associated with mathematical discourse. They are able to bridge students' intuitive understandings with the understandings sanctioned by the larger mathematical community.

Providing cognitive structure for students' ideas

Scaffolding students' mathematical thinking enhances the immediate construction of knowledge. But it does more than this: it also provides a foundation for future independent learning (Holton & Clarke, 2006). It seems that students can benefit from explicit instructional support. For example, Lowrie and Diezmann (2005) found that students working with graphic and textual information needed specific scaffolding to retrieve information from diagrams and graphs.

Assessing students' thinking is a key step in the scaffolding process (Groves & Doig, 2004; Hunter, 2007; Owens, 2005; Prescott & Mitchelmore, 2005; Pritchard & McDiarmid, 2006; Smith, 2006). The assessments teachers make influence the kinds of feedback they give to students. Rawlins (2007) showed that students value teachers' feedback, because it points students 'in the right direction.' Teachers who provide quality structuring and feedback draw on a range of pedagogical content knowledge skills that provide them with the knowledge about when to intervene and when not to interfere. It entails listening more and telling less and coming to terms with learning when and what to let go (Smith, 2006).

Some classroom research has shown that much of the teacher feedback that students receive is not particularly constructive. In addition, too much feedback is counterproductive to learning. This point is illustrated by Woodward and Irwin (2005). In their comparative study on the talk and feedback provided by two teachers of Years 5/6 the researchers recorded, for one of the project teachers, numerous occasions of too much feedback or 'teacher lust'. Whilst this teacher had created a positive learning environment and revealed a keen desire for talk to occur in the classroom, the level of engagement in actual mathematical talk was minimal. Cognitive space was limited as pause times for thinking were rarely offered and students were occasionally 'talked over'.

Exploring how teachers provide cognitive space was the core interest of a study undertaken by Wood, Williams, and McNeal (2006). In their investigation into the association between social interaction and students' development of mathematical thinking within five classes of 7-8 year old students, Wood and colleagues found that increased complexity in students' verbalised mathematical thinking was closely related to the interaction patterns that differentiated class discussions. They make the point that it is "the social cognitive process of joint attention and understanding of others' communicative intentionality that is the medium by which mathematical thought develops through meaning making with others" (p. 250).

Communicating with others, however, is a learned strategy and can be taught. In an intervention study with a composite Years 6-8 class, Rowe and Bicknell (2004) focused on promoting students' quality engagement in verbal interactions. Working individually on a problem initially, the partners shared answers and arrived at an agreed solution which they then presented and justified to another pair. As a result of the intervention, the researcher reported a higher level of task related talk and higher verbalised cognitive engagement. Groves and Doig (2004) investigated the specific teacher actions that contributed to productive discourse within a classroom in Japan and a classroom in Australia. Although the lessons differed in a range of ways, in both classrooms the teacher focused clearly on the conceptual aspects of the curriculum and used tasks that were able to problematise the curriculum. Both teachers orchestrated the reporting of students' solutions, focusing on mathematical thinking and dialogue.

Quality teaching familiarises students with the discourse patterns that relate to mathematical argumentation. Reporting on a year-long programme aimed at enhancing teachers' pedagogical strategies and the classroom's collective argumentation, Brown and Renshaw (2006) note that effective teachers use collective argumentation as a formalised pedagogical tool. They propose two key constructs that extend students' participation in mathematical discourse: the *replacement* pattern of discourse emphasises precise vocabulary and *interweaving* occurs when students weave mathematical ideas into a discourse that is representative of their specific circumstances.

Hunter (2005) examined the pedagogical strategies that support effective discourse patterns within a Year 5 classroom. The teacher shaped students' higher level thinking by encouraging critical feedback on students' responses and by skilfully fostering their involvement in taking and defending a particular position against the claims of other students. Tracking the communication skills developed by one primary school teacher, Hunter (2006) found that as the year progressed, the teacher modelled the communicative processes and pressed students for understanding, and as a result she moved towards a focus on student challenge, justification and defence of ideas. Classrooms cultures that provide opportunities for mathematical argumentation on a regular basis, lead to enhanced students' mathematical understandings (Brown & Redmond, 2007; Hunter, 2007).

Whilst most teachers are familiar with forms of dialogue centred on initiation-response-feedback exchanges, many have grappled with the challenge of how to use classroom discourse to enhance learning more productively. Muir (2006) designed an observational schedule to raise the teacher's awareness of the discursive patterns within the classroom and to enhance the effectiveness of that discourse. Sawyer (2006) used video and audio recordings to explore how a teacher of a First Grade classroom explicitly developed a culture of thinking so that very young students could test hypotheses and explain and justify reasoning to others. Sullivan et al. (2006a) also identified the value of explicit pedagogies which encompass routines, communication modes, responses valued, legitimacy of knowledge produced, and individual responsibilities within the community of learning.

Goos (2004) described how a secondary school mathematics teacher developed his students' mathematical thinking through scaffolding the processes of inquiry. The teacher's practice was focused on probing and interpreting students' understanding and on generating new knowledge through a respectful exchange of ideas. He scaffolded thinking and as the year progressed he gradually withdrew his support to push students towards more independent engagement with mathematical ideas. For their part, the students responded by completing tasks with decreasing teacher assistance but also by proposing and evaluating alternative solutions.

In summary, research has shown that effective teachers provide quality structuring and feedback involving a respectful exchange of ideas. They model the kinds of discursive exchanges desirable and legitimate the types of communication processes that lead to higher levels of cognitive growth.

MATHEMATICAL TASKS AND TOOLS

In this section we explore how students are afforded opportunities to learn mathematics through tasks and tools. Tasks and their associated activities influence how learners come to think about, develop, use, and make sense of mathematics. We begin by investigating how teachers provide students with access to important mathematical ideas through a range of task formats. We see that effective teachers use tasks that are aligned with students' current mathematical proficiencies, and selected with careful consideration to students' prior knowledge, and experiences and interests. We also explore how teachers' instructional activities associated with tasks take account of appropriate student challenge and use of artefacts/tools and representations.

Attending to the mathematics

Tasks should share some commonality: they should be problematic for the learner and leave a mathematical 'learning residue' (Davis, 1992). Most importantly, this residue should consist of insight into the structure of mathematics and strategies or methods for solving problems. The importance of attending to the structure of mathematics is emphasised in the research of Mulligan, Mitchelmore, and Prescott (2005). Their research found that early school mathematics achievement was strongly linked with the child's development and perception of mathematical structure. Building on these findings, Mulligan, Prescott, Papic, and Mitchelmore (2006) conducted a school-based numeracy initiative involving 683 low-achieving students. They found that tasks that focused on improving students' visual memory, on enhancing their ability to identify and apply patterns, and on structure in mathematical ideas and representations, resulted in a marked improvement in students' *Pattern and Structure Assessment* scores and school-based and system-wide measures of mathematical achievement.

Research studies have also found that children's exploration of mathematical structure within early numeracy provides an effective bridge from numerical to early algebraic thinking. In a teaching experiment investigating children's patterning skills, Warren (2005) noted the importance of providing multiple representations of patterns. Through explicit questioning and focus on 'across patterns' students in the study were more likely to express generalisations. Stephens (2006) explored the incidence of relational versus computational thinking used by students in Years 5 to 7. He argued that explicit teaching of relational approaches provided greater opportunities for students to engage in quasi-variable thinking. Similarly, Irwin and Britt (2005) found that those teachers who encouraged children to use a "flexible array of skills for manipulating arithmetical relations in ways that exhibit number sense as well as operational sense" (p. 182) were more likely to facilitate students' acquisition of algebraic thinking. Following the student cohort as they moved from Year 8 to Year 9 and 10 levels, Irwin and

Britt (2007) found that those students who exhibited flexible use of computational strategies continued to make marked progress in algebraic skills.

Providing a mathematical focus can occur through a range of task formats. Practice tasks that help children develop fluency and automaticity has been the focus of research by Pegg, Graham, and Bellert (2005). Instructional features of their *QuickSmart* intervention programme included an emphasis on self-regulation, metacognition and self-esteem, use of extended practice in the application of understanding and strategy use, use of feedback, and positive reinforcement.

Game playing activities are also a common task activity – especially in primary classes. Asplin, Frid, and Sparrow (2006) investigated the use of game playing for mental computation in a Year 6 class. They found that where peer assistance was an integral component of participation the "game playing environment provided opportunities to learn about as well as practice and consolidate mental computations" (p. 51). In addition to evidence that students develop automaticity of facts, benefits of game playing included the exposure to new ideas, and the challenge to practise and learn new skills. Bragg (2006) also reported the positive effects of games involving conceptual feedback and dialogue about the mathematics involved. Like Asplin and colleagues, Bragg (2007) noted that games needed to provide appropriate levels of challenge and purpose for all students involved.

The use of the example – a task format that is traditionally taken-for-granted within the mathematics classroom (Watson & Mason, 2005) – has been the focus of several research studies. Chick (2007) and Muir (2007) both explored the affordances offered by particular examples within primary lessons, noting the critical role of the teachers' pedagogical content knowledge (PCK) in both selecting and responding to examples. Chick noted that those teachers who were able to identify the dimensions of possible variation for examples were in a stronger position to be able to "make them conceptually harder or easier, to produce counterexamples, or to emphasise a different principle" (p. 18). Counter-examples, too, have been shown to enhance learning, particularly at the senior level (Klymchuk, 2005).

Mathematical modelling has received considerable research attention with advocates arguing modelling introduces learners to the "complex systems that are an inherent part of our 21st century world" (English, 2006a, p. 195). Whilst modelling activities have traditionally resided in senior classes, research studies have increasingly focused on mathematical modelling at the primary school level, with calls for the use of modelling activities at the pre-school level (Fox, 2006). A three year longitudinal study undertaken by English and colleagues (English, 2006b; English, 2007; English & Fox, 2005; English & Watters, 2005) demonstrated that mathematical modelling tasks can provide young children with rich opportunities to engage in a range of mathematical practices – interpreting and re-interpreting the problem information, posing hypotheses and problems, presenting arguments and counter-arguments, applying previous learning, and acting metacognitively as well as developing important mathematical constructs.

Mitchelmore, White, and McMaster's (2007) research on the *Teaching for Abstraction* method highlighted the difficulties some teachers have in drawing out the mathematics within contextualised tasks. Teachers were unsure when to let a "discussion ramble, when to cut it off to draw out a mathematical point" (p. 509) and matching teacher interest with student interest was also a challenge. Doyle (2006) and Hurst (2007) both investigated the teaching of specific strategies to assist children make sense of contextually-based written material. In Hurst's study, students used a *Mathematical Search* strategy to search for mathematical ideas embedded within text and then used the mathematical ideas to interrogate the text. The teaching strategy proved effective in helping students connect their classroom mathematics to various other learning areas and contexts and situations. In Doyle's study children utilised a top-level structuring (TLS) activity to aid the comprehension of the textual material involved in mathematical-modelling tasks.

At the junior secondary level, Galbraith, Stillman, and Brown's (2006) research on learning and applying modelling skills in a technology-rich teaching and learning environment (TRTLE) explored the nature of the real world modelling process. They found that effective teachers design tasks for which "the basic elements required to initiate and support such activity should be within [students'] experience" (p. 244). In particular, "designing problems where data are generated from a real context, before being subjected to analyses, reduces the potential for mindless manipulation driven by available calculator or computer menus" (p. 243). Galbraith, Stillman, Brown, and Edwards (2007) highlight that quality task design and implementation requires teacher awareness of the modelling process and of potential blockages (e.g., lack of mathematical knowledge, lack of technological skills, technical expertise, or accuracy) in the students' modelling process.

In summary, these studies highlight how effective teachers use tasks involving modelling, game playing, examples, and practice activities that allow students to access important mathematical concepts and relationships and investigate mathematical structure. Engagement in quality tasks involves students in a range of mathematical practices and mathematical thinking.

Linking to learners' knowledge, experiences and interests

Effective teachers focus planning for instruction on their students' current mathematical competencies, experiences and interests. Teachers' attempts to make mathematics interesting are, however, sometimes made at the expense of accuracy and meaning (Christensen, 2004). The danger of marginalising mathematics is evident in Diezmann's (2004) investigation of a story-activity. In Teacher I's classroom, this involved students faithfully acting out the story with assigned roles and prompts. Focused on the physical enjoyment, the mathematical purpose and challenge of the task was incidental. In contrast, Teacher U directed the actions whilst simultaneously questioning the students about mathematical aspects of the unfolding story. In particular, her provision of cues for children to think about the mathematical situations, and her use of appropriate representations that focused on the mathematics – all within a supportive community orientation – ensured that mathematical thinking was "woven into the daily fabric of instruction" (Blanton & Kaput, 2005, p. 441).

In providing productive learning opportunities for diverse students, Sullivan et al. (2006a, 2006b) cautioned that teachers need to consider students' prior experience and knowledge, including their language and reading or listening skills; mathematics concepts; contextual knowledge; mathematics skills; and the ability to cope with complexity. O'Toole's (2006) case study also showed the importance of a teacher's recognition and respect for students' informal knowledge. In developing an understanding of area measure the teacher listened to and talked with the student so that connections could be made between the student's informal language with the language of mathematics, as well as connections between measurement, number, space, data and pattern that the student's informal language made possible.

Teachers who make valuable connections have sound mathematical knowledge. Watson, Beswick, and Brown (2006) found in their investigation of 42 middle school teachers that sound content knowledge was a prerequisite for accessing what conceptual understandings students are articulating in their methods, deciding how those understandings might have come about and where those understandings might be heading. Research evaluations of large-scale numeracy programmes (Clarke, Stephens, Lewis, & Downton, 2005; Higgins, Irwin, Thomas, Trinick, & Young-Loveridge, 2005) noted that teachers' use of diagnostic interviewing supported a more responsive pedagogy that used students' existing mathematical knowledge as a basis for task design and instructional direction. In a teaching experiment designed to foster conceptual understanding of multiplication, Worlley and Jamieson-Proctor (2005) identified the teacher's use of the SOLO taxonomy (Biggs & Collis, 1989) as an effective tool to frame the teaching and learning activities to match students' needs.

The importance of teachers noticing and attending to the mathematics inherent in their students' solution strategies is highlighted in numerous classroom studies. It is at the listening and noticing stages of their work with students (Davies & Walker, 2007), that "teachers have the most responsibility for doing the cognitive work" (Meaney, Fairhall, & Trinick, 2007, p. 501) necessary for advancing students' understanding. Davies and Walker (2005) explored how teacher knowledge and pedagogical practices changed through collaboration between teachers and two researchers. As a result of their more finely tuned classroom listening and questioning skills developed during the meetings, the teachers began to notice changes in their teaching behaviours, including changes in their planning and an increased sensitivity to value and utilise students' responses.

Teachers who understand student thinking as 'understanding in progress' are able to use misconceptions as building blocks, creating tasks that challenge misconceptions. Watson (2007) demonstrated the potential of cognitive conflict tasks involving understanding of average. Whilst this study was not situated in a classroom, Watson offered possible ways to move the study of cognitive conflict into the classroom by working with the activities in groups. Prescott and Mitchelmore's (2005) teaching experiment also used activities to challenge misconceptions about projectile motion. In comparing four teachers, effective teaching was strongly linked to teacher knowledge. Only with strong awareness of their own misconceptions or those of their students, were the teachers able to provide appropriate activities and discussion that sufficiently challenged students' misconceptions.

Cooper, Baturo, and Grant (2006) more directly investigated teacher knowledge as part of a larger professional development project of schools. Findings from the study revealed three distinct aspects of pedagogy: technical, domain, and generic. The technical level refers to the practical content know-how that allows lessons to function. The domain aspect "ensured that the particular topic was covered fully and in an appropriate learning sequence" (p. 366) and the generic level operated to provide deep insights about mathematics. The researchers found that planning and instruction that incorporates all three pedagogies required sound mathematical and pedagogical knowledge. Notably, they found that "even limited increase in teacher knowledge can lead to student improvements" (p. 366). Burgess (2007) developed a framework to investigate the nature of teacher knowledge that is needed for teaching statistics through investigations. Applying the framework to classroom episodes, Burgess tracked the impact of teaching knowledge on teacher/student discussion and task enactment. Lack of content knowledge, he concluded, led to missed opportunities for student/teacher engagement in statistical thinking.

In summary, these studies provide evidence that effective pedagogy requires that teachers adapt tasks to their students' learning needs. In attending to students' thinking these studies confirm the importance of teacher knowledge in all its forms.

Creating appropriate task challenge

Creating appropriate challenge is important for all learners. In their *Overcoming Barriers Project* Mousley, Sullivan, and Zevenbergen (2004) provided students who were experiencing difficulty alternative pathways leading to the learning goal. Ways to reduce task complexity, while still maintaining the mathematical integrity of the task, included 'enabling prompts' and resources as appropriate that reduced the number of steps or variables, simplified the mode of representing results, reduced the written elements in recording, and made the task more concrete. Appropriately adjusting the task demands allowed all students to participate in class discussion and reviews and to move together to the next stage of the learning.

In order to make mathematics more meaningful and accessible for all learners, mathematics curricula frequently advocate the use of contexts (see Arnold, 2005; and curriculum integration studies: Nettleship, 2006; Still & Bobis, 2005; Vale, 2006). Researchers however, caution that the link between contextual task and improved learning outcomes is fragile. Certain contexts that are taken as realistic and relevant may in fact restrict the mathematical development of students from socially and culturally divergent backgrounds (Meaney & Irwin, 2005; Sullivan et al. 2006a; Taylor & Cowie, 2006). Clarke (2005) suggested that dichotomisation of real-world and school mathematics can be avoided by "viewing students as simultaneously members of complementary communities of practice within a broader integrative socio-cultural context" (p. 9). Whilst such links are common in early years settings (see Clarke & Robbins, 2004), Lowrie's (2005) study on children learning mathematics at a distance is a timely reminder of the "overlapping spheres of influence in children's education and development" (Goos & Jolly, 2004, p. 280). Parent supervisors in Lowrie's study established strong connections between in and out of school engagement through (re)constructing the learning activities to include life-experience activities that were both practical and meaningful.

Research concerning task-based practices that are appropriate for students with special needs is scarce. In a study of young children with Down's syndrome, Rietveld (2005) found that their opportunities to learn were masked by 'busy work' and inappropriate tasks and feedback. In her study, Ian, a case study child, was frequently praised for obtaining the 'right' answer when observations indicated that he had not developed the thinking processes that would produce those 'right' answers. Working with gifted and talented children, Diezmann and Watters (2004) reported ways to successfully increase task challenge by methods such as inserting obstacles to the solutions, removing some information, or requiring students to use particular representations or develop generalisations. It was noted that problematising, adapting, and enriching regular curriculum tasks provided underachieving gifted students in the study with the opportunity to oscillate between regular activities and more challenging activities according to their capability, confidence, and motivation.

Whilst the provision of mathematical challenge is integrally linked to productive learning communities, not all high quality tasks achieve their desired purpose. Inappropriate challenge, a shift in focus from understanding to correctness or completeness, inappropriate allocation of time, relaxing of accountability requirements, and a lack of alignment between the task and students' prior knowledge, interest, and motivation, can all contribute to the lowering of task demands (Anthony & Hunter, 2005). Walls (2005) found that teacher instructions for problem-solving activities were dominated by procedural compliance and task-related protocols. Teachers in Cavanagh's (2006) research reported that attempts to incorporate *Working Mathematically* activities with their classes were met with "resistance from students who saw little value in tasks that neither conformed to their usual lesson structures nor appeared to have any direct bearing on their formal assessment for the subject" (p. 119).

In summary, these research studies address issues of creating appropriate challenge for diverse learners in the classroom. Effective teachers, when planning for instruction, focus on students' current mathematical competencies and interests, using these as a resource on which they can build. These studies provide evidence of ways that task challenge can be maintained within the learning community of the classroom.

Tools and representations

An important way in which the teacher can take account both of students' current competencies and interest and their long term learning goals is by introducing judiciously chosen tools and representations. Used effectively, tools and representations offer spaces to help organise mathematical thinking. For example, Chick, Pfannkuch, and Watson (2005) illustrated how telling stories with graphical representations supported Year 7 and 8 students' development of statistical thinking. These researchers argued that the effectiveness of standard representations may be more apparent if the students first grappled with their own representations.

Several studies found that the teacher use of multiple representations was critical to students' learning. Warren (2004) noted that the use of different representations appeared to influence students' ability to generalise. In Heirdsfield's (2005) study, the success of an instructional sequence promoting young children's understanding of mental computation, was in part attributed to the use of multiple models (hundred chart, ninety-nine chart, empty number line), and the freedom of students to choose any (or no model) to solve the examples. Multiple representations lightened the cognitive load, assisted as a means of communication, and encouraged students to develop flexibility in computation. Similarly, O'Toole's (2006) showed how the use of multiple representations prompted students to move "back and forth between different interrelated conceptual ideas of mathematics, within and across different strands" (p. 390).

Numerous studies demonstrate that the tools students access makes a difference to learning outcomes. The use of authentic artefacts to provide a bridge between the mathematics and the situation was the focus of a planning activity in Lowrie's (2004) study. Children assisted to 'make sense' of the task through the use of brochures, menus, bus timetables, and photographs readily established their own sense of authenticity by aligning the problem with their personal experiences and understandings. Significantly, some of the children who were not considered 'mathematically capable' invented more powerful ideas than those who did not see the task as an open-ended challenge. Artefact use was also found to mediate emotional responses. McDonald, Le, Higgins, and Podmore (2005) detail a pair of 6 year-olds' frustrating encounter with a jigsaw puzzle in which the use of the artefact was appropriated for self-appraisal and judgment.

The mathematical textbook is an example of an often taken-for-granted tool. Ewing's (2004) analysis of interviews with senior students highlighted the potential of the text to marginalise some learners. For some students in her study "despite their efforts to try, learning mathematics from a text, with little or no support from the teacher, just became too difficult" (p. 14).

Technological tools can act as catalysts for classroom collaboration (Ball & Stacey, 2005; Yelland, 2005), independent enquiry (Williams, 2006), shared knowledge (Arnold, 2005), and mathematical engagement (Brown, 2006). Brown (2006) investigated the conditions within technology-rich teaching and learning environments that facilitate students' understanding of, and working with, functions. She concluded that effective teaching provided students with opportunities to "enact multiple manifestations of a range of affordances and develop an understanding of when given manifestations are appropriate" (p. 248). Extended tasks that provided multiple solution pathways proved "particularly valuable in providing students with opportunities to develop and consolidate understanding of essential notions related to function" (Brown, 2007, p. 161).

Providing opportunities for mathematical exploration, technological tools can also serve to increase the relevance and accessibility of mathematical practices for learners. Herbert and Pierce (2005) investigated Year 10 students' use of *MathWorlds* to explore multiple representations of the simulated movement of a lift in a multi-storey building. The experientially real context for students provided a bridge between their informal and formal knowledge, leading to a concept image of rate of change. The researcher claimed that the use of the motion context to develop numerous 'models of' rate of change "facilitated the development of an emergent model for rate of change in other contexts" (p. 441).

Vale (2006) reports on how a mathematics teacher used digital technology for real applications within an integrated curriculum. The teacher's use of projection technology to model digital technology skills and provide visual displays proved particularly effective in providing access to the mathematical ideas. This practice ensured that students' access was not restricted by their initial inequitable knowledge and confidence with digital technological skills. In another study with senior mathematics students, McHardy (2006) reported that students found email contact and availability of online resources increased their access to information and gave them a more flexible work environment and greater opportunities to practise.

Geiger (2006) illustrated how technology can be used to involve students who are initially reluctant to engage in, or in some cases resist, the social and cultural norms of a community of learners. Geiger's conclusion that mathematically enabled technologies and applications (META) should be afforded "even greater attention in terms of its pedagogical power than has been previously assigned" (p. 252) signals an important and ongoing area of research for quality pedagogical practice. However, integrating new technologies effectively is challenging and without monitoring may have unintended effects on student participation practices. For example, Zevenbergen and Lerman (2007) investigated the use of Interactive White Boards in middle school classrooms and reported their potential to "close down further the possibility of rich communications and interactions" (p. 861).

Whatever the technologies in use, research has found that messages conveyed by teachers' words and actions are of paramount importance in influencing the way in which technological tools are used by students. For example, Pierce, Herbert, and Giri (2004) found that where teachers privileged the high value of done-by-hand algebraic manipulations, students more likely perceived that CAS offered insufficient advantages to warrant the time and cognitive effort required to become effective users of this new technology. Ball and Stacey (2005) suggest that teachers share the decision making about technology-based approaches with their classes and require students to monitor their own underuse or overuse of technology.

In summary, research has shown the various ways tools and artefacts are implicated in the developing mathematical practices evident within classrooms. The studies provide evidence of ways that teachers can enhance learning opportunities for their students through the effective selection and use of tools and representations.

CONCLUSION

This chapter has explored how effective mathematics teachers create the conditions and provide activities for learning in their classroom communities. The research adds to our knowledge about the kinds of pedagogical approaches that facilitate desirable student outcomes. While most studies focused on specific aspects of effective pedagogical practice, collectively these studies allow us to envisage what quality teaching might look like.

The research has contributed to our knowledge of how effective teachers establish classroom communities that open up spaces for developing students' mathematical and cultural identities. Rather than conforming to one way to organise learning, effective teachers use a range of organisational processes and tasks and tools to enhance students' thinking and to engage them more fully in the creation of mathematical knowledge. When teachers develop structural arrangements, and continually monitor their effectiveness for all students, they ensure that the ideas put forward are, or become, commensurate with local and national curricular goals.

Addressing the challenges of our diverse learning community has been a central focus of many classroom based studies. Many studies have shown that disadvantaged students can benefit from pedagogic strategies that involve inclusiveness and engagement in meaningful mathematics tasks. These studies represent a change in focus to an agenda of proficiency rather than deficiency within classroom-based research over the review period. Such studies confirm that effective teachers truly care about their students. They have high yet realistic expectations for their students and provide them with opportunities to ask why the class is doing certain things and with what effect. In particular, the effective teacher is able to sensitise students, irrespective of their home language, to the nuances of mathematical language. Classroom work involves the respectful exchange of ideas, during which the teacher orchestrates discussion and argumentation and facilitates dialogue. In this way the teacher not only contributes to the development of mathematical competencies and identities but also ensures important social outcomes.

Facilitating the development of students' thinking and classroom mathematical practices towards those compatible with the discipline of mathematics is the hallmark of those studies that focus on the integration of modelling and contextually based tasks and activities. Studies reinforce the need for instructional planning that incorporates students' prior experiences and provides an appropriate level of challenge and a focus on mathematics. It is important that task contexts are accessible to all students, and that teachers capitalise on students' thinking to support their development of increasingly sophisticated forms of mathematical reasoning. It is clear from the research that these forms of reasoning can be supported by a range of judiciously chosen tools and representations. Teachers play a critical role, particularly in understanding how the tools can act as a springboard for discussion and for structuring mathematical knowledge. However, despite positive signposts with effective tool use the research community needs to continue to monitor the affective and social aspects of student use of new technologies.

Sound teacher knowledge underwrites many of the reviewed studies. This knowledge is a prerequisite for accessing students' conceptual understandings and for deciding where those understandings might be heading. Knowing what knowledge teachers need in order to be effective and how teacher knowledge can continue to be strengthened in response to interactions within the classroom and within a professional community of learners, is signalled as an important area of ongoing research.

The studies we reviewed in this chapter highlighted increased attention to specific learner outcomes, within specific contexts. The diverse range of research studies – be they a single case, or a multi-site longitudinal study – combine to offer a developing picture of what it looks like for a teacher's practice to cultivate students' proficiency. Design type research involving collaborative partnerships with researchers and teachers, alongside studies that combine professional development with a research focus, provided a promising approach to access "how [teachers'] actions played out in the classroom and how students were involved in this" (Blanton & Kaput, 2005, p. 435).

In an era of increased accountability (Walshaw, 2007) it is critical that future research on effective teaching focuses not just on "educational means but should at the same time inquire into the desirability of educational ends" (Biesta, 2007, p. 18). For mathematics, these educational ends include not only mathematical proficiency, but also a range of social, affective, and participatory outcomes. Attending to those pedagogical practices that prioritise desirable outcomes for the learner will generate new kinds of evidence about effective teaching.

REFERENCES

Anthony, G., & Hunter, R. (2005). A window into mathematics classrooms: Traditional to reform. *New Zealand Journal of Educational Studies, 40*(1 & 2), 25-43.

Anthony, G., & Walshaw, M. (2007). *Effective pedagogy in Mathematics/Pangarau: Best evidence synthesis iteration [BES]*. Wellington: Ministry of Education.

Arnold, S. (2005). Investigation functions: Exploring calculus concepts. *Australian Senior Mathematics Journal, 19*(1), 50-56.

Asplin, P., Frid, S., & Sparrow, L. (2006). Game playing to develop mental computation: A case study. In P. Grootenboer, R. Zevenbergen, & M. Chinnappan (Eds.), *Identities, cultures and learning spaces* (Proceedings of the 29th annual conference of the Mathematics Education Research Group of Australia, Vol. 1, pp. 46-53). Canberra: MERGA.

Averill, R., & Clark, M. (2006). "If they don't care, then I won't": The importance of caring about our students' mathematics learning. *SET 3,* 15-20.

Ball, D. L., & Cohen, D. K. (1997). Developing practice, developing practitioners: Toward a practice-based theory of professional education. In L. Darling-Hammond & G. Sykes (Eds.), *Teaching as the learning profession: Handbook of policy and practice* (pp. 3-32). San Francisco, CA: Jossey-Bass.

Ball, L., & Stacey, K. (2005). Teaching strategies for developing judicious technology use. In W. Masalski & P. Elliot (Eds.), *Technology-supported mathematics learning environments* (pp. 3-15). Reston: National Council of Teachers of Mathematics.

Barnes, M. (2005). 'Outsiders' in a collaborative learning classroom. In M. Goos, C. Kanes, & R. Brown (Eds.), *Mathematics education and society* (Proceedings of the 4th International Mathematics Education and Society conference, pp. 58-68). Brisbane: Griffith University.

Bartholomew, H. (2005). Top set identities and the marginalisation of girls. In M. Goos, C. Kanes, & R. Brown (Eds.), *Mathematics education and society* (Proceedings of the 4th International Mathematics Education and Society Conference, pp. 69-77). Brisbane: Mathematics Education and Society.

Barton, B., Paterson, J., Kensington-Miller, B., & Bartholomew, H. (2005). Dodging the dragon. Strategies for mathematics professional development in low socio-economic areas. In M. Goos, C. Kanes, & R. Brown (Eds.), *Mathematics education and society* (Proceedings of the 4th International Mathematics Education and Society Conference, pp. 78-87). Brisbane: Mathematics Education and Society.

Barwell, R., Barton, B., & Setati, M. (2006). Multilingual issues on mathematic education: Introduction. *Special Issue of Educational Studies in Mathematics, 64*, 1-7.

Bicknell, B. (2006). Investigating parental roles of mathematically gifted students. In P. Grootenboer, R. Zevenbergen, & M. Chinnappan (Eds.), *Identities, cultures and learning spaces* (Proceedings of the 29th annual conference of the Mathematics Education Research Group of Australasia, Canberra, Vol. 1, pp.76-83). Adelaide: MERGA.

Bicknell, B., & Riley, T. (2005). Addressing mathematical promise. *SET: Research Information for Teachers, 1*, 38-42.

Biesta, G. (2007). Why "what works" won't work: Evidence-based practice and the democratic deficit in educational research. *Educational Theory, 57*(1), 1-22.

Biggs, J., & Collis, K. (1989). Towards a model of school-based curriculum development and assessment using the SOLO taxonomy. *Australian Journal of Education, 33*(2), 151-163.

Blanton, M., & Kaput, J. (2005). Characterizing a classroom practice that promotes algebraic reasoning. *Journal for Research in Mathematics Education, 36*(5), 412-446.

Bobis, J., & Anderson, J. (2006). Reform-oriented teaching practices and the influence of school context. In P. Grootenboer, R. Zevenbergen, & M. Chinnappan (Eds.), *Identities, cultures and learning spaces* (Proceedings of the 29th annual conference of the Mathematics Education Research Group of Australasia, Canberra, Vol. 1, pp. 92-98). Adelaide: MERGA.

Bower, M. L. W. (2005). The mathematics talk of a secondary school teacher of mathematics and physics. In P. Clarkson, A. Downton, D. Gronn, M. Horne, A. McDonough, R. Pierce, & A. Roche (Eds.), *Building connections: Theory, research and practice* (Proceedings of the 28th annual conference of the Mathematics Education Research Group of Australasia, Melbourne, Vol. 1, pp. 161-168). Sydney: MERGA.

Bragg, L. (2006). Students' impressions of the value of games for the learning of mathematics. In J. Novotná, H. Moraová, M. Krátká, & N. Stehliková (Eds.), *Proceedings of the 30th annual conference for the Psychology of Mathematics Education* (Vol. 2, pp. 217-224). Prague: PME.

Bragg, L. (2007). Students' conflicting attitudes towards games as a vehicle for learning mathematics: A methodological dilemma. *Mathematics Education Research Journal, 19*(1), 29-44.

Brown, J. (2006). Manifestations of affordances of a technology-rich teaching and learning environment (TRTLE). In J. Novotná, H. Moraová, M. Krátká, & N. Stehliková (Eds.), *Proceedings of the 30th annual conference for the Psychology of Mathematics Education* (Vol. 2, pp. 242-248). Prague: Charles University.

Brown, J. (2007). Early notions of functions in a technology-rich teaching and learning environment (TRTLE). In J. Watson & K. Beswick (Eds.), *Mathematics: Essential research, essential practice* (Proceedings of the 30th annual conference of the Mathematics Education Research Group of Australasia, Hobart, Vol. 1, pp. 153-162). Adelaide: MERGA.

Brown, R., & Redmond, T. (2007). Collective argumentation and modeling mathematics practices outside the classroom. In J. Watson & K. Beswick (Eds.), *Mathematics: Essential research, essential practice* (Proceedings of the 30th Annual Conference of the Mathematics Education Research Group, Hobart, Vol. 1, pp. 163-171). Adelaide: MERGA.

Brown, R., & Renshaw, P. (2006). Transforming practice: Using collective argumentation to bring about teacher change in a Year 7 mathematics classroom. In P. Grootenboer, R. Zevenbergen, & M. Chinnappan (Eds.), *Identities, cultures and learning spaces* (Proceedings of the 29th annual conference of the Mathematics Education Research Group of Australasia, Canberra, Vol. 1, pp. 99-106). Adelaide: MERGA.

Burgess, T. (2007). *Investigating the nature of teacher knowledge needed and used in teaching statistics.* Unpublished doctoral thesis, Massey University, Palmerston North, New Zealand.

Callingham, R. (2005). A whole-school approach to developing mental computation strategies. In H. L. Chick & J. L. Vincent (Eds.), *Proceedings of the 29th annual conference for the Psychology of Mathematics Education* (Vol. 2, pp. 201-208). Melbourne: University of Melbourne.

Carroll, J. (2005). Developing effective teachers of mathematics: Factors contributing to development in mathematics education for primary school teachers. In P. Clarkson, A. Downton, D. Gronn, M. Horne, A. McDonough, R. Pierce, & A. Roche (Eds.), *Building connections: Theory, research and practice* (Proceedings of the 28th annual conference of the Mathematics Education Research Group of Australasia, Melbourne, Vol. 1, pp. 201-208). Sydney: MERGA.

Cavanagh, M. (2006). Mathematics teachers and working mathematically: Response to curriculum change. In P. Grootenboer, R. Zevenbergen, & M. Chinnappan (Eds.), *Identities, cultures and learning spaces* (Proceedings of the 29th annual conference of the Mathematics Education Research Group of Australia, Canberra, Vol. 1, pp. 115-122). Adelaide: MERGA.

Chick, H. (2007). Teaching and learning by example. In J. Watson & K. Beswick (Eds.), *Mathematics: Essential research, essential practice* (Proceedings of the 30th annual conference of the Mathematics Education Research Group of Australasia, Canberra, Vol. 1, pp. 3-19). Adelaide: MERGA.

Chick, H., Pfannkuch, M., & Watson, J. (2005). Transnumerative thinking: Finding and telling stories within data. *Curriculum Matters, 1*, 86-107.

Christensen, I. (2004). *An evaluation of Te Poutama Tau 2003: Exploring issues in mathematical education.* Wellington: Ministry of Education.

Clarke, B., & Robbins, J. (2004). Numeracy enacted: Preschool families conceptions of their children's engagements with numeracy. In I. Putt, R. Faragher, & M. McLean (Eds.), *Mathematics education for the third millennium: Towards 2010* (Proceedings of the 27th annual conference of the Mathematics Education Research Group of Australasia, Townsville, Vol. 1, pp. 175-182). Sydney: MERGA.

Clarke, D. (2004). Patterns of participation in the mathematics classroom. In M. J. Høines & A. B. Fuglestad (Eds.), *Proceedings of the 28th annual conference for the Psychology of Mathematics Education* (Vol. 2, pp. 231-238). Bergen: Bergen University College.

Clarke, D. (2005). Essential complementarities: Arguing for an integrative approach to research in mathematics classrooms. In P. Clarkson, A. Downton, D. Gronn, M. Horne, A. McDonough, R. Pierce, & A. Roche (Eds.), *Building connections: Theory, research and practice* (Proceedings of the 28th annual conference of the Mathematics Education Research Group of Australasia, Melbourne, Vol. 1, pp. 3-17). Sydney: MERGA.

Clarke, D., Stephens, M., Lewis, G., & Downton, A. (2005). The evaluation of the success in numeracy education program. In P. Clarkson, A. Downton, D. Gronn, M. Horne, A. McDonough, R. Pierce, & A. Roche (Eds.), *Building connections: Theory, research and practice* (Proceedings of the 28th annual conference of the Mathematics Education Research Group of Australasia, Melbourne, Vol. 1, pp. 257-264). Sydney: MERGA.

Clarkson, P. C. (2006). Australian Vietnamese students learning mathematics: High ability bilinguals and their use of their languages. *Educational Studies in Mathematics, 64,* 191-215.

Cooper, T., Batauro, A., & Grant, E. (2006). Collaboration with teachers to improve mathematics learning: Pedagogy at three levels. In J. Novotná, H. Moraová, M. Krátká, & N. Stehliková (Eds.), *Proceedings of the 30th annual conference for the Psychology of Mathematics* (Vol. 2, pp. 361-368). Prague: Charles University.

Cooper, T., Baturo, A., & Warren, E. (2005). Indigenous and non-indigenous teaching relationships in three remote mathematics classrooms in remote Queensland. In H. L. Chick & J. L. Vincent (Eds.), *Proceedings of the 29th annual conference for the Psychology of Mathematics Education* (Vol. 2, pp.265-272). Melbourne: University of Melbourne.

Davies, N., & Walker, K. (2005). Learning to notice: One aspect of teachers' content knowledge in numeracy classrooms. In P. Clarkson, A. Downton, D. Gronn, M. Horne, A. McDonough, R. Pierce, & A. Roche (Eds.), *Building connections: Theory, research and practice* (Proceedings of the 28th annual conference of the Mathematics Education Research Group of Australasia, Melbourne, Vol. 1, pp. 273-280). Sydney: MERGA.

Davies, N., & Walker, K. (2007). Teaching as listening: Another aspect of teachers' content knowledge in the numeracy classroom. In J. Watson & K. Beswick (Eds.), *Mathematics: Essential research, essential practice* (Proceedings of the 30th annual conference of the Mathematics Education Research Group, Hobart, Vol. 1, pp. 217-225). Adelaide: MERGA.

Davis, R. (1992). Understanding "understanding". *Journal of Mathematical Behavior, 11*, 225-241.

Diezmann, C. (2004). Teachers' characteristics: One lesson – two learning environments. In I. Putt, R. Faragher & M. McLean (Eds.), *Mathematics education for the third millennium: Towards 2010* (Proceedings of the 27th annual conference of the Mathematics Education Research Group, Townsville, Vol. 1, pp. 191-197). Sydney: MERGA.

Diezmann, C., & Watters, J. J. (2004). *Challenges and connectedness in the mathematics classroom: Using lateral strategies with gifted elementary students.* Paper presented at the 10th International Congress on Mathematical Education, Copenhagen, Denmark.

Doig, B., McCrae, B., & Rowe, K. J. (2003). *A good start to numeracy: Effective numeracy strategies from research and practice in early childhood* (IMR): Australian Council of Educational Research.

Doyle, K. (2006). Organisational structure for mathematical modelling. In P. Grootenboer, R. Zevenbergen, & M. Chinnappan (Eds.), *Identities, cultures and learning spaces* (Proceedings of the 29th annual conference of the Mathematics Education Research Group of Australia, Canberra, Vol. 1, pp. 187-194). Adelaide: MERGA.

Doyle, K. (2007). The teacher, the task: Their role in students' mathematical literacy. In J. Watson & K. Beswick (Eds.), *Mathematics: Essential research, essential practice* (Proceedings of the 30th annual conference of the Mathematics Education Research Group of Australasia, Canberra, Vol. 1, pp. 246-254). Adelaide: MERGA.

English, L. D. (2006a). Introducing young children to complex systems through modelling. In P. Grootenboer, R. Zevenbergen, & M. Chinnappan (Eds.), *Identities, cultures and learning spaces* (Proceedings of the 29th annual conference of the Mathematics Education Research Group of Australia, Canberra, Vol. 1, pp. 195-202). Adelaide: MERGA.

English, L. D. (2006b). Mathematical modelling in the primary school: Children's construction of a consumer guide. *Educational Studies in Mathematics, 63*, 303-323.

English, L. D. (2007). Interdisciplinary modelling in the primary mathematics curriculum. In J. Watson & K. Beswick (Eds.), *Mathematics: Essential research, essential practice* (Proceedings of the 30th annual conference of the Mathematics Education Research Group of Australasia, Hobart, Vol. 1, pp. 275-283). Adelaide: MERGA.

English, L. D., & Fox, J. (2005). Seventh-graders' mathematical modelling on completion of a three-year program. In P. Clarkson, A. Downton, D. Gronn, M. Horne, A. McDonough, R. Pierce, & A. Roche (Eds.), *Building connections: Research, theory and practices* (Proceedings of the 28th annual conference of the Mathematics Education Research Group of Australasia, Melbourne, Vol. 1, pp. 321-328). Sydney: MERGA.

English, L. D., & Watters, J. (2005). Mathematical modelling in the early school years. *Mathematics Education Research Journal, 16*(3), 59-79.

Ewing, B. (2004). "Go to the page and work it from there": Young people's experiences of learning mathematics from a text. *Australian Senior Mathematics Journal, 20*(1), 8-14.

Fox, J. (2006). A justification for mathematical modelling experiences in the preparatory classroom. In P. Grootenboer, R. Zevenbergen, & M. Chinnappan (Eds.), *Identities, cultures and learning spaces*

(Proceedings of the 29th annual conference of the Mathematics Education Research Group of Australasia, Canberra, Vol. 1, pp. 221-228). Adelaide: MERGA.

Galbraith, P., Stillman, G., & Brown, J. (2006). Identifying key transition activities for enhanced engagement in mathematical modelling. In P. Grootenboer, R. Zevenbergen, & M. Chinnappan (Eds.), *Identities, cultures and learning spaces* (Proceedings of the 29th annual conference of the Mathematics Education Research Group of Australasia, Canberra, Vol. 1, pp. 237-245). Adelaide: MERGA.

Galbraith, P., Stillman, G., Brown, J., & Edwards, I. (2007). Facilitating middle secondary modelling competencies. In C. Haines, P. Galbraith, W. Blum, & S. Khan (Eds.), *Mathematical modelling: Education, engineering and economics* (pp. 130-140). Chichester, UK: Horwood.

Geiger, V. (2006). Standing on the outside: A tale of how technology can engage those working on the margins of a community of inquiry. In P. Grootenboer, R. Zevenbergen, & M. Chinnappan (Eds.), *Identities, cultures and learning spaces* (Proceedings of the 29th annual conference of the Mathematics Education Research Group of Australasia, Canberra, Vol. 1, pp. 246-253). Adelaide: MERGA.

Goos, M., & Bennison, A. (2006). An online community of practice for pre-service and beginning teachers of secondary mathematics. In J. Novotná, H. Moraová, M. Krátká, & N. Stehliková (Eds.), *Proceedings of the 30th annual conference for the Psychology of Mathematics* (Vol. 3, pp. 209-216). Prague: Charles University.

Goos, M. (2004). Learning mathematics in a classroom community of inquiry. *Mathematics Education Research Journal, 25*(4), 258-291.

Goos, M., & Jolly, L. (2004). Investigating partnerships with families and communities to support children's numeracy learning. In I. Putt, R. Faragher, & M. McLean (Eds.), *Mathematics education for the third millennium: Towards 2010* (Proceedings of the 27th annual conference of the Mathematics Education Research Group of Australasia, Townsville, Vol. 1, pp. 279-286). Sydney: MERGA.

Groves, S., & Doig, B. (2004). Progressive discourse in mathematics classes – The task of the teacher. In M. Høines & A. Fuglestad (Eds.), *Proceedings of the 28th annual conference for the Psychology of Mathematics Education* (Vol. 2, pp. 485-502). Bergen, Norway: Bergen University College.

Heirdsfield, A. (2005). One teacher's role in promoting understanding in mental computation. In H. L. Chick & J. L. Vincent (Eds.), *Proceedings of the 29th annual conference for the Psychology of Mathematics Education* (Vol. 3, pp. 113-120). Melbourne: University of Melbourne.

Herbert, S., & Pierce, R. (2005). Potential of technology and a familiar context to enhance students' concept of rate of change. In P. Clarkson, A. Downton, D. Gronn, M. Horne, A. McDonough, R. Pierce &, A. Roche (Eds.), *Building connections: Research, theory and practices* (Proceedings of the 28th annual conference of the Mathematics Education Research Group of Australasia, Melbourne, Vol. 2, pp. 435-442). Sydney: MERGA.

Higgins, J., Irwin, K., Thomas, G., Trinick, T., & Young-Loveridge, J. (2005). *Findings from the New Zealand Numeracy Development Project 2004*. Wellington: Ministry of Education.

Higgins, R. (2005). Effective teaching strategies for Māori students in an English-medium numeracy classroom. In J. Higgins, K. C. Irwin, G. Thomas, T. Trinick, & J. Young-Loveridge (Eds.), *Findings from the New Zealand numeracy development Project 2004* (pp. 74-79). Wellington: Learning Media.

Holton, D., & Clarke, D. (2006). Scaffolding and metacognition. *International Journal of Mathematical Education in Science and Technology, 37*(2), 127-143.

Horne, M. (2004). Class grouping for mathematics: What do we know? In B. Clarke, D. Clarke, G. Emanuelsson, B. Johansson, D. Lambdin, F. Lester, A. Wallby, & K. Wallby (Eds.), *International perspectives on learning and teaching mathematics* (pp. 395-410). Göteborg, Sweden: National Centre for Mathematics Education.

Howard, P., & Perry, B. (2005). Learning mathematics: Perspectives of Australian aboriginal children and their teachers. In H. L. Chick & J. L. Vincent (Eds.), *Proceedings of the 29th annual conference for the Psychology of Mathematics Education* (Vol. 3, pp. 153-160). Melbourne: University of Melbourne.

Hunter, R. (2005). Reforming communication in the classroom: One teacher's journey of change. In P. Clarkson, A. Downton, D. Gronn, M. Horne, A. McDonough, R. Pierce, & A. Roche (Eds.), *Building connections: Research, theory and practice* (Proceedings of the 28th annual conference of the Mathematics Education Research Group of Australasia, Melbourne, pp. 451-458). Sydney: MERGA.

Hunter, R. (2006). Structuring the talk towards mathematical inquiry. In P. Grootenboer, R. Zevenbergen, & M. Chinnappan (Eds.), *Identities, cultures and learning spaces* (Proceedings of the 29th annual conference of the Mathematics Education Research Group of Australasia, Canberra, Vol. 1, pp. 309-318). Adelaide: MERGA.

Hunter, R. (2007). Can you convince me: Learning to use mathematical argumentation. In J. H. Woo, H. C. Lew, K. S. Park, & D. Y. Seo (Eds.), *Proceedings of the 31st annual conference for the Psychology of Mathematics Education* (Vol., 3, pp. 81-88). Seoul, Korea: The Korea Society of Educational Studies in Mathematics.

Hurst, C. (2007). Numeracy in action: Students connecting mathematical knowledge to a range of contexts. In J. Watson & K. Beswick (Eds.), *Mathematics: Essential research, essential practice* (Proceedings of the 30th annual conference of the Mathematics Education Research Group of Australasia, Hobart, Vol. 1, pp. 440-449). Adelaide: MERGA.

Hyett, M. (2005). *Interactions and feedback with diverse students in mathematics classrooms.* Unpublished Master of Education project, Massey University, New Zealand.

Ingvarson, L., Beavis, A., Bishop, A., Peck, R., & Elsworth, G. (2004). *Investigation of effective mathematics teaching and learning in Australian secondary schools,* Canberra: Australian Council for Educational Research.

Irwin, K., & Britt, M. (2005). The algebraic nature of students' numerical manipulation in the New Zealand Number Project. *Educational Studies in Mathematics, 58*, 169-188.

Irwin, K., & Britt, M. (2007). The development of algebraic thinking: Results of a three-year study. In *Findings from the New Zealand Numeracy Development Project 2006* (pp. 33-43). Wellington: Ministry of Education.

Klymchuk, S. (2005). *Counter-examples in calculus.* Auckland: Maths Press.

Latu, V. F. (2005). Language factors that affect mathematics teaching and learning of Pasifika students. In P. Clarkson, A. Downton, D. Gronn, M. Horne, A. McDonough, R. Pierce, & A. Roche (Eds.), *Building connections: Research, theory and practice* (Proceedings of the 28th annual conference of the Mathematics Education Research Group of Australasia, Melbourne, pp. 483-490). Sydney: MERGA.

Lobato, J., Clarke, D., & Ellis, A. B. (2005). Initiating and eliciting in teaching: A reformulation of telling. *Journal for Research in Mathematics Education, 36*(2), 101-136.

Lowrie, T. (2004). Authentic artefacts: Influencing practice and supporting problem solving in the mathematics classroom. In I. Putt, R. Faragher, & M. McLean (Eds.), *Mathematics education for the third millennium: Towards 2010* (Proceedings of the 27th annual conference of the Mathematics Education Research Group of Australasia, Townsville, Vol. 2, pp. 351-358). Sydney: MERGA.

Lowrie, T. (2005). Establishing a numeracy culture in a distance education learning environment: A case study. In P. Clarkson, A. Downton, D. Gronn, M. Horne, A. McDonough, R. Pierce, & A. Roche (Eds.), *Building connections: Research, theory and practice* (Proceedings of the 28th annual conference of the Mathematics Education Research Group of Australasia, Melbourne, Vol., 2, pp. 498-504). Sydney: MERGA.

Lowrie, T., & Diezmann, C. (2005). Fourth-grade students' performance on graphical languages in mathematics. In H. L. Chick & J. L. Vincent (Eds.), *Proceedings of the 29th annual conference for the Psychology of Mathematics Education* (Vol. 3, pp. 265-272). Melbourne: University of Melbourne.

McChesney, J. (2005). *Student-initiated interactions: Student agency in mathematics classrooms.* Paper presented at the AERA annual meeting, Montreal, Canada, 11-15 April 2005 as part of the NZARE symposium, Making meaning: Students agency in four curriculum contexts.

McDonald, G., Le, H., Higgins, J., & Podmore, V. (2005). Artifacts, tools, and classrooms. *Mind, Culture, and Activity, 12*(2), 113-127.

McHardy, J. (2006). *Investigating the impact of a technology integrated flexible programme on senior mathematics students' learning.* Unpublished Master of Science thesis, The University of Auckland, Auckland, New Zealand.

Meaney, T. (2006). Acquiring the mathematics register in classrooms. *SET, 3*, 39-43.

Meaney, T., Fairhall, U., Trinick, T. (2007). Acquiring the mathematics register in te reo Maori. In J. Watson & K. Beswick (Eds.), *Mathematics: Essential research, essential practice* (Proceedings of the 30th annual conference of the Mathematics Education Research Group, Hobart, Vol. 2, pp. 493-502). Adelaide: MERGA.

Meaney, T., & Flett, K. (2006). Learning to read in mathematics classrooms. *Australian Mathematics Teacher, 62*(2), 10-16.

Meaney, T., & Irwin, K. (2005). *Language used by students in mathematics for quantitative and numerical comparisons* (NEMP Probe study). Dunedin: EARU, University of Otago.

Mitchelmore, M., White, P., & McMaster, H. (2007). Teaching ratio and rates for abstraction. In J. Watson & K. Beswick (Eds.), *Mathematics: Essential research, essential practice* (Proceedings of the 30th annual conference of the Mathematics Education Research Group of Australasia, Hobart, Vol. 2, pp. 503-512). Adelaide: MERGA.

Mousley, J., Sullivan, P., & Zevenbergen, R. (2004). Alternative learning trajectories. In I. Putt, R. Faragher, & M. McLean (Eds.), *Mathematics education for the third millennium: Towards 2010.* (Proceedings of the 21st conference of the Mathematics Education Research Group of Australasia, Townsville, Vol. 2, pp. 374-381). Sydney: MERGA.

Muir, T. (2006). What does effective teaching for numeracy look like? The design of an observation schedule. In P. Grootenboer, R. Zevenbergen, & M. Chinnappan (Eds.), *Identities, cultures and learning spaces* (Proceedings of the 29th annual conference of the Mathematics Education Research Group of Australasia, Canberra, pp. 368-375). Adelaide: MERGA.

Muir, T. (2007). Setting a good example: Teachers' choice of examples and their contribution to effective teaching of numeracy. In J. Watson & K. Beswick (Eds.), *Mathematics: Essential research, essential practice* (Proceedings of the 30th annual conference of the Mathematics Education Research Group of Australasia, Hobart, Vol. 2, pp. 513-522). Adelaide: MERGA.

Mulligan, J., & Busatto, S. (2004). The numeracy research in NSW primary schools project 2001-2003. In I. Putt, R. Faragher, & M. McLean (Eds.), *Mathematics education for the third millennium: Towards 2010* (Proceedings of the 27th annual conference of the Mathematics Education Research Group of Australasia, Townsville, Vol. 2, pp. 382-389). Sydney: MERGA.

Mulligan, J., Mitchelmore, M., & Prescott, A. (2005). Case studies of children's development of structure in early mathematics: A two-year longitudinal study. In H. L. Chick & J. L. Vincent (Eds.), *Proceedings of the 29th annual conference for the Psychology of Mathematics Education* (Vol. 4, pp. 1-8). Melbourne: University of Melbourne.

Mulligan, J., Prescott, A., Papic, M., & Mitchelmore, M. (2006). Improving early numeracy through a pattern and structure mathematics awareness program (PASMAP). In P. Grootenboer, R. Zevenbergen & M. Chinnappan (Eds.), *Identities, cultures and learning space* (Proceedings of the 29th annual conference of the Mathematics Education Research Group of Australasia, Canberra, Vol. 2, pp. 376-383). Adelaide: MERGA.

Nettleship, E. (2006). *Students' perspective of a mathematics extension programme designed with special interest in history.* Unpublished Master of Educational Studies thesis, Massey University, Palmerston North, New Zealand.

Neville-Barton, P., & Barton, B. (2005). *The relationship between English-language and mathematics learning for non-native speakers.* Final Report. Retrieved September 7, 2005 from http://www.tlri.org.nz/index.html

Norton, S., & Irvin, J. (2007). Developing positive attitudes towards algebra. In J. Watson & K. Beswick (Eds.), *Mathematics: Essential research, essential practice* (Proceedings of the 30th annual conference of the Mathematics Education Research Group, Hobart, Vol. 2, pp. 561-570). Adelaide: MERGA.

O'Keefe, C. A., Hua, X. L., & Clarke, D. (2006). Kikan-Shido: Through the lens of guiding student activity. In J. Novotná, H. Moraová, M. Krátká, & N. Stehliková (Eds.), *Proceedings of the 30th annual conference for the Psychology of Mathematics* (Vol. 4, pp. 265-272). Prague: Charles University.

O'Toole, T. (2006). Building powerful understanding by connecting informal and formal knowledge. In P. Grootenboer, R. Zevenbergen, & M. Chinnappan (Eds.), *Identities, cultures and learning spaces* (Proceedings of the 29th annual conference of the Mathematics Education Research Group of Australasia, Canberra, Vol 2., pp. 384-391). Adelaide: MERGA.

Owens, K. (2005). Teachers' development of substantive communication about mathematics. In P. Clarkson, A. Downton, D. Gronn, M. Horne, & A. McDonough, R. Pierce, & A. Roche (Eds.), *Building connections: Research, theory and practice* (Proceedings of the 28th annual conference of the Mathematics Education Research Group of Australasia, Melbourne, Vol. 2, pp. 601-608). Sydney: MERGA.

Pegg, J., Graham, L., & Bellert, A. (2005). The effect of improved automaticity of basic number skills on persistently low-achieving pupils. In H. L. Chick & J. L. Vincent (Eds.), *Proceedings of the 29th annual conference for the Psychology of Mathematics Education* (Vol. 4, pp. 49-56). Melbourne: PME.

Pegg, J., Panizzon, D., & Lynch, T. (2007). Identifying and analyzing processes in NSW public schooling producing outstanding educational outcomes in mathematics. In J. Watson & K. Beswick (Eds.), *Mathematics: Essential research, essential practice* (Proceedings of the 30th annual conference of the Mathematics Education Research Group, Hobart, Vol. 2, pp. 611-620). Adelaide: MERGA.

Pierce, R., Herbert, S., & Giri, J. (2004). CAS: Student engagement requires unambiguous advantages. In I. Putt, R. Faragher, & M. McLean (Eds.), *Mathematics education for the third millennium: Towards 2010.* (Proceedings of the 27th annual conference of the Mathematics Education Research Group, Townsville, Vol. 2, pp. 462-469). Sydney: MERGA.

Prescott, A., & Mitchelmore, M. (2005). Teaching projectile motion to eliminate misconceptions. In H. L. Chick & J. L. Vincent (Eds.), *Proceedings of the 29th annual conference for the Psychology of Mathematics Education* (Vol. 4, pp. 97-104). Melbourne: University of Melbourne.

Pritchard, R., & McDiarmid, F. (2006). Promoting change in teacher practices: Investigating factors which contribute to sustainability. In P. Grootenboer, R. Zevenbergen, & M. Chinnappan (Eds.), *Identities, cultures and learning spaces* (Proceedings of the 29th annual conference of the Mathematics Education Research Group of Australasia, Canberra, Vol. 2, pp. 432-439). Adelaide: MERGA.

Rawlins, P. (2007). *Students' perceptions of the formative potential of NCEA.* Unpublished doctoral thesis, Massey University, Palmerston North.

Rietveld, C. (2005). Classroom learning experiences of mathematics by new entrant children with Down Syndrome. *Journal of Intellectual & Developmental Disability, 30*(3), 127-138.

Rogers, P. (2006). Mathematical teacher professional development incorporating an external critical friend. In P. Grootenboer, R. Zevenbergen, & M. Chinnappan (Eds.), *Identities, cultures and learning spaces* (Proceedings of the 29th annual conference of the Mathematics Education Research Group of Australasia, Canberra, Vol. 2, pp. 447-454). Adelaide: MERGA.

Rowe, K., & Bicknell, B. (2004). Structured peer interactions to enhance learning in mathematics. In I. Putt, R. Faragher, & M. McLean (Eds.), *Mathematics education for the third millennium: Towards 2010* (Proceedings of the 27th annual conference of the Mathematics Education Research Group of Australasia, Townsville, Vol. 2, pp. 494-500). Sydney: MERGA.

Sawyer, A. (2006). Unpacking the rules of class discussion: Young children learning mathematics within a community of inquiry. In P. Grootenboer, R. Zevenbergen, & M. Chinnappan (Eds.),

Identities, cultures and learning spaces (Proceedings of the 29th annual conference of the Mathematics Education Research Group of Australasia, Canberra, Vol. 2, pp. 455-462). Adelaide: MERGA.

Seah, W. T. (2005). Negotiating about perceived value differences in mathematics teaching: The case of immigrant teachers in Australia. In H. L. Chick & J. L. Vincent (Eds.), *Proceedings of the 29th annual conference for the Psychology of Mathematics Education* (Vol. 4, pp. 145-152). Melbourne: University of Melbourne.

Smith, T. (2006). Exploring connections between teaching, learning and assessing mathematics for understanding. In J. O. Masingila (Ed.), *Teachers engaged in research: Inquiry into mathematics practice grades 6-8* (pp. 185-202). Greenwich, CT: Information Age/National Council of Teachers of Mathematics.

Stephens, M. (2006). Describing and exploring the power of relational thinking. In P. Grootenboer, R. Zevenbergen, & M. Chinnappan (Eds.), *Identities, cultures and learning spaces* (Proceedings of the 29th annual conference of the Mathematics Education Research Group of Australasia, Canberra, Vol. 2, pp. 479-486). Adelaide: MERGA.

Still, K., & Bobis, J. (2005). The integration of mathematics and music in the primary school classroom. In P. Clarkson, A. Downton, D. Gronn, M. Horne, A. McDonough, R. Pierce, & A. Roche (Eds.), *Building connections: Theory, research and practice* (Proceedings of the 28th annual conference of the Mathematics Education Research Group of Australasia, Melbourne, Vol. 2, pp. 712-719). Sydney: MERGA.

Sullivan, P., Mousley, J., & Zevenbergen, R. (2006a). Developing guidelines for teachers helping students experiencing difficulty in learning mathematics. In P. Grootenboer, R. Zevenbergen, & M. Chinnappan (Eds.), *Identities, cultures and learning spaces* (Proceedings of the 29th annual conference of the Mathematics Education Research Group of Australasia, Canberra, Vol. 2, pp. 496-503). Adelaide: MERGA.

Sullivan, P., Mousley, J., & Zevenbergen, R. (2006b). Teacher actions to maximize mathematics learning opportunities in heterogeneous classrooms. *International Journal of Science and Mathematics Education, 4*(1), 117-143.

Taylor, M., & Cowie, B. (2006). Fast horses, slow cows: Context and mathematical tasks. *SET: Research Information for Teachers, 3*, 7-10.

Trinick, T. (2005). Te Poutama Tau: A case study of two schools. In J. Higgins, K. C. Irwin, G. Thomas, T. Trinick, & J. Young-Loveridge (Eds.), *Findings from the New Zealand numeracy development project 2004* (pp. 80-88). Wellington: Learning Media.

Vale, C. (2006). Equity and technology: A case study. In P. Grootenboer, R. Zevenbergen, & M. Chinnappan (Eds.), *Identities, cultures and learning space* (Proceedings of the 29th annual conference of the Mathematics Education Research Group of Australasia, Canberra, Vol. 2, pp. 512-518). Adelaide: MERGA.

Walls, F. (2005). Challenging task-driven pedagogies of mathematics. In P. Clarkson, A. Downton, D. Gronn, M. Horne, A. McDonough, R. Pierce, & A. Roche (Eds.), *Building connections: Research, theory and practices* (Proceedings of the 28th annual conference of the Mathematics Education Research Group of Australasia, Melbourne, Vol. 2, pp. 751-758). Sydney: MERGA.

Walls, F. (2006). Discursive practices as cultural cargo. *For the Learning of Mathematics, 26*(3), 2-8.

Walshaw, M. (2005). Getting political and unravelling layers of gendered mathematical identifications. *Cambridge Journal of Education, 35*(1), 19-34.

Walshaw, M. (2007). Responding to calls for greater accountability. *Mathematics Education Research Journal, 19*(1), 1-2.

Walshaw, M., & Anthony, G. (2006). Numeracy reforms in New Zealand: Factors that influence classroom enactment. In J. Novotná, H. Moraová, M. Krátká, & N. Stehliková (Eds.), *Proceedings of the 30th annual conference for the Psychology of Mathematics Education* (Vol. 5, pp. 36-368). Prague: Charles University.

Warren, E. (2004). Generalising arithmetic: Supporting the process in the early years. In In M. Johnsen Høines & A. B. Fuglestad (Eds.), *Proceedings of the 28th annual conference for the Psychology of Mathematics Education* (Vol. 4, pp. 417-424). Bergen, Norway: Bergen University College.

Warren, E. (2005). Patterns supporting the development of early algebraic thinking. In P. Clarkson, A. Downton, D. Gronn, M. Horne, & A. McDonough, R. Pierce, & A. Roche (Eds.), *Building connections: Research, theory and practice* (Proceedings of the 28th annual conference of the Mathematics Education Research Group of Australasia, Melbourne, Vol. 2, pp. 759-766). Sydney: MERGA.

Warren, E., Young, J., & de Vries, E. (2007). Australian indigenous students: The role of oral language and representations in the negotiation of mathematical understanding. In J. Watson & K. Beswick (Eds.), *Mathematics: Essential research, essential practice* (Proceedings of the 30th annual conference of the Mathematics Education Research Group, Hobart, Vol. 2, pp. 775-784). Adelaide: MERGA.

Watson, A., & Mason, J. (2005). *Mathematics as a constructive activity*. Mahwah: Lawrence Erlbaum Associates.

Watson, J. M. (2007). The role of cognitive conflict in developing students' understanding of average. *Educational Studies in Mathematics, 65*, 21-47.

Watson, J., Beswick, K., & Brown, N. (2006). Teachers' knowledge of their students as learners and how to intervene. In P. Grootenboer, R. Zevenbergen, & M. Chinnappan (Eds.), *Identities, cultures and learning spaces* (Proceedings of the 29th annual conference of the Mathematics Education Research Group of Australasia, Canberra, Vol. 2, pp. 551-558). Adelaide: MERGA.

Williams, G. (2006). Student-engineered 'space to think'. In P. Grootenboer, R. Zevenbergen, & M. Chinnappan (Eds.), *Identities, cultures and learning spaces* (Proceedings of the 29th annual conference of the Mathematics Education Research Group of Australasia, Canberra, Vol. 2, pp. 567-574). Adelaide: MERGA.

Wood, T., Williams, G., & McNeal, B. (2006). Children's mathematical thinking in different classroom cultures. *Journal for Research in Mathematics Education, 37*(3), 222-255.

Woodward, J., & Irwin, K. (2005). Language appropriate for the New Zealand Numeracy Project. In P. Clarkson, A. Downton, D. Gronn, M. Horne, A. McDonough, R. Pierce, & A. Roche (Eds.), *Building connections: Research, theory and practice* (Proceedings of the 28th annual conference of the Mathematics Education Research Group of Australasia, Melbourne, Vol. 2, pp. 799-806). Sydney: MERGA.

Woodward, J., & Pfannkuch, M. (2007). Statistics teachers as scientific lawyers. In J. Watson & K. Beswick (Eds.), *Mathematics: Essential research, essential practice* (Proceedings of the 30th annual conference of the Mathematics Education Research Group, Hobart, Vol. 2, pp. 834-842). Adelaide: MERGA.

Worlley, A., & Jamieson-Proctor, R. (2005). Results of a teaching experiment to foster the conceptual understanding of multiplication based on children's literature. In P. Clarkson, A. Downton, D. Gronn, M. Horne, A. McDonough, R. Pierce, & A. Roche (Eds.), *Building connections: Theory, research and practice* (Proceedings of the 28th annual conference of the Mathematics Education Research Group of Australasia, Melbourne, Vol., 2, pp. 807-814). Sydney: MERGA.

Yelland, N. (2005). Curriculum, pedagogies and practice with ICT in the information age. In N. Yelland (Ed.), *Critical issues in early childhood education* (pp. 224-242). Maidenhead: Open University.

Zevenbergen, R. (2005). The construction of a mathematical habitus: Implications of ability grouping in the middle years. *Journal of Curriculum Studies, 37*(5), 607-619.

Zevenbergen, R., & Lerman, S. (2007). Pedagogy and interactive whiteboards: Using an activity theory approach to understand tensions in practice. In J. Watson & K. Beswick (Eds.), *Mathematics: Essential research, essential practice* (Proceedings of the 30th annual conference of the Mathematics Education Research Group of Australasia, Hobart, Vol. 2, pp. 853-862). Adelaide: MERGA.

AFFILIATIONS

Glenda Anthony
Massey University, New Zealand

Margaret Walshaw
Massey University, New Zealand

WEE TIONG SEAH, BILL ATWEH, PHILIP CLARKSON, & NERIDA ELLERTON

SOCIOCULTURAL PERSPECTIVES IN MATHEMATICS TEACHING AND LEARNING

In this chapter, Australasian mathematics education research conducted over the period 2004 to 2007 and adopting the sociocultural perspective is reviewed and critiqued. Two main approaches within this perspective have been noted. The first approach acknowledges the variety of socioeconomic factors that mediate mathematics teaching and learning. These may be usefully grouped into three major threads, namely student background factors, language and discourse, and attitudes, beliefs and values. The second approach in mathematics education research incorporates the sociocultural perspective into its theoretical conceptions and methodologies, and is associated with poststructuralist, feminist, or community of practice stances. Critiques of research that has been conducted are discussed within the various sociocultural perspectives, as well as more globally.

Key words: attitudes, beliefs, class, discourse, ethnicity, language, race, socioeconomic status, values

INTRODUCTION

Since the 1992 four-yearly review published by the Mathematics Education Research Group of Australasia [MERGA], research into the social-cultural context has been discussed under different banners, signifying the multidimensionality of issues related to this perspective. In the 1992 review, Atweh, Cooper, and Kanes (1992) reviewed research on background factors affecting mathematics learning such as indigeniety, bilingual and socioeconomic backgrounds, and the classroom as a social context of communication. In the 1996 review, Zevenbergen, Atweh, Kanes, and Cooper (1996) discussed several theoretical constructs used by researchers to study social and cultural contexts including social constructivism, Habermas, poststructuralist theory, and Wittgenstein. In the last review, Atweh, Meaney, McMurchy-Pilkington, Neyland, and Trinick (2004) highlighted an increase in the use of sociocultural theories including Vygotsky, and theories of ethics and social justice.

It is becoming increasingly difficult to positively specify the type of research that can be classified under the social context in mathematics education. Here we posit four stances that can be taken to understand this construct. First, there has been continual interest in the social factors that affect students' learning of

H. Forgasz et al. (eds.), Research in Mathematics Education in Australasia 2004-2007, 223-253.

mathematics including: gender, socioeconomic and cultural backgrounds of students, as well as the attitudes, beliefs and values they bring with them into the mathematics classroom.[1] Second, many researchers have looked at the classroom as a social context with its own culture and norms of communication where language and discourse play a central role. Third, there has been a steady increase and diversification of sociocultural theories of teaching and learning that have been utilised by researchers. These include the use of Vygotskian analysis and the associated activity theory (see, for example, Leont'ev, 1978), as well as poststructuralist/postmodern and feminist perspectives. Fourth, several researchers have addressed the role of mathematics in society including topics such as numeracy and modelling.

Regardless of the different conceptions of social context, there is evidence relating sociocultural factors to students' mathematics achievement (e.g. Powell-Mikle, 2003), thus affording an approach to understanding and optimising learning outcomes which complements cognitive and affective factors. Highlighting the sociocultural factors influencing mathematics teaching and learning recognises the day-to-day attempts by students and teachers to make sense of mathematical ideas and concepts by making what Aikenhead (2001) called *cultural border crossings* from their respective cultures and subcultures. It should be noted that within the science education literature in which Aikenhead writes, the notions of borderlands and the like have been cogently critiqued (Carter, 2004, 2006).

These alternative conceptions of the social context pose dilemmas for any review of this area of research in terms of the focus and structure to adopt. Given the focus on research on sociocultural *factors*, in this chapter we consider the first two interpretations of research into the sociocultural theories identified above, reflecting the focus of research by Australasian researchers during the period of the review. In particular, in the following section, research on the effects of cultural and socioeconomic factors is considered. In the section that follows the focus is on research on language factors and classroom communication. Research on attitudes, beliefs, and values comprises the fourth section in which sociocultural, theoretical perspectives identified by various researchers in the literature are also considered. Finally, there are some overall comments by way of a critique of sociocultural research in mathematics education.

In compiling this review we have undertaken an extensive search of published literature in the MERGA conference proceedings and journals, the *Mathematics Education Research Journal* and the *Mathematics Teacher Education and Development.* Similarly, the proceedings of conferences frequented by Australasian researchers, such as the Australian Association for Research in Education, the International Conference of the Psychology of Mathematics Education and the annual International Conferences organised by the Department of Science and

[1] As in previous MERGA reviews, gender is the topic of a separate chapter in this volume, indicating its centrality for many researchers; issues related to Indigenous students are also discussed in a separate chapter.

Mathematics Education at Universiti Brunei Darussalam in the period 2004-2007, were perused. We also examined publications in international journals such as *For the Learning of Mathematics*, *Educational Studies in Mathematics* and the *Journal for Research in Mathematics Education*. In addition, scholarly articles and chapters in books, as well as higher degree theses, were examined.

STUDENT BACKGROUND FACTORS

The contribution of Australasian mathematics educators to ethnomathematics has a long history, as demonstrated at the Third International Conference on Ethnomathematics held in Auckland in 2006. Incidentally, at least one third of the participants there had come from the Asia Pacific region (Barton, Poisard, & Do Carmo Domite, 2006).

Ethnomathematics was used as the theoretical stance of three further studies. Matang (2005) reported on a longitudinal study on the effects of teaching early arithmetic strategies using the Kâte counting system within the Papua New Guinea school curriculum, to demonstrate that the use of counting and arithmetic strategies embedded in the children's own traditional counting systems is an added advantage in their learning of formal English arithmetic strategies normally taught in schools. Similarly, Adam (2004) discussed the development of a unit of study in two primary schools that involved teaching grade 5 students an ethnomathematical unit of work on measurement in the Maldives. The author argued that the ethnomathematical approach was appreciated and understood by teachers and students. Finally, Walls (2006) discussed her experiences with teaching mathematics in the Pacific islands. A collection of photographs was used to tell a narrative and to raise questions about "political and ontological implications of globalised mathematics curriculum" (Walls, 2006, p. 3). It was concluded that:

> if children and communities of the region are fully to benefit from education in the 21st century, new models of mathematics education for the Pacific must be developed that go beyond an uncritical adoption of the discursive practices of Western mathematics education. (Walls, 2006, p. 8)

Two further studies that represent cross-cultural comparisons in mathematics education are worth mentioning here. Ya-Amphan and Bishop (2004) compared Thai and Japanese students' error patterns in solving equations. They concluded that:

> rather than just using mathematics tests in international comparisons to focus on total right answer scores, it might be much more productive to develop more cross-cultural studies of student errors and their possible causes. Researchers already know that one learns more from students' errors than from their right answers, and linking these errors to possible causes such as teachers' teaching styles or, as in this case, textbook features … will increase our knowledge enormously. (Ya-Amphan & Bishop, 2004, pp. 603-604)

Finally Cao, Bishop, and Forgasz (2007) demonstrated differences in perceived parental influences on mathematics learning among Chinese students in Australia and in China. Cultural factors were posited as possible explanations of these differences. The authors suggested that such differences may explain some of the variations in international comparisons on achievement.

Socioeconomic status/class

Politicians' beliefs that the analysis of data from ubiquitous system-wide testing of literacy and numeracy is the best way of judging the worth of education systems must be contested. This stance ignores key issues, including the impact that structural societal variables such as socio-economic status [SES] and social class have on numeracy scores and reflects a major change from 25 years ago. The 'renewed' importance attributed to testing was highlighted when the then Australian Federal Minister for Education quoted an economist on what makes effective teachers. The economist's report used correlational analyses of system wide numeracy test data, allowing the Minister to dismiss the impact of any societal structural variables (Leigh, 2007). And yet in the week prior to Leigh's paper being publicised, a report decrying the quality of government schools in 'poor' areas of the city of Melbourne, and the impact this was having on the students' numeracy abilities also gained wide publicity (e.g. Tomazin, 2007). It is time that the mathematics education research community looked again at key issues such as SES and social class, if the inequalities inherent in society, as they impact on educational opportunities, are to be made transparent.

In the introduction or literature review sections of the many studies scanned for this review SES was briefly mentioned; social class less often. Thus Ayers and McCormick (2006) suggested that teachers in their study assumed that students from low SES suburbs would have low performances in mathematics, and hence the implementation of the NSW 'reforms' would be difficult in such suburbs. When discussing gifted students in high SES suburbs, Bicknell (2006) noted the assumptions made about the willingness and ability of parents to be involved in their students' education. But assumptions such as these were rarely returned to in any meaningful manner in these and other studies. At most there was a cursory statement in the discussion or conclusion concerning them. Such near silences are worrying in high quality research. Hidden assumptions need to be dealt with explicitly and made transparent. Vale (2006) argued that as members of a research community, we must avoid stereotyping if better understanding of the impact of such societal structural variables is to be achieved. We should not simply accept commonly held assumptions, but should contest them to establish whether they do indeed hold in the particular context of particular research studies.

The direct impact of SES was discovered in one study, but none was found on students' beliefs about computers for mathematics learning (Forgasz, 2004). Clarke and Robbins (2004) challenged various SES assumptions. They worked with pre-school families in low SES areas, believing that if students from such areas were going to be disadvantaged because of where they lived, then early intervention to alleviate this disadvantage was crucial. The findings suggested that many parents actively supported their children's learning of school mathematics in the home. Much informal mathematical activity that was enacted quite purposefully in students' homes was identified. The results clearly surprised, challenged, and delighted the teachers, who had expected little active support for mathematics learning from the low SES parents.

Another Australasian study considered here confirmed some of the understandings regarding SES held by the education community. The New Zealand Numeracy Project used SES as a key variable. Analyses showed that 11-year-old students from higher SES areas outperformed comparable students from lower SES areas, after various co-variables were taken into account (Irwin & Irwin, 2005). During the years that students were followed, the lower SES students' performances hardly increased. It appeared that these students had not gained sufficient foundation in earlier school years to deal with new concepts encountered in later years (Clarke & Robbins, 2004). The project also addressed the relationship between teachers' professional development (PD) and the SES area in which their schools were located. The key finding from the study was that although teachers in low SES area schools wanted to engage in PD, there were overwhelming demands on their time and emotional energy that prevented real and sustained engagement. This key factor would need to be addressed if mentoring endeavours were to be successful (Kensington-Miller, 2004, 2005).

The final two publications considered here are based on the Programme for International Student Assessment's [PISA] international comparison of achievement in mathematics education. McGaw (2004) discussed major findings about Australia's performance in the 2000 PISA round in literacy and mathematics. A highly publicised finding was the relatively high performance of Australian students on these tests. In mathematical literacy, Australia ranked third along with five other countries. Perhaps less publicised were the data on the variation between schools and the relatively high inequalities within the Australian student population. For example, gender differences in Australia to the benefit of boys represented the 11th largest difference among 27 countries, although it was not clear if this difference was statistically significant. Similarly, McGaw demonstrated that the regression line of mathematics as a function of social disadvantage is curved at the higher levels of disadvantage, indicating:

> social advantage being associated with smaller increments in mean mathematical literacy level at lower levels of social advantage and larger increments in mean mathematical literacy level at higher levels of social advantage. (McGaw, 2004, p. 20)

McGaw (2004) concluded that:

> There is a tendency in Australia to dismiss attempts to redress inequities in education by declaring that they will result in 'dumbing down'. The PISA 2000 evidence that some countries achieve high quality and at the same time more equitable results than Australia makes it clear that 'dumbing down' is

> no necessary consequence of the pursuit of equity. 'Levelling up' is clearly an alternative. (p. 24)

Using the same PISA data set, Marks (2006) considered the between-schools and within-schools differences in performance to investigate the overall effect of socioeconomic factors on performance and the practice of tracking students into different types of mathematics courses at the secondary level. The author concluded that:

> differences in student performance between and within schools cannot be accounted for by socioeconomic background. The study lends some support to the official rationale for tracking and other forms of educational differentiation, that the allocation of students to different school types and academic locations within schools is largely based on student ability. (p. 21)

However, there is a danger in early differentiation. Marks (2006) mitigates to some degree his earlier claim by explaining:

> Higher-status parents have the resolve and resources to place their children in more academic locations than lower-status parents with equally intelligent children. However, once in different academic locations, differences in student performance arise and increase because students from higher socio-economic backgrounds are in pedagogically richer and academically more challenging environments. Students in higher-ability groupings tend to show greater gains in learning than students in lower-ability groupings … Countries with more highly differentiated systems tend to have more variation in student performance … This is the process by which socioeconomic inequalities tend to be magnified by educational differentiation. (p. 38)

The dearth of studies that directly address societal structural variables in mathematics education research in recent times should be addressed. There are many silences that need to be given voice if we are to promote equity and justice in our societies. In particular, it was noted that in all the studies reported here the effects of socioeconomic background of the students were considered, rather than research on how to deal with it. Considering the consistent evidence of the importance of this factor in determining achievement in mathematics, more research is needed in this area.

LANGUAGE/DISCOURSE

Language factors and discourse patterns are inherently associated with the teaching and learning of mathematics. Skalicky (2007) included the language of mathematics as the first of five dimensions of numeracy, and Goos (2006) claimed that all mental processes are mediated by tools and signs, which include multiple forms of language. As Booker, Bond, Sparrow, and Swan (2004) noted, "materials by themselves do not literally carry meaning … it is language that communicated ideas, not only in describing concepts but also helping them take shape in each learner's mind" (p. 20). Such views are consistent with the concept of apperception, within the theoretical stance of Johann Herbart, the great German educator (Ellerton & Clements, 2005).

In this section, four main headings, taken from the framework originally developed by Ellerton (1989), will be used: mathematics curricula and classrooms, classroom discourse, mathematics and problem solving, and sociolinguistics. This framework was also adopted in the chapters on language factors in earlier four-yearly MERGA reviews (published in 1996 and 2000), and in the chapter (Ellerton & Clarkson, 1996) on language factors in the first *International Handbook of Mathematics Education.*

Language factors in mathematics curricula and classrooms

Over the past four years much attention has been given to how teachers use language in the classroom. The Japanese Lesson Study emphasised the need for teachers to communicate effectively with each other within structured environments that focus specifically on lesson improvement (see, e.g., White, 2006a). Bobis et al. (2005) reported how teachers assisted participating students in the Victorian Early Numeracy Project to become better at explaining their reasoning and strategies. Cooper, Baturo, and Warren (2005) found everyday communication between teachers and their Indigenous aides to be crucial for effective teaching. Davies and Walker (2005) noted positive outcomes generated by teachers changing their questioning modes so that their questions became more open-ended. Woodward and Irwin (2005) suggested a language model based on having teachers listen and question students' justifications and explanations.

The *Victorian Essential Learning Standards* [VELS] (Victorian Curriculum and Assessment Authority, 2005) maintained that "communication was central to the capacity to construct meaning and to convey information to others in a range of ways and in a variety of settings" (p. 4). However, ways in which domains such as 'Communication' and 'Thinking' manifest themselves in the classroom remain unclear (Gough, 2006). The VELS makes frequent reference to both receptive and expressive forms of language. For example, students are involved in an expressive sense as they "describe", "represent", "give travel directions", and "justify". In the Communication domain, however, descriptions start by referring to more receptive forms of language such as "being attentive listeners and viewers", and suggest that students develop their "understanding of basic communication conventions" through guidance, presumably given by the teacher. In the Working Mathematically dimension for Level 4, students are expected to develop and test semantically complex conjectures.

The term 'structure' has been introduced into the language of VELS. "Structure includes logic and sets as well as number patterns leading to algebra" (Lowe, 2006, p. 22). Teachers will need to consider carefully the implications of the concept of 'structure' as they implement their mathematics curricula.

Caswell and Nisbet (2005) described meta-language as the domain-specific language of mathematics. Students appeared to value the fact that their teacher "asked specific questions about the value and use of meta-language" (p. 211). In other studies, teachers modelled the use of domain-specific language: research reports made reference, for example, to "the language of graphs" (Fitzallen, 2006), "graphical languages" (Diezmann, Lowrie, & Kozak, 2007), "diagram literacy" (Diezmann & English, 2001; Diezmann, 2005), "fractional language" (Pearn & Stephens, 2007), "language of inference" (Woodward & Pfannuch, 2007), "language of probability" (Chick & Baker, 2005), "statistical literacy" (Watson, Kelly, & Izard, 2005), and "language of percentages" (White, Wilson, Faragher, & Mitchelmore, 2007). The terms "graphical languages" and "axis language" were used by Diezmann and Lowrie (2006) and by Lowrie and Diezmann (2005) in their studies of primary students' understanding of number lines. Contrasting forms of language used between researchers and between teachers and researchers were described by Geiger and Goos (2006). The need for high quality teacher-researcher communication was also emphasised by Thomas and Hong (2005).

Several researchers have studied how students can develop mathematical thinking through written reflections. Scott (2007), for example, used three questions to provide scaffolding for Grade 5/6 students as they wrote reflections on space-based mathematics tasks; and Ellerton (2005) described pre-service education students' metacognitive growth as shown through successive written reflections.

Language aspects of mathematics problem solving and strategy development

Language has many forms – for example, spoken, written, symbolic, diagrammatic, graphical, tabular – and all of these inevitably enter into the domain of problem solving in mathematics (Anderson, 2005). Falle (2005) emphasised that understanding the language of mathematics "does not involve simply the acquisition and use of the appropriate vocabulary" (p. 345), and that students often struggle because mathematical language uses natural language and borrows words or ascribes specific but different meanings to commonly used words. In her study of Grades 8 and 9 students learning algebra, Falle (2007) concluded that students needed "to experience, and use, a variety of mathematical language and symbols in order to explore and express mathematical meanings," and maintained that "without this, their algebraic development must be inhibited" (p. 352). The effectiveness of having pre-service education students express algebraic meanings in a variety of ways was explored by Clements and Ellerton (2006). In a study with Grade 9 students in which there was explicit teaching of the links between algebraic materials, symbols, and language, Norton and Irvin (2007a, 2007b) reported students' progress in developing algebraic understanding, and Warren (2005) found that Grade 4 students had difficulty expressing pattern generalizations.

Research findings indicate that teachers need consciously to assist students to develop the formal vocabulary and linguistic structures that characterise mathematics text (see, e.g., Clements, 2006). Chick and Baker (2005) noted that a teacher in their study introduced a lesson through class discussion "during which the concept of likelihood was developed with rich and rigorous use of language" (p. 238). Another teacher did not help the students with language, and these students struggled to understand the meaning of the very small difference between the chances of winning a game. Mulligan, Prescott, and Mitchelmore (2004) investigated students' use of pattern and structure across several different strands of the mathematics curriculum and found that some children focused on aspects of a mathematics problem which were not always conducive to formal mathematical development. According to Mulligan (2006), if a student's attention were refocused on essential elements of mathematical pattern and structure, then the teacher could assist the student "to make better sense of what may seem a confusing and artificial system of symbols and language" (p. 29).

Sometimes, however, teacher assistance can be unproductive, especially if students are not linguistically ready for what their teacher is trying to teach them. Students' appropriate use of language involving comparative terms such as "more" and "most" develops slowly from early childhood to the middle years, and generally occurs after students have developed an intuitive appreciation of variation (Watson, 2005). Callingham (2005) noted the articulate manner with which primary students were able to describe their mental computation strategies. In a detailed study of Grade 6 children's development of mathematical modeling strategies, English (2006) demonstrated how properties of the problems themselves invited children "to communicate and share their mathematical ideas" (p. 319). In addition, children needed to use different written formats as they documented their progress, developed skills in explaining, justifying, and refining their mathematical models (English & Fox, 2005; English & Watters, 2005).

Ellerton and Olson (2005), applying a modified Newman error approach, found that students often make errors on written mathematical tasks even though they fully understand the associated mathematics, but paradoxically students who do not understand the mathematics often give correct answers.

Classroom discourse

Coming from a poststructuralist perspective, Letts (2006) urged that the principles of feminist postructuralism be applied to discursive practices relating to the education of prospective mathematics teachers. Thornton (2006) considered students' "funds of knowledge" as comprised of "three spaces," with the overt and covert languages of mathematics playing crucial roles in the second of these spaces. Walshaw and Anthony (2007) explored, through analyses of classroom discourse, the forms of pedagogy that contribute to students' active engagement with mathematics. They noted that it was difficult to plan for and implement effective classroom environments in which students explain, make predictions, debate, clarify, and justify. The importance of finding new ways to analyse mathematics classroom discourse has been explored by Barnes (2005).

Teachers involved in a study conducted by Doyle (2006, 2007) became leaders of functional discourse communities who promoted mathematical literacy in students through mathematical modeling contexts. Discourse, of course, also involves listening and responding to the language used by others, and J. Hunter (2006) and R. Hunter (2006) investigated ways of structuring classroom interactions to encourage student participation and increase cognitive demand.

Attempts by teachers to assist their students to link real-life knowledge and abstract mathematical knowledge can lead to the formation of "invisible pedagogies" (Sawyer, 2006) in which learners may not be aware of, or comfortable with, associated norms. An example of finding ways to connect informal and formal knowledge, thereby stimulating students' use of visual representation and verbal and written language, was reported by O'Toole (2006).

Clarke (2005) developed a method for analyzing teacher and student discourse by examining the distribution of mathematics-related expressions which were classified in three ways: terms related to the teacher's stated instructional goals; terms which were subordinate or supportive of these stated goals; and terms which appeared infrequently during lessons. This analytic approach was used to distinguish between classrooms in different cultural contexts. Argumentation profile charts have been developed by Vincent, Chick, and McCrae (2005) to facilitate analysis of student-student and student-teacher discourses during argumentation.

A re-examination of teachers' communications with their students in terms of the *function* rather than *form* of the communication led Clarke and Lobato (2002) and, more recently, Lobato, Clarke, and Ellis (2005), and Clarke and Hoon (2005), to distinguish between "eliciting" and "initiating". Initiation can occur, for example, when a teacher poses a question whose answer initiates subsequent action by the teacher. If, however, a teacher's communicative act functions to facilitate action or expression on the students' part, then this is described as eliciting. "Cognitive emptying" (Brousseau, 1997) could be an outcome of teachers' acts of initiating, if students are consistently pushed towards answers which the teachers have in mind (Clements, 2005).

But, are teachers aware of the meanings of technical terms such as *cognitive emptying*? Nally (2005) has reported on the fundamental importance of developing a shared language between teachers and researchers – a language which would not only allow informed conversations to take place, but which would help to move the research process forward in productive ways.

Averill and Harvey (2005) reported positive results after they told stories to their pre-service education students about beginning teachers. Students felt that these stories (scenario-based learning) had supported their personal growth as mathematics teachers; through the stories, learners were able to identify with the beginning teacher in the story.

The role of discourse in interpreting the links between what teachers/researchers and students say and do is an important tool in the analysis of classroom or interview data (e.g., Stillman, 2004; Vaiyavutjamai & Clements, 2006). Calder (2005), for example, investigated grade 6 students' initial approaches to working with spreadsheets. The students first needed to become familiar with reading and interpreting spreadsheets. Then, they quickly developed a visual referent and were able to make conjectures based on observed patterns. In the case of spreadsheets, being able to read, interpret and set up spreadsheets are all part of a specialised language – without this language, students would not be able to make effective use of spreadsheets (Calder, Brown, Hanley, & Darby, 2006). Clarkson's (2004a, 2004b, 2005b) research has revealed that if there is a mix of English and non-English languages in a mathematics classroom, this can be a confounding factor for mediating understanding. But when the students are encouraged to be competent in all their languages, this can become a powerful learning strategy (Clarkson, 2006; Clements, 2006).

Sociolinguistics

After conducting a pilot study with Indigenous students in North Queensland, Warren, Young, and de Vries (2007) concluded that strategies needed to be developed by which precise mathematical language could be used to link school mathematics to home environments, and to use questioning to help establish effective classroom discourse. Thornton (2007) employed sociolinguistic approaches to discuss the mathematical discourses observed in a Grade 7 classroom, and Joseph (2005) took a critical theory stance when analyzing mathematics discourses within Malaysian secondary schools.

Goos (2004) reported details of a Grade 12 mathematics classroom in which the teacher established social norms and called on students to clarify, explain, critique, and justify their ideas. The teacher also made explicit reference to (and expected students to apply) appropriate mathematical language, conventions, and symbolism. But discourse is not always in face-to-face environments and, indeed, on-line discussion introduces forms of social discourse which contrast with face-to-face classroom interactions (Goos & Bennison, 2005).

In a major international study, Clarke, Emanuelsson, Jablonka, and Mok (2006), and Clarke, Keitel, and Shimizu (2006) reported and interpreted video and interview data from a range of mathematics classrooms around the world. In each classroom, mathematical ideas were explored in different social contexts, through different linguistic forms. The study design required several video cameras to be in operation in each participating class, raising the issue of whether data gathered under such circumstances can be regarded as indicative of what happens "normally". The researchers believed that the quality of data was not affected, but that view could be challenged.

Studies of communication and discourse in various information and communication technology environments by authors such as White (2006b), and Robertson, Fluck, Webb, and Loechel (2004) have identified both the potential for developing cultural literacy and the complexity of the issues involved.

ATTITUDES, BELIEFS AND VALUES

Attitudes

In reviewing relevant research, it was evident that the educational outcomes of attitudes, beliefs, and values (Krathwohl, Bloom, & Masia, 1964) constitute another broad category of sociocultural perspectives of mathematics education, although relatively little ongoing deep research into attitudes has been reported in the publications reviewed (a view shared as well by Grootenboer et al. in this volume). In the main the notion of attitudes was mentioned as important, and at times was included in the rationale for the particular study, but rarely was attitude itself isolated as an issue for which data were collected, nor was it mentioned in the conclusion of such studies. The impression is gained that although students' and teachers' attitudes are of some importance, other constructs take centre stage. In the reporting of research results, attitudes are at best an afterthought to ensure that there is some roundedness to the results. One wonders whether central expressions that sum up schools of thought that seem to drive the research on student learning and mathematics teaching, such as notions of *community of practice* and the *zone of proximal development*, are really thought of as more than theoretical notions. If researchers really considered what it takes to have a community of practice that actually functions well, or a situation where students' zones become key teaching parameters, then for the participants to possess attitudes to each other and their mathematical activities that are positive becomes self evidentially crucial. But the review of the literature on attitudes presented here suggested that researchers are not seeing this crucial assumption that they are making, and leaving it under-examined. However given this rather sombre opening, there have been snippets of research in the last four years that are well worth noting.

Leder and Grootenboer (2005) explored the relationships between attitudes, beliefs, values, feelings, and the like. They suggest that for them attitudes are an intermediary category, more stable than emotions and feelings, but not as stable as beliefs and/or values. However, is there a potential developmental path from mere feelings towards beliefs? Clearly some feelings do transform into deeper beliefs, but whether there is a set path, that may or may not include attitudes, would be interesting to research in the mathematics context. Further it might be that researchers need to take more care in this area when engaging in self-reporting studies. How the issue of whether respondents are reporting their relatively superficial emotions and feelings, or whether the deeper attitudes and indeed beliefs and values are influencing responses in self reports is rarely noted as an issue. Even for observational and interviewing data, it seems to take careful preparation, insight and persistence by the researchers to discern whether respondents are going beyond feelings and emotions. These notions are complex in themselves, as is the interplay between them. In these comments we have to some degree downplayed feelings and emotions in preference to attitudes, beliefs, and values. However this may not be appropriate, since many of us in everyday circumstances act on our surface perceptions, rather than deep well thought through beliefs. This may mean an undervaluing of emotions and feelings, as seems to be happening at present in the research community. Hence we may not be capturing all that is needed. Thus the interplay, and perhaps transition, between all of these notions needs further research.

There were some research reports that dealt with students' attitudes to mathematics and its teaching. Attitude was a key component of judging the success of the thematic teaching of mathematics in one study. Handal and Bobis (2004) noted that past research studies have claimed that students' motivational levels and attitudes have increased with the use of this approach, but in their view the results were both inconsistent and inconclusive. It was found that the initial peak of positive attitudes from the students with the introduction of this approach fell away over time. Games have also been a perennial teaching approach for which claims for increased motivation in student learning have been made, and hence ensuring an enduring positive attitude. Bragg (2006) found otherwise. She concluded that the use of games was at the best problematic as far as impacting on students' attitudes to mathematics. Certainly they enjoyed playing the games she used, but they did not always comprehend the connection between the games and the mathematics the games were embodying.

Other research studies focussed on teachers' attitudes to mathematics and its teaching. Pre-service education students' attitudes were probed, since researchers felt that if young teachers start out with problematic attitudes to both mathematics and mathematics teaching, then trying to change these at a later time will be difficult. Grootenboer (2006) noted research that documented the negative attitudes often held by pre-service teachers. To address this issue, Wilson and Thornton (2006) developed a technique based on young teachers developing, and then reflecting on, their own bibliography of their mathematics learning. In particular the authors led their students to confront any negative anxieties they might hold, but also celebrate the positives.

The attitudes of pre- and in-service teachers also came in for some research attention, although few studies appeared to regard attitudes as sociocultural constructs. Nisbet and Grimbeek (2004) commented that it is often assumed, naively by many, that teachers change their attitudes in a simple linear fashion. Hence if professional development is given to teachers, this will change their ways of thinking and their attitudes, which will lead to changes in their teaching practice, and result in student change, always for the better. But the authors noted that change is always complex, multi-faceted, and indeed takes a long time. The sociocultural dimension of this change process should not be ignored. Nisbet and Grimbeek also noted that their Queensland teachers had negative views of the imposed system-wide tests they have to administer. Hence to find that these tests had little influence on the teachers' practices in the classroom comes as no surprise. It is strange that governments continually claim the exact opposite. On the other hand, Cooper, Baturo, Warren, and Grant (2006) remind us that Indigenous conceptions of personal attributes such as attitudes and motivation vary considerably to western notions, and hence the way they operate in classrooms with high numbers of Indigenous students and/or with an Indigenous teacher may well be very different. It seems that this is a crucial area that needs much further and urgent attention.

In looking forward, we note one important issue that as yet has not been explored extensively. One of the enduring relationships in the literature has been how attitudes impact on understanding and performance. However there are other relationships that might be as important. At present there is much discussion in the literature and in public why few students continue with the serious (undertaken for its own sake and not to gain bonus points for tertiary entry, as happens in some education systems), non-compulsory study of mathematics in the last years of secondary school and in university. In this discussion, much has been made of the mathematics curriculum, students' performance in mathematics, and the lack of adequately qualified mathematics teachers throughout secondary schools. However a little considered variable in this important debate is students' attitudes to mathematics and its teaching, and how sociocultural factors impact on these attitudes. Khoo and Ainley (2005) have shown that students' general attitude to schooling does impact on whether they do continue on with tertiary education, but students' performances in literacy and numeracy have little impact. An analogous longitudinal study of students' attitudes to mathematics may well be useful.

Beliefs

The next sociocultural factor to be reviewed is that of beliefs. Recent projects conducted in various places, both big- and smaller-scale, have suggested the possibility that beliefs underlying the pedagogic methods, rather than the pedagogic methods themselves (e.g., Askew, Brown, Rhodes, Johnson, & William, 1997; Beswick, 2007), regulate successful learning. Beliefs are discussed in another chapter in this book (see Grootenboer et al.), as constituting an affective variable. In this chapter, Australasian mathematics education research in which beliefs were investigated as a sociocultural construct, or in which the research design featured beliefs as a sociocultural variable, will be the focus. Thus, while beliefs can be regarded as an (internal) affective variable which regulate the cognitive development of mathematical knowledge and skills, in this chapter the roles of beliefs as a (external) sociocultural factor in serving similar functions in mathematics education will be examined in the context of Australasian research.

Beliefs espoused in the mathematics classroom are shaped by the various sociocultural forces prevalent in it. Anderson, White, and Sullivan's (2005) research with 162 primary school teachers produced a model relating teachers' professed beliefs to their practices. In this model, the constraints and opportunities arising from the social context of teaching were recognised as impacting on teacher beliefs and thus affecting subsequent teacher practices. Researching with 465 secondary school mathematics teachers, principals and consultants in Greece, Australia-based researchers Barkatsas and Malone (2005) found that teachers' negotiation of beliefs that were in conflict in the mathematics classroom was typically regulated by the social and cultural climate within it.

Students' beliefs

Australasian mathematics education research into beliefs as a sociocultural factor can be broadly categorised into those which focussed on students' beliefs, and others on the beliefs of teachers. In terms of the former, the PISA study considered two aspects of student beliefs, namely, mathematics self-efficacy and self-concept. Amongst the Australian students surveyed in this comparative study, both these two aspects of beliefs were found to correlate positively with mathematical literacy performance (Thomson, Cresswell, & de Bortoli, 2004). Multilevel analysis "suggests that the factors that may have the greatest influence on Australian students' mathematical literacy, as assessed in PISA, are the attitudes and beliefs of students, which stand out above any of the other factors incorporated into the model" (Thomson et al., 2004, p. 203). It should be noted, however, that below-average self-efficacy means were reported by both higher-performing countries (compared to Australia) such as Finland, the Netherlands, and Korea and lower-performing nations like Brazil and Thailand (Thomson et al., 2004, Figure 7.12). That is, the direction of beliefs may not be a determinant of performance. Interestingly, Asian students (from Japan, Korea, Hong Kong, and Macau) reported the four lowest mathematics self-concept scores amongst the nations surveyed. From a sociocultural standpoint, this result may have implications for Australasian nations which are experiencing high numbers of Asian immigrants.

Teachers' beliefs

Teacher beliefs have also been found to be influenced by school size. Nisbet and Grimbeek (2004) surveyed 121 primary school teachers across 56 schools in Queensland for their beliefs and views about statewide numeracy tests. The data indicated that teachers from large schools (where there were more than 400 students) were more likely to modify their practices based on the test results of their respective students. The researchers postulated that teachers in these large schools were more competitive, aiming to maximise student scores as a means of comparing the performance of their respective schools with other schools.

School location also appears to affect the beliefs held by different teachers. Nisbet and Grimbeek (2004) reported that teachers from rural or remote settings were more likely than their peers elsewhere to hold beliefs that statewide tests were valid indicators of teacher ability and school performance, and to fine tune their professional practices in light of the test scores received.

Within schools, student characteristics constitute another category of factors impacting on teacher beliefs. Beswick's (2005a) survey of 22 early childhood, primary and middle years teachers found that teacher beliefs for students with mathematics learning difficulties were different from those they held for students in general. For the former group of students, the teacher participants were more likely to employ teaching approaches reflective of a skills-based curriculum focussed on 'real world' survival (versus one which emphasised deep conceptual understanding), and to use concrete materials for answer-getting (versus using concrete materials to develop conceptual understanding). Beswick concluded that "attention will need to be paid to teachers' relevant beliefs if inclusive policy is to have a real impact on students with MLD [mathematics learning difficulties]" (p. 143).

The extent of teachers' knowledge also appears to shape teachers' pedagogical beliefs. Beswick's (2005b) quantitative analysis of data collected from 25 secondary school teachers in Tasmania revealed that teachers who have studied tertiary mathematics and/or who possessed a Masters degree tended to subscribe to beliefs aligned with constructivist ideas. It is not clear from the data collected, however, whether it was mathematical content knowledge or pedagogical ideas which are really associated with constructivist beliefs, as well as how these beliefs might correlate with the chronological periods in which the teachers did their respective studies, as noted by Beswick herself.

Values

Values in mathematics education have been categorised as mathematical, mathematics educational, and general educational (Bishop, 1996), and organisational (Seah, 2004b). The affordances arising from investigating across different disciplines in the Australian *Values in Mathematics and Science Education* project in 2004 led to the renaming of one of the six mathematical values (see Bishop, 1988), *objectism*, as *empiricism,* "in order to accommodate the scientist's approach" (Bishop & Seah, in press).

The constructs of values and beliefs might be related, but they are by no means similar. Bishop and Seah (in press) asserted that they:

> have based … [their] current research on the theoretical notion that values operate, and are revealed, when choices are made …. The essential difference between values and beliefs is that one may hold various beliefs but it is when one must make choices that one's values come into play.

In this sense, if a belief represents what one considers to be true, then a value represents the emphasis and importance one accords to related belief(s).

Mathematics through values, or values through mathematics education?

There are two ways in which values might be discussed in mathematics education research. In one, values emphasised and portrayed in mathematics lessons represents one way in which broader educational goals of cultivating civic, ethically and morally responsible citizens might be realised. This perspective does not appear to have been targeted much by Australasian (mathematics) education research, except for Seah and Kalogeropoulos (2004). Written at the time of release of a federal Values Education Study report (Department of Education, Science and Training, 2003a) and of a new curriculum reform statement in Victoria, Australia, this paper identified ways in which the mathematics teacher might finetune his/her classroom practice and discourse to more explicitly inculcate socially 'desirable' values to students.

There have been relatively more investigations into how more effective lessons in mathematics and numeracy might be facilitated in schools through the concept of values. This perspective underlies much of the research reviewed below.

Researching values

In terms of exploring what teachers value, a 2005 survey of 2924 teachers of mathematics and 612 heads of mathematics in Australia found that "irrespective of their tertiary background or years of teaching experience, teachers [of mathematics] valued ongoing professional development" (Harris & Jensz, 2006, p. 20), and that "in general, teachers [of mathematics] valued their disciplinary grounding and the practical aspects of their education studies" (p. 28). Another study researching with a smaller group of teachers echoed this observation: "teachers' values in the classroom are shaped to some extent by the values embedded in each subject, as perceived by them" (Bishop, Clarke, Corrigan, & Gunstone, 2005, p. 158). Yet, teachers' values were also found to be shaped in some ways by official curriculum statements. In the Values in Science and Mathematics Education project, it was found that features of mathematics pedagogy that were valued by the primary and secondary teachers reflected the ideals of constructivist, reform-oriented mathematics lesson planning, as opposed to those of more "traditional" pedagogical practices.

Teachers' values also constituted one of the variables in Galligan's (2005) ethnographic research with Australian lecturers teaching university preparatory mathematics overseas. Specifically, the values that were in conflict were grounded in the respective Hong Kong and Australian cultures. As such, these constitute the sociocultural factors that regulated the practices and discourses of the mathematics lessons offered offshore.

The impact of migration on mathematics teachers' values was a subject of Seah's (2004b) study. The ways in which these teachers negotiated perceived differences of values in the secondary mathematics classroom as they made their respective cultural border crossings (Aikenhead & Jegede, 1999) between professional practices in different cultures were summarised as a range of responsive approaches that spans across degrees of assimilating the values of the host cultures (Seah, 2005a, 2005b).

Contexts of valuing have also been researched. Clarkson (2005a), for example, sought to build on the earlier observation that the mathematics education community lacked a shared language with which to discuss values. Taking the example of *rationalism* (which values logical thinking and deductive reasoning (see Bishop, 1988)), Clarkson highlighted the role played by logical connectives in facilitating associated language. His analysis revealed, however, that very few logical connectives had been identified at what Robitaille (1993) called the intended curriculum level. Student facility with the relevant language of the values being inculcated, however, is important, such that Clarkson called for the "curriculum documents … [to be] broaden[ed] to take this type of language seriously, and in turn make the values that they want taught, including rationalism, more explicit" (p. 16). Acknowledging the role of social interaction within the mathematics classroom, Seah's (2007a, 2007b) preliminary study identified what teachers and students of effective primary mathematics lessons co-valued, namely *fun*, (teacher) *experience*, *boardwork*, *instruction/explanation*, and *interestingness*.

DEVELOPMENTS IN SOCIOCULTURAL THEORETICAL PERSPECTIVES

Thus far, Australasian research in the period 2004-2007 and involving sociocultural factors for school mathematics education has been reviewed. In this section, we will identify some of the main theoretical perspectives that are increasingly used by researchers investigating sociocultural aspects of mathematics education.

Poststructuralist/postmodern perspectives

A significant contribution to the use of postmodern analysis in mathematics education is the publication of the book *Mathematics Education Within the Postmodern*, edited by Walshaw (2004). This collection of papers "embracing uncertainty and the instability of language and identity, opens up potential 'lines of flight' in mathematics education and research" (Klein, 2005, p. 101).

Kline and Saunders (2004) argued that pre-service teachers often engage in binary thinking in classifying students as good/bad, as learning/not learning or engaged/not engaged. Using a poststructuralist analysis, the authors demonstrated that good students are not essentially so, but are so constituted through how they are positioned in the classroom, and demonstrated how a more "quality" (p. 337) relationship between students and teachers that does not construct ability as "rock solid" (p. 337) is more capable of providing a better mathematics education to all students. Similarly, Kline (2004) critiqued the *Australia's Teachers: Australia's Future* (Department of Education, Science and Training, 2003b) document, where humanist, psychological notions of teachers and students prevail. The instrumental construction of teaching constructs the teacher as a technician rather than a professional. The author concluded:

> To me, the inherent worth of poststructuralist theorising is its interruption of taken-for-granted notions of rational, autonomous individuals able to understand the importance of, in this case, innovative classroom practice in mathematics, and act on this understanding. In making the agency to act innovatively or 'outside the square' problematic, given that everything we do is discursively constituted, poststructuralist thought forces educators and researchers to attend to the social as well as the intellectual quality of teaching/learning interactions and relationships. (p. 334)

Finally, Kline (2006) argued how a poststructuralist perspective used to analyse the difficulties of developing one-year postgraduate teacher education courses raises the question of not what to leave out, but how to structure the course to produce competent, inquiry-oriented teachers of mathematics in less time. Walls (2005) used postmodernism to argue that the task-oriented approach to mathematics education leads to a lack of student engagement. The author argues that a reframing of educational discourse as *participant-determined pedagogy of mathematics* more appropriately reflects the discourse of enhanced empowerment for students. The argument for an increased student-centered approach is given in terms of human rights, governance and learning.

Feminist perspectives

In mathematics education feminist discourses are no longer restricted to research into gender issues. We hasten to add, however, that feminism is not a unitary mode of analysis. Rather, it consists of a wide variety of perspectives.

Letts (2006) used poststructuralist-feminist analysis to discuss a case study of one mathematics teacher. Rather than constructing the identity of the teacher as coherent, rational, and knowable, this perspective sees identity as multiple, shifting and reflecting conflict. Similarly, it understands power and agency not as hegemony that some people have over others, or as something that is lost or gained, but rather as a productive power.

On the other hand, Atweh (2007), and Atweh and Keitel (2007) employed the theorisation of social justice presented by Fraser (2005) to present a multidimensional model of social justice and used it to study issues in international collaborations in mathematics education. In both publications the authors argued that even though equity concerns are central to mathematics education research, and increasingly policy and curriculum documents, the term remains undertheorised. While accepting some of the critique of poststructuralist authors, in particular the politics of difference, the model of social justice presented does not follow poststructuralism to the limit, but focuses on a socialist emphasis on distribution models of social justice.

Community of practice

Increasingly in the literature reviewed here, researchers in mathematics education in the Australasian region have been employing the concept of community of practice, as developed by Wenger, in their research. However, we note that rarely do authors give a definition of the term. Frequently the meaning of the term is more intuitive than analytic, and often different understandings of the term are presented. For example, Smith (2006) used the term as a "space of learning" (p. 621). In a keynote address to the 27th MERGA conference, Anthony (2004) defined community of practice as *learning* communities in which the discourses and practices of that community are negotiated by all the participants; members become informally bound by the value that they find in learning together. Goos and Bennison (2005) discussed three defining characteristics of communities of practice as "mutual engagement of participants, negotiation of a joint enterprise, and development of a shared repertoire of resources for creating meaning" (p. 385). Finally, Walshaw and Anthony (2006) understood the term as "a set of relations among persons, activity, and world, over time and in relation with other tangential and overlapping communities of practice" (p. 528).

REFLECTING ON AUSTRALASIAN RESEARCH IN 2004-2007

Australasian researchers investigating sociocultural factors influencing mathematics teaching and learning have often experienced external constraints that have limited their modes of operation. Heavy teaching workloads, demanding ethical requirements for research designs, limited access to adequate funding resources, and political pressures have all combined to shape and often limit research investigations. This has placed researchers in difficult and potentially controversial situations as they attempt to undertake research. Researchers need to resolve questions such as: What are the sources of funding? What are the pressures on researchers to complete and publish research in the minimum time, at minimum cost with maximum visibility, readability and appeal? What ethics approval hurdles have to be overcome in order to undertake research (Sullivan, Tobias, & McDonough, 2006)? Has the norm for mathematics education publications become papers with lengthy scholarly reviews, followed by a research question which can do little more than map one or two fine details of mathematics classrooms in which the participants are only those who feel comfortable enough to give (or for their parents to give) permission for the research to be done? In the resulting relatively small-scale studies interesting aspects of teaching and learning mathematics within narrowly defined domains are discussed. Substantive studies of students in low socio-economic areas have been difficult to find. Where are the innovative theories which step beyond the narrow confines of ideas expounded by others in different contexts? The literature abounds with interesting "mapping" exercises, but studies which peel back the layers of respectability and accepted theories, to lay bare realities and fundamental ways of finding research-based strategies which make a difference in the mathematics classroom, are few and far between.

Institutional and political constraints which have operated during these past four years have certainly affected mathematics education research in the Australasian region, and many of these constraints do not show signs of going away. It appears then that Australasian (mathematics) educational researchers will need to be more creative and innovative in finding ways of collaborating and theorizing in order to extend what is known, to influence policy formulations, *and* to bring about pedagogical change.

At the same time, we acknowledge the multi-facetted challenges associated with researching the evolution and development of particular socio-cultural factors that have an impact on mathematics learning and teaching. The challenge for researchers is to find ways to make the hidden transparent, to work in classrooms and with teachers where others have not gone, and to disseminate their findings. They need to provide convincing evidence that deep, research-based strategies for the teaching and learning of mathematics which take into account important socio-cultural factors have the power to illuminate and change what happens in all places, and at all levels, of mathematics learning.

REFERENCES

Adam, S. (2004). Ethnomathematical ideas in the curriculum. *Mathematics Education Research Journal, 16*(2), 49-68.

Aikenhead, G. S., & Jegede, O. J. (1999). Cross-cultural science education: A cognitive explanation of a cultural phenomenon. *Journal of Research in Science Teaching, 36*, 269-287.

Aikenhead, G. S. (2001). Students' ease in crossing cultural borders into school science. *Science Education, 85*(2), 180-188.

Anderson, J. (2005). Implementing problem solving in mathematics classrooms: What support do teachers want? In P. Clarkson, A. Downton, D. Gronn, M. Horne, A. McDonough, R. Pierce, & A. Roche (Eds.), *Building connections: Theory, research and practice* (Proceedings of the 28th annual conference of the Mathematics Education Research Group of Australasia, Melbourne, Vol. 1, pp. 89-96). Sydney: MERGA.

Anderson, J., White, P., & Sullivan, P. (2005). Using a schematic model to represent influences on, and relationships between, teachers' problem-solving beliefs and practices. *Mathematics Education Research Journal, 17*(2), 9-38.

Anthony, G. (2004). MERGA: A community of practice. In I. Putt, R. Faragher, &, M. McLean (Eds.), *Mathematics education for the third millennium: Towards 2010* (Proceedings of the 27th annual conference of the Mathematics Education Research Group of Australasia, Townsville, Vol. 1, pp. 2-15). Sydney: MERGA.

Askew, M., Brown, M., Rhodes, V., Johnson, D., & Wiliam, D. (1997). *Effective teachers of numeracy. Final report.* London: King's College.

Atweh, B. (2007). International interactions in mathematics education. In U. Gillbert, & E. Jablonka (Eds.), *Mathematisation and demathematisations* (pp. 171-186). Rotterdam, The Netherlands: Sense Publishers.

Atweh, B., Cooper, T., & Kanes, C. (1992). Social and cultural context in mathematics education. In W. Atweh, & J. Watson (Eds.), *Research in mathematics education in Australasia: 1988-1991* (pp. 43-66). Sydney: MERGA.

Atweh, B., & Keitel, C. (2007). Social (in)justice in international collaborations in mathematics education. In B. Atweh, M. Borba, A. Calabrese Barton, N. Gough, C. Keitel, C. Vistro-Yu, & R. Vithal (Eds.), *Internationalisation and globalisation in mathematics and science education* (pp. 95-112). Dordrecht, The Netherlands: Springer.

Atweh, B., Meaney, T., McMurchy-Pilkington, C., Neyland, J., & Trinick, T. (2004). Social justice and sociocultural perspectives in mathematics education. In B. Perry, G. Anthony, & C. Diezmann (Eds.), *Research in mathematics education in Australasia 2000-2003*. Sydney: MERGA.

Averill, R., & Harvey, R. (2005). Using Jamie's experiences: An investigation into using teachers' stories in pre-service mathematics teacher education. In P. Clarkson, A. Downton, D. Gronn, M. Horne, A. McDonough, R. Pierce, & A. Roche (Eds.), *Building connections: Theory, research and practice* (Proceedings of the 28th annual conference of the Mathematics Education Research Group of Australasia, Melbourne, Vol. 1, pp. 113-120). Sydney: MERGA.

Ayers, P., & McCormick, J. (2006). Grade 12 mathematics teachers' views on curriculum reform in New South Wales. In P. Grootenboer, R. Zevenbergen, & M. Chinnappan (Eds.), *Identities, cultures and learning spaces* (Proceedings of the 29th annual conference of the Mathematics Education Research Group of Australasia, Canberra, Vol. 1, pp. 54-59). Adelaide: MERGA.

Barkatsas, A., & Malone, J. (2005). A typology of mathematics teachers' beliefs about teaching and learning mathematics and instructional practices. *Mathematics Education Research Journal, 17*(2), 69-90.

Barnes, M. (2005). Exploring how power is enacted in small groups. In H. L. Chick & J. L. Vincent (Eds.), *Proceedings of the 29th annual conference for the Psychology of Mathematics Education* (Vol. 2, pp. 137-144). Melbourne: University of Melbourne.

Barton, B., Poisard, C., & Do Carmo Domite, M. (2006). An overview of the third international conference on ethnomathematics: Cultural connections and mathematical manipulations. *For the Learning of Mathematics, 26*(2), 21.

Beswick, K. (2005a). It depends on the students: Influencing teachers' beliefs about the ends and means of numeracy teaching. In P. Clarkson, A. Downton, D. Gronn, M. Horne, A. McDonough, R. Pierce, & A. Roche (Eds.), *Building connections: Research, theory and practice* (Proceedings of the 28th annual conference of the Mathematics Education Research Group of Australasia, Melbourne, Vol. 1, pp. 137-144). Sydney: MERGA.

Beswick, K. (2005b). The beliefs/practice connection in broadly defined contexts. *Mathematics Education Research Journal, 17*(2), 39-68.

Beswick, K. (2007). Teachers' beliefs that matter in secondary mathematics classroom. *Educational Studies in Mathematics, 65*, 95-120.

Bicknell, B. (2006). Investigating parental roles of mathematically gifted students. In P. Grootenboer, R. Zevenbergen, & M. Chinnappan (Eds.), *Identities, cultures and learning spaces* (Proceedings of the 29th annual conference of the Mathematics Education Research Group of Australasia, Vol. 1, pp. 76-83). Sydney: MERGA.

Bishop, A. (1988). *Mathematical enculturation: A cultural perspective on mathematics education.* Dordrecht, The Netherlands: Kluwer Academic Publishers.

Bishop, A. J. (1996). *How should mathematics teaching in modern societies relate to cultural values – Some preliminary questions.* Paper presented at the Seventh Southeast Asian Conference on Mathematics Education, June 3-7, Hanoi, Vietnam.

Bishop, A., Clarke, B., Corrigan, D., & Gunstone, D. (2005). Teachers' preferences and practices regarding values in teaching mathematics and science. In P. Clarkson, A. Downton, D. Gronn, M. Horne, A. McDonough, R. Pierce, & A. Roche (Eds.), *Building connections: Research, theory and practice* (Proceedings of the 28th annual conference of the Mathematics Education Research Group of Australasia, Melbourne, Vol. 1, pp. 153-160). Sydney: MERGA.

Bishop, A. J., & Seah, W. T. (in press). Educating values through mathematics teaching: Possibilities and challenges. In M. H. Chau, & T. Kerry (Eds.), *International perspectives on education.* London: Continuum.

Bobis, J., Clarke, B., Clarke, D., Thomas, G., Wright, R., Young-Loveridge, J., & Gould, P. (2005). Supporting teachers in the development of young children's mathematical thinking: Three large scale cases. *Mathematics Education Research Journal, 16*(3), 27-57.

Booker, G., Bond, D., Sparrow, L., & Swan, P. (2004). *Teaching primary mathematics.* French's Forest, Australia?: Pearson Education.

Bragg, L. (2006). Students' impressions of the value of games for the learning of mathematics. In J. Novetná, H. Moraová, M. Krátká, & N. Stehlíková (Eds.), *Proceedings of the 30th annual conference for the Psychology of Mathematics Education* (Vol. 2, pp. 217-224). Prague: Charles University.

Brousseau, G. (1997). *Theory of didactical situations in mathematics (Didactique des mathématiques – Guy Brousseau)* (translated by M. Balacheff, M. Cooper, & R. Sutherland). Dordrecht, The Netherlands: Kluwer Academic Publishers.

Calder, N. (2005). "I type what I think and try it": Children's initial approaches to investigation through spreadsheets. In P. Clarkson, A. Downton, D. Gronn, M. Horne, A. McDonough, R. Pierce, & A. Roche (Eds.), *Building connections: Theory, research and practice* (Proceedings of the 28th annual conference of the Mathematics Education Research Group of Australasia, Melbourne, Vol.1, pp. 185-192). Sydney: MERGA.

Calder, N., Brown, T., Hanley, U., & Darby, S. (2006). Forming conjectures within a spreadsheet environment. *Mathematics Education Research Journal, 18*(3), 100-116.

Callingham, R. (2005). Primary students' mental computation: Strategies and achievement. In P. Clarkson, A. Downton, D. Gronn, M. Horne, A. McDonough, R. Pierce, & A. Roche (Eds.), *Building connections: Theory, research and practice* (Proceedings of the 28th annual conference of the Mathematics Education Research Group of Australasia, Melbourne, Vol.1, pp. 193-200). Sydney: MERGA.

Cao, Z., Bishop, A., & Forgasz, H. (2007). Perceived parental influence on mathematics learning: A comparison among students in China and Australia. *Educational Studies in Mathematics, 64*(1), 85-106.

Carter, L. (2004). Thinking differently about cultural diversity: Using postcolonial theory to (re)read science education. *Science Education, 88*(6), 819-836.

Carter, L. (2006). The challenges of postcolonialism to science education. *Educational Philosophy and Theory, 38*(5), 677-692.

Caswell, R., & Nisbet, S. (2005). Enhancing mathematical understanding through self-assessment and self-regulation: The value of meta-awareness. In P. Clarkson, A. Downton, D. Gronn, M. Horne, A. McDonough, R. Pierce, & A. Roche (Eds.), *Building connections: Theory, research and practice* (Proceedings of the 28th annual conference of the Mathematics Education Research Group of Australasia, Melbourne, Vol.1, pp. 209-216). Sydney: MERGA.

Chick, H., & Baker, M. (2005). Teaching elementary probability: Not leaving it to chance. In P. Clarkson, A. Downton, D. Gronn, M. Horne, A. McDonough, R. Pierce, & A. Roche (Eds.), *Building connections: Theory, research and practice* (Proceedings of the 28th annual conference of the Mathematics Education Research Group of Australasia, Melbourne, Vol.1, pp. 233-240). Sydney: MERGA.

Clarke, B., & Robbins, J. (2004). Numeracy enacted: Preschool families conceptions of their children's engagements with numeracy. In I. Putt, R. Faragher, & M. McLean (Eds.), *Mathematics education for the third millennium: Towards 2010* (Proceedings of the 27th annual conference of the Mathematics Education Research Group of Australasia, Townsville, Vol.1, pp. 173-182). Sydney: MERGA.

Clarke, D. (2005). Essential complementarities: Arguing for an integrative approach to research in mathematics classrooms. In P. Clarkson, A. Downton, D. Gronn, M. Horne, A. McDonough, R. Pierce, & A. Roche (Eds.), *Building connections: Theory, research and practice* (Proceedings of the 28th annual conference of the Mathematics Education Research Group of Australasia, Melbourne, Vol.1, pp. 3-17). Sydney: MERGA.

Clarke, D., Emanuelsson, J., Jablonka, E., & Ah Chee Mok, I. (Eds.), (2006). *Making connections: Comparing mathematics classrooms around the world.* Rotterdam, The Netherlands: Sense Publishers.

Clarke, D., & Hoon, S. L. (2005). Studying the distribution of responsibility for the generation of knowledge in mathematics classrooms in Hong Kong, Melbourne, San Diego and Shanghai. In H. L. Chick & J. L. Vincent (Eds.), *Proceedings of the 29th annual conference for the Psychology of Mathematics Education* (Vol. 2, pp. 250-264). Melbourne: University of Melbourne.

Clarke, D., Keitel, C., & Shimizu, Y. (Eds.), (2006). *Mathematics classrooms in twelve countries: The insider's perspective.* Rotterdam, The Netherlands: Sense Publishers.

Clarke, D., & Lobato, J. (2002). To tell or not to tell: A reformulation of telling and the development of an initiating/eliciting model of teaching. In C. Malcolm, & C. Lubisi (Eds.), *Proceedings of the tenth annual meeting of the Southern African Association for Research in Mathematics, Science and Technology Education* (pp. 15-22). Durban, South Africa: University of Natal.

Clarkson, P. C. (2004a). Multilingual contexts for teaching mathematics. In M. J. Høines & A. B. Fuglestad (Eds.), *Proceedings of the 28th annual conference for the Psychology of Mathematics Education* (pp. 236-239). Bergen, Norway: Bergen University College.

Clarkson, P. C. (2004b). Teaching mathematics in multilingual classrooms: The global importance of contexts In I. Cheong, H. Dhindsa, I. Kyeleve, & O. Chukwu (Eds.), *Globalisation trends in science, mathematics and technical education* (pp. 9-23). Brunei Darussalam: Universiti Brunei Darussalam.

Clarkson, P. C. (2005a). *Researching the language for rational explanations in mathematics teaching and learning.* Paper presented at the Australian Association for Research in Education Conference, Melbourne, Australia. Retrieved March 5, 2007 from http://www.aare.edu.au/04pap/cla04891.pdf

Clarkson, P. C. (2005b). Two perspectives of bilingual students learning mathematics in Australia: A discussion. In D. Hewitt & A. Noyes (Eds.), *British Congress of Mathematics Education conference proceedings* (CD). Warwick, UK: BSRLM.

Clarkson, P.C. (2006). Australian Vietnamese students learning mathematics: High ability bilinguals and their use of their languages. *Educational Studies in Education, 64,* 191-215.

Clements, M. A. (2005). A synthesis of selected Universiti Brunei Darussalam mathematics education research studies, 1998-2004. In H. S. Dhindsa, I. J. Kyeleve, O. Chukwu, & J. S. H. Quintus Perera (Eds.), *Future directions in science, mathematics and technical education* (pp. 196-205). Gadong, Brunei Darussalam: Universiti Brunei Darussalam.

Clements, M. A. (2006). *Active mathematics in classrooms.* Bandar Seri Begawan, Brunei Darussalam: Brunei Darussalam Ministry of Education.

Clements, M. A., & Ellerton, N. F. (2006). Improving prospective mathematics teachers' understanding of the solution process for quadratic equations in the form $(x \quad a)(x \quad b) = 0$. In H. S. Dhindsa, I. J. Kyeleve, O. Chukwu, L. S. Bee, Hjh Z. Bte Hj Taha, A. Bamba, & S. Upex (Eds.), *Shaping the future of science, mathematics and technical education* (pp. 173-182). Gadong, Brunei Darussalam: Universiti Brunei Darussalam.

Cooper, T. J., Baturo, A. R., & Warren, E. (2005). Indigenous and non-indigenous teaching relationships in three mathematics classrooms in remote Queensland. In H. L. Chick & J. L. Vincent (Eds.), *Proceedings of the 29th annual conference for the Psychology of Mathematics Education* (Vol. 2, pp. 265-272). Melbourne: University of Melbourne.

Cooper, T., Baturo, A., Warren, L., & Grant, E. (2006). Aim high – beat yourself: Effective mathematics teaching in a remote indigenous community. In J. Novetná, H. Moraová, M. Krátká, & N. Stehliková (Eds.), *Proceedings of the 30th annual conference for the Psychology of Mathematics Education* (Vol. 2, pp. 369-376). Prague: Charles University.

Davies, N., & Walker, K. (2005). Learning to notice: One aspect of teachers' content knowledge in the numeracy classroom. In P. Clarkson, A. Downton, D. Gronn, M. Horne, A. McDonough, R. Pierce, & A. Roche (Eds.), *Building connections: Theory, research and practice* (Proceedings of the 28th annual conference of the Mathematics Education Research Group of Australasia, Melbourne, Vol.1, pp. 273-280). Sydney: MERGA.

Department of Education, Science and Training (2003a). *Values education study* (Final report). Melbourne: Curriculum Corporation.

Department of Education, Science and Training (DEST) (2003b). *Australia's teachers: Australia's future.* Canberra: DEST.

Diezmann, C. (2005). Assessing primary students' knowledge of networks, hierarchies and matrices using scenario-based tasks. In P. Clarkson, A. Downton, D. Gronn, M. Horne, A. McDonough, R. Pierce, & A. Roche (Eds.), *Building connections: Theory, research and practice* (Proceedings of the 28th annual conference of the Mathematics Education Research Group of Australasia, Melbourne, Vol.1, pp. 289-296). Sydney: MERGA.

Diezmann, C. M., & English, L. D. (2001). Promoting the use of diagrams as tools for thinking. In A. A. Cuoco (Ed.), *The role of representation in school mathematics* (pp. 77-89). Reston, VA: National Council of Teachers of Mathematics.

Diezmann, C. M., & Lowrie, T. J. (2006). Primary students' knowledge of and errors on number lines. In P. Grootenboer, R. Zevenbergen, & M. Chinnappan (Eds.), *Identities, cultures and learning spaces* (Proceedings of the 29th annual conference of the Mathematics Education Research Group of Australasia, Canberra, Vol. 1, pp. 171-178). Adelaide: MERGA.

Diezmann, C. M., Lowrie, T. J., & Kozak, N. (2007). Essential differences between high and low performers' thinking about graphically-oriented numeracy items. In J. Watson & K. Beswick (Eds.), *Mathematics: Essential research, essential practice* (Proceedings of the 30th annual conference of the Mathematics Education Research Group of Australasia, Hobart, Vol. 1, pp. 226-235). Adelaide: MERGA.

Doyle, K. (2006). Organisational structure for mathematical modeling. In P. Grootenboer, R. Zevenbergen, & M. Chinnappan (Eds.), *Identities, cultures and learning spaces* (Proceedings of the

29th annual conference of the Mathematics Education Research Group of Australasia, Canberra, Vol. 1, pp. 187-194). Adelaide: MERGA.

Doyle, K. (2007). The teacher, the tasks: Their role in students' mathematical literacy. In J. Watson & K. Beswick (Eds.), *Mathematics: Essential research, essential practice* (Proceedings of the 30th annual conference of the Mathematics Education Research Group of Australasia, Hobart, Vol. 1, pp. 246-254). Adelaide: MERGA.

Ellerton, N. F. (1989). The interface between mathematics and language. *Australian Journal of Reading, 12* (2), 92–102.

Ellerton, N. F. (2005). Establishing reflective practice with preservice mathematics teachers. In H. S. Dhindsa, I. J. Kyeleve, O. Chukwu, & J. S. H. Quintus Perera (Eds.), *Future directions in science, mathematics and technical education* (pp. 216-225). Gadong, Brunei Darussalam: Universiti Brunei Darussalam.

Ellerton, N., & Clarkson, P.C. (1996). Language factors in mathematics teaching and learning. In A. Bishop, K. Clements, C. Keitel, J. Kilpatrick, & C. Laborde (Eds.), *International handbook of mathematics education* (pp.987-1034). Dordrecht: Kluwer Academic.

Ellerton, N. F., & Clements, M. A. (2005). A mathematics education ghost story: Herbartianism and school mathematics. In P. Clarkson, A. Downton, D. Gronn, M. Horne, A. McDonagh, R. Pierce, & A. Roche (Eds.), *Building connections: Research, theory and practice* (Proceedings of the 28th annual conference of the Mathematics Education Research Group of Australasia, Melbourne, Vol. 1, pp. 313-320). Sydney, MERGA.

Ellerton, N. F., & Olson, J. (2005). The assessment dilemma: Correct answers with no understanding and incorrect answers with some understanding. In H. S. Dhindsa, I. J. Kyeleve, O. Chukwu, & J. S. H. Quintus Perera (Eds.), *Future directions in science, mathematics and technical education* (pp. 226-235). Gadong, Brunei Darussalam: Universiti Brunei Darussalam.

English, L. (2006). Mathematical modeling in the primary school: Children's construction of a consumer guide. *Educational Studies in Mathematics, 63*(3), 303-323.

English, L. D., & Fox, J. I. (2005). Seventh-graders mathematical modeling on completion of a three-year program. In P. Clarkson, A. Downton, D. Gronn, M. Horne, A. McDonough, R. Pierce, & A. Roche (Eds.), *Building connections: Theory, research and practice* (Proceedings of the 28th annual conference of the Mathematics Education Research Group of Australasia, Melbourne, Vol.1, pp. 321-328). Sydney: MERGA.

English, L. D., & Watters, J. J. (2005). Mathematical modeling with 9-year-olds. In H. L. Chick & J. L. Vincent (Eds.), *Proceedings of the 29th annual conference for the Psychology of Mathematics Education* (Vol. 2, pp. 297-304). Melbourne: University of Melbourne.

Falle, J. (2005). Towards a language-based model of students' early algebraic understandings: Some preliminary findings. In P. Clarkson, A. Downton, D. Gronn, M. Horne, A. McDonough, R. Pierce, & A. Roche (Eds.), *Building connections: Theory, research and practice* (Proceedings of the 28th annual conference of the Mathematics Education Research Group of Australasia, Melbourne, Vol.1, pp. 345-352). Sydney: MERGA.

Falle, J. (2007). Students' tendency to cojoin terms: An inhibition to their development of algebra. In J. Watson & K. Beswick (Eds.), *Mathematics: Essential research, essential practice* (Proceedings of the 30th annual conference of the Mathematics Education Research Group of Australasia, Hobart, Vol. 1, pp. 285-294). Adelaide: MERGA.

Fitzallen, N. (2006). A model of students' statistical thinking and reasoning about graphs in an ICT environment. In P. Grootenboer, R. Zevenbergen, & M. Chinnappan (Eds.), *Identities, cultures and learning spaces* (Proceedings of the 29th annual conference of the Mathematics Education Research Group of Australasia, Canberra, Vol. 1, pp. 203-210). Adelaide: MERGA.

Forgasz, H. (2004). *Year 11 students' beliefs about computers for mathematics learning*. Paper presented at the International Congress of Mathematics Education, Copenhagen. Retrieved February 23, 2008 from: http://www.icme-organisers.dk/tsg26/1Forgasz.doc

Fraser, N. (2005). *Reframing justice*. Assen, The Netherlands: Royal Van Gorcum.

Galligan, L. (2005). Conflicts in offshore learning environments of a university preparatory mathematics course. In H. L. Chick & J. L. Vincent (Eds.), *Proceedings of the 29th annual conference for the Psychology of Mathematics Education* (Vol. 3, pp. 25-32). Melbourne: University of Melbourne.

Geiger, V., & Goos, M. (2006). Living in the gap: A tale of two different types of researchers. In P. Grootenboer, R. Zevenbergen, & M. Chinnappan (Eds.), *Identities, cultures and learning spaces* (Proceedings of the 29th annual conference of the Mathematics Education Research Group of Australasia, Canberra, Vol. 1, pp. 254-261). Adelaide: MERGA.

Goos, M. (2004). Learning mathematics in a classroom community of inquiry. *Journal of Research in Mathematics Education, 35*(4), 258-291.

Goos, M. (2006). Creating learning spaces. In P. Grootenboer, R. Zevenbergen, & M. Chinnappan (Eds.), *Identities, cultures and learning spaces* (Proceedings of the 29th annual conference of the Mathematics Education Research Group of Australasia, Canberra, Vol. 1, pp. 3-14). Adelaide: MERGA.

Goos, M., & Bennison, A. (2005). The role of online discussion in building a community of practice for beginning teachers of secondary mathematics. In P. Clarkson, A. Downton, D. Gronn, M. Horne, A. McDonough, R. Pierce, & A. Roche (Eds.), *Building connections: Theory, research and practice* (Proceedings of the 28th annual conference of the Mathematics Education Research Group of Australasia, Melbourne, Vol.1, pp. 385-392). Sydney: MERGA.

Grootenboer, P. (2006). Mathematics educators: Identity, beliefs, roles and ethical dilemmas. In P. Grootenboer, R. Zevenbergen, & M. Chinnappan (Eds.), *Identities, cultures and learning spaces* (Proceedings of the 29th annual conference of the Mathematics Education Research Group of Australasia, Canberra, Vol. 1, pp. 270-277). Adelaide: MERGA.

Gough, J. (2006). VELS versus CSF; Replacement, conflict or harmony? *Vinculum, 43*(3), 3-6.

Handal, B., & Bobis, J. (2004). Teaching mathematics thematically: Teachers' perspectives. *Mathematics Education Research Journal, 16*(1), 3-18.

Harris, K. L., & Jensz, F. (2006). *The preparation of mathematics teachers in Australia: Meeting the demand for suitably qualified mathematics teachers in secondary schools.* Melbourne: Australian Council of Deans of Science.

Hunter, J. (2006). Students in a mathematical community of inquiry: What do they think? In P. Grootenboer, R. Zevenbergen, & M. Chinnappan (Eds.), *Identities, cultures and learning spaces* (Proceedings of the 29th annual conference of the Mathematics Education Research Group of Australasia, Canberra, Vol. 1, pp. 301-308). Adelaide: MERGA.

Hunter, R. (2006). Towards mathematical inquiry. In P. Grootenboer, R. Zevenbergen, & M. Chinnappan (Eds.), *Identities, cultures and learning spaces* (Proceedings of the 29th annual conference of the Mathematics Education Research Group of Australasia, Canberra, Vol. 1, pp. 309-317). Adelaide: MERGA.

Irwin, K., & Irwin, J. (2005). Assessing development in numeracy of students. *Educational Studies in Mathematics, 58*(3), 283-298.

Joseph, C. (2005). Discourses of schooling in contemporary Malaysia: Pedagogical practices and ethnic policies. *Australian Journal of Education, 49*(1), 28-45.

Kensington-Miller, B. (2004). Professional development of mathematics teachers in low socio-economic secondary schools in New Zealand. In I. Putt, R. Faragher, & M. McLean (Eds.), *Mathematics education for the third millennium: Towards 2010* (Proceedings of the 27th annual conference of the Mathematics Education Research Group of Australasia, Townsville, pp. 320-327). Sydney: MERGA.

Kensington-Miller, B. (2005). Mentoring mathematics teachers in low socio-economic secondary schools in New Zealand. In P. Clarkson, A. Downton, D. Gronn, M. Horne, A. McDonough, R. Piece, & A. Roche (Eds.), *Building connections: Theory, research and practice* (Proceedings of the 28th annual conference of the Mathematics Education Research Group of Australasia, Melbourne, pp. 459-466). Sydney: MERGA.

Khoo, S., & Ainley, J. (2005). *Attitudes, intentions and participation.* Camberwell, Victoria: Australian Council for Educational Research.

Klein, M. (2005). Lines of flight in mathematics education and research. *Mathematics Education Research Journal, 17*(1), 101-106.

Kline, M. (2004). Preservice teacher education in mathematics: Thinking innovatively about innovation. In I. Putt, R. Faragher, & M. McLean (Eds.), *Mathematics education for the third millennium: Towards 2010* (Proceedings of the 27th annual conference of the Mathematics Education Research Group of Australasia, Townsville, Vol. 1, pp. 328-335). Sydney: MERGA.

Kline, M. (2006). What to leave out when preservice mathematics education goes from four years to one: A poststructural account. In P. Grootenboer, R. Zevenbergen, & M. Chinnappan (Eds.), *Identities, cultures and learning spaces* (Proceedings of the 29th annual conference of the Mathematics Education Research Group of Australasia, Canberra, Vol. 1, pp. 328-335). Adelaide: MERGA.

Kline, M., & Saunders, L. (2004). How binary thinking in school mathematics masks and maintains its coercive and conservative effects: A poststructuralist analysis. In I. Putt, R. Faragher, & M. McLean (Eds.), *Mathematics education for the third millennium: Towards 2010* (Proceedings of the 27th annual conference of the Mathematics Education Research Group of Australasia, Townsville, Vol. 2, pp. 336-342). Sydney: MERGA.

Krathwohl, D. R., Bloom, B. S., & Masia, B. B. (1964). *Taxonomy of educational objectives: The classification of educational goals (Handbook II: Affective domain).* New York: David McKay.

Leder, G., & Grootenboer, P. (2005). Affect and mathematics education. *Mathematics Education Research Journal, 17*(2), 1-8.

Leigh, A. (2007). Estimating teacher effectiveness from two-year changes in students' test scores. Retrieved May 22, 2007 from: http://econrsss.anu.edu.au/~aleigh/pdf/TQPanel.pdf

Leont'ev, A. N. (1978). *Activity, consciousness, and personality.* Englewood Cliffs: Prentice Hall.

Letts, W. (2006). Bringing feminist poststucturalism to bear on [mathematics] teacher education. In P. Grootenboer, R. Zevenbergen, & M. Chinnappan (Eds.), *Identities, cultures and learning spaces* (Proceedings of the 29th annual conference of the Mathematics Education Research Group of Australasia, Canberra, Vol. 2, pp. 623-625). Adelaide: MERGA.

Lobato, J., Clarke, D., & Ellis, A. B. (2005). Initiating and eliciting in teaching: A reformulation of telling. *Journal for Research in Mathematics Education, 36*(2), 101-136.

Lowe, I. (2006). Making VELS work in mathematics. *Prime Number, 21*(3), 21-25.

Lowrie, T., & Diezmann, C. (2005). Fourth-grade students' performance in graphical languages in mathematics. In H. L. Chick & J. L. Vincent (Eds.), *Proceedings of the 29th annual conference for the Psychology of Mathematics Education* (Vol. 3, pp. 265-272). Melbourne: University of Melbourne.

Marks, G. (2006). Are between- and within-school differences in student performance largely due to socio-economic background? Evidence from 30 countries. *Educational Research, 48*(1), 21-40.

Matang, R. (2005). Formalising the role of Indigenous counting systems in teaching the formal English arithmetic strategies through local vernaculars: An example from Papua New Guinea. In P. Clarkson, A. Downton, D. Gronn, M. Horne, A. McDonough, R. Pierce, & A. Roche (Eds.), *Building connections: Research, theory and practice* (Proceedings of the 28th annual conference of the Mathematics Education Research Group of Australasia, Melbourne, Vol. 2, pp. 505-512). Sydney: MERGA.

McGaw, B. (2004). Australian mathematics learning in an international context. In I. Putt, R. Faragher, & M. McLean (Eds.), *Mathematics education for the third millennium: Towards 2010* (Proceedings of the 27th annual conference of the Mathematics Education Research Group of Australasia, Townsville, Vol. 1, pp. 29). Sydney: MERGA.

Mulligan, J. (2006). Using structure in young children's mathematics learning. *Prime Number, 21*(2), 25-29.

Mulligan, J., Prescott, A., & Mitchelmore, M. C. (2004). Children's development of structure in early mathematics. In M. J. Høines & A. B. Fuglestad (Eds.), *Proceedings of the 28th annual conference for the Psychology of Mathematics Education* (Vol. 3, pp. 393-400). Bergen, Norway: Bergen University College.

Nally, V. (2005). Building connections: Research, theory and practice: A view from a practitioner. In P. Clarkson, A. Downton, D. Gronn, M. Horne, A. McDonough, R. Pierce, & A. Roche (Eds.), *Building connections: Theory, research and practice* (Proceedings of the 28th annual conference of the Mathematics Education Research Group of Australasia, Melbourne, Vol.1, pp. 58-62). Sydney: MERGA.

Nisbet, S., & Grimbeek, P. (2004). Primary teachers' beliefs and practices with respect to compulsory numeracy testing. In I. Putt, R. Faragher, & M. McLean (Eds.), *Mathematics education for the third millennium: Towards 2010* (Proceedings of the 27th annual conference of the Mathematics Education Research Group of Australasia, Townsville, Vol. 2, pp. 406-413). Sydney: MERGA.

Norton, S., & Irvin, J. (2007a). A concrete approach to teaching symbolic algebra. In J. Watson & K. Beswick (Eds.), *Mathematics: Essential research, essential practice* (Proceedings of the 30th annual conference of the Mathematics Education Research Group of Australasia, Hobart, Vol. 2, pp. 551-560). Adelaide: MERGA.

Norton, S., & Irvin, J. (2007b). Developing positive attitudes towards algebra. In J. Watson & K. Beswick (Eds.), *Mathematics: Essential research, essential practice* (Proceedings of the 30th annual conference of the Mathematics Education Research Group of Australasia, Hobart, Vol. 2, pp. 561-570). Adelaide: MERGA.

O'Toole, T. (2006). Building powerful understanding by connecting informal and formal knowledge. In P. Grootenboer, R. Zevenbergen, & M. Chinnappan (Eds.), *Identities, cultures and learning spaces* (Proceedings of the 29th annual conference of the Mathematics Education Research Group of Australasia, Canberra, Vol. 2, pp. 384-391). Adelaide: MERGA.

Pearn, C., & Stephens, M. (2007). Whole number knowledge and number lines help to develop fraction concepts. In J. Watson & K. Beswick (Eds.), *Mathematics: Essential research, essential practice* (Proceedings of the 30th annual conference of the Mathematics Education Research Group of Australasia, Hobart, Vol. 2, pp. 601-610). Adelaide: MERGA.

Powell-Mikle, A. (2003). We are more than numbers: African American students' perceptions of exemplary mathematics teachers. *International Journal of Educational Reform, 12*(2), 83-96.

Robertson, M., Fluck, A., Webb, I., & Loechel, B. (2004). Classroom computer climate, teacher reflections and re-envisioning pedagogy in Australian schools. *Australasian Journal of Educational Technology, 20*(3), 351-370.

Robitaille, D. F. (Ed.), (1993). *Curriculum frameworks for mathematics and science*. Vancouver, Canada: Pacific Educational Press.

Sawyer, A. (2006). Unpacking the rules of class discussion: Young children learning mathematics within a community of inquiry. In P. Grootenboer, R. Zevenbergen, & M. Chinnappan (Eds.), *Identities, cultures and learning spaces* (Proceedings of the 29th annual conference of the Mathematics Education Research Group of Australasia, Canberra, Vol. 2, pp. 455-462). Adelaide: MERGA.

Scott, A. (2007). Seeking evidence of thinking and mathematical understandings in students' writing. In J. Watson & K. Beswick (Eds.), *Mathematics: Essential research, essential practice* (Proceedings of the 30th annual conference of the Mathematics Education Research Group of Australasia, Hobart, Vol. 2, pp. 641-650). Adelaide: MERGA.

Seah, W. T. (2004a). Shifting the lens of inquiry into the socialisation of mathematics teachers: Nature of value differences. In I. Putt, R. Faragher, & M. McLean (Eds.), *Mathematics education for the third millennium: Towards 2010* (Proceedings of the 27th annual conference of the Mathematics Education Research Group of Australasia, Townsville, Vol. 2, pp. 501-508). Sydney: MERGA.

Seah, W. T. (2004b). *The negotiation of perceived value differences by immigrant teachers of mathematics in Australia.* Unpublished PhD dissertation, Monash University, Australia.

Seah, W. T. (2005a). Negotiating about perceived value differences in mathematics teaching: The case of immigrant teachers in Australia. In H. L. Chick & J. L. Vincent (Eds.), *Proceedings of the 29th annual conference for the Psychology of Mathematics Education* (Vol. 4, pp. 145-152). Melbourne: University of Melbourne.

Seah, W. T. (2005b). *Teacher approaches to cultural differences in mathematics pedagogy between Asia and the 'West'.* Paper presented at the 3rd East Asia Regional Conference on Mathematics Education, Shanghai, China.

Seah, W. T. (2007a). Qualities co-valued in effective mathematics lessons in Australia: Preliminary findings. In J. H. Woo, H. C. Lew, K. S. Park, & D. Y. Seo (Eds.), *Proceedings of the 31st annual conference for the Psychology of Mathematics Education* (Vol. 4, pp. 161-168). Seoul, Korea: The Korea Society of Educational Studies in Mathematics.

Seah, W. T. (2007b). What do effective mathematics lessons value: Preliminary findings from Australian primary schools. In C. S. Lim, S. Fatimah, S. Hajar, M. Y. Hashimah, W. L. Gan, & T. Y. Hwa (Eds.), *Proceedings of the Fourth East Asia Regional Conference on Mathematics Education* (pp. 417-424). Penang, Malaysia: Universiti Sains Malaysia.

Seah, W. T., & Kalogeropoulos, P. (2004). Victorian curriculum reform 2004 and values: The role for mathematics education. In B. Tadich, S. Tobias, C. Brew, B. Beatty, & P. Sullivan (Eds.), *Towards excellence in mathematics* (pp. 441-448). Brunswick, Vic.: Mathematical Association of Victoria.

Skalicky, J. (2007). Exploring teachers' numeracy pedagogies and subsequent student learning across five dimensions of numeracy. In J. Watson & K. Beswick (Eds.), *Mathematics: Essential research, essential practice* (Proceedings of the 30th annual conference of the Mathematics Education Research Group of Australasia, Hobart, Vol. 2, pp. 661-670). Adelaide: MERGA.

Smith, T. (2006). Becoming a teacher of mathematics: Wenger's social theory of learning perspective. In P. Grootenboer, R. Zevenbergen, & M. Chinnappan (Eds.), *Identities, cultures and learning spaces* (Proceedings of the 29th annual conference of the Mathematics Education Research Group of Australasia, Canberra, Vol. 2, pp. 619-622). Adelaide: MERGA.

Stillman, G. (2004). Strategies employed by upper secondary students for overcoming or exploiting conditions affecting accessibility of applications tasks. *Mathematics Education Research Journal, 16*(1), 41-70.

Sullivan, P., Tobias, S., & McDonough, A. (2006). Perhaps the decision of some students not to engage in learning mathematics in school is deliberate. *Educational Studies in Mathematics, 62*(1), 81-99.

Thomas, O. J., & Hong, Y. Y. (2005). Teacher factors in integration of graphic calculators into mathematics learning. In H. L. Chick & J. L. Vincent (Eds.), *Proceedings of the 29th annual conference for the Psychology of Mathematics Education* (Vol. 4, pp. 257-264). Melbourne: University of Melbourne.

Thomson, S., Cresswell, J., & de Bortoli, L. D. (2004). *Facing the future: A focus on mathematical literacy among Australian 15-year-old students in PISA 2003.* Victoria, Australia: Australian Council for Educational Research.

Thornton, S. (2006). Moving into third space – high school students' funds of knowledge in the mathematics classroom. In P. Grootenboer, R. Zevenbergen, & M. Chinnappan (Eds.), *Identities, cultures and learning spaces* (Proceedings of the 29th annual conference of the Mathematics Education Research Group of Australasia, Canberra, Vol. 2, pp. 504-511). Adelaide: MERGA.

Thornton, S. (2007). Dependency and objectification in a year 7 mathematics classroom: Insights from sociolinguistics. In J. Watson & K. Beswick (Eds.), *Mathematics: Essential research, essential practice* (Proceedings of the 30th annual conference of the Mathematics Education Research Group of Australasia, Hobart, Vol. 2, pp. 717-726). Adelaide: MERGA.

Tomazin, F. (2007, May 16). State fails school in poor areas. *The Age*. Retrieved May 22, 2007 from: http://www.theage.com.au/news/national/state-failing-schools-in-poor-areas/2007/05/15/1178995158668.html

Vaiyavutjamai, P., & Clements, M. A. (2006). Effects of classroom instruction on student performance on, and understanding of, linear equations and linear inequalities. *Mathematical Thinking and Learning, 8*(2), 113-147.

Vale, C. (2006). Equity and technology: A case study. In P. Grootenboer, R. Zevenbergen, & M. Chinnappan (Eds.), *Identities, cultures and learning spaces* (Proceedings of the 29th annual conference of the Mathematics Education Research Group of Australasia, Canberra, Vol. 2, pp. 510-518). Adelaide: MERGA.

Victorian Curriculum and Assessment Authority (2005). *Victorian Essential Learning Standards: Overview*. East Melbourne, Victoria: Victorian Curriculum and Assessment Authority.

Vincent, J., Chick, H., & McCrae, B. (2005). Argumentation profile charts as tools for analyzing students' argumentations. In H. L. Chick & J. L. Vincent (Eds.), *Proceedings of the 29th annual conference for the Psychology of Mathematics Education* (Vol. 4, pp. 281-288). Melbourne: University of Melbourne.

Walls, F. (2005). Challenging task-driven pedagogies of mathematics. In P. Clarkson, A. Downton, D. Gronn, M. Horne, A. McDonough, R. Pierce, & A. Roche (Eds.), *Building connections: Theory, research and practice* (Proceedings of the 28th annual conference of the Mathematics Education Research Group of Australasia, Melbourne, Vol. 2, pp. 751-758). Sydney: MERGA.

Walls, F. (2006) Discursive practices as cultural cargo. *For the Learning of Mathematics, 26*(3), 2-8.

Walshaw, M. (Ed.), (2004). *Mathematics education within the postmodern*. Greenwich, Connecticut: Information Age Publishing.

Walshaw, M., & Anthony, G. (2006). Classroom arrangements that benefit students. In P. Grootenboer, R. Zevenbergen, & M. Chinnappan (Eds.), *Identities, cultures and learning spaces* (Proceedings of the 29th annual conference of the Mathematics Education Research Group of Australasia, Canberra, Vol. 2, pp. 527-534). Adelaide: MERGA.

Walshaw, M., & Anthony, G. (2007). The role of pedagogy on classroom discourse. In J. Watson & K. Beswick (Eds.), *Mathematics: Essential research, essential practice* (Proceedings of the 30th annual conference of the Mathematics Education Research Group of Australasia, Hobart, Vol. 2, pp. 765-774). Adelaide: MERGA.

Warren, E. (2005). Young children's ability to generalise the pattern rule for growing patterns. In H. L. Chick & J. L. Vincent (Eds.), *Proceedings of the 29th annual conference for the Psychology of Mathematics Education* (Vol. 4, pp. 305-312). Melbourne: University of Melbourne.

Warren, E., Young, J., & de Vries, E. (2007). Australian indigenous students: The role of oral language and representations in the negotiation of mathematical understanding. In J. Watson & K. Beswick (Eds.), *Mathematics: Essential research, essential practice* (Proceedings of the 30th annual conference of the Mathematics Education Research Group of Australasia, Hobart, Vol. 2, pp. 775-784). Adelaide: MERGA.

Watson, J., Kelly, B. A., & Izard, J. F. (2005). Statistical literacy over a decade. In P. Clarkson, A. Downton, D. Gronn, M. Horne, A. McDonough, R. Pierce, & A. Roche (Eds.), *Building connections: Theory, research and practice* (Proceedings of the 28th annual conference of the Mathematics Education Research Group of Australasia, Melbourne, Vol.2, pp. 775-782). Sydney: MERGA.

Watson, J. M. (2005). Variation and expectation as foundations for the chance and data curriculum. In P. Clarkson, A. Downton, D. Gronn, M. Horne, A. McDonough, R. Pierce, & A. Roche (Eds.), *Building connections: Theory, research and practice* (Proceedings of the 28th annual conference of the Mathematics Education Research Group of Australasia, Melbourne, Vol.2, pp. 35-42). Sydney: MERGA.

White, A. L. (2006a). An examination of the international manifestations of lesson study that began as a Japanese model of teacher professional learning. In H. S. Dhindsa, I. J. Kyeleve, O. Chukwu, L. S. Bee, Hjh Z. Bte Hj Taha, A. Bamba, & S. Upex (Eds.), *Shaping the future of science, mathematics and technical education* (pp. 129-138). Gadong, Brunei Darussalam: Universiti Brunei Darussalam.

White, A. L. (2006b). Shaping the future of secondary school mathematics through modeling with the graphics calculator. In H. S. Dhindsa, I. J. Kyeleve, O. Chukwu, L. S. Bee, Hjh Z. Bte Hj Taha, A. Bamba, & S. Upex (Eds.), *Shaping the future of science, mathematics and technical education* (pp. 139-146). Gadong, Brunei Darussalam: Universiti Brunei Darussalam.

White, P., Wilson, S., Faragher, R., & Mitchelmore, M. (2007). Percentages as part whole relationships. In J. Watson & K. Beswick (Eds.), *Mathematics: Essential research, essential practice* (Proceedings of the 30th annual conference of the Mathematics Education Research Group of Australasia, Hobart, Vol. 2, pp. 805-814). Adelaide: MERGA.

Wilson, S., & Thornton, S. (2006). To heal and enthuse: Developmental bibliography and pre-service primary teachers' reflections on learning and teaching mathematics. In P. Grootenboer, R. Zevenbergen, & M. Chinnappan (Eds.), *Identities, cultures and learning spaces* (Proceedings of the 29th annual conference of the Mathematics Education Research Group of Australasia, Canberra, pp. 36-44). Adelaide: MERGA.

Woodward, J., & Irwin, K. C. (2005). Language appropriate for the New Zealand Literacy Project. In P. Clarkson, A. Downton, D. Gronn, M. Horne, A. McDonough, R. Pierce, & A. Roche (Eds.), *Building connections: Theory, research and practice* (Proceedings of the 28th annual conference of the Mathematics Education Research Group of Australasia, Melbourne, Vol. 2, pp. 799-806). Sydney: MERGA.

Woodward, J., & Pfannkuch, M. (2007). Statistics teachers as scientific lawyers. In J. Watson & K. Beswick (Eds.), *Mathematics: Essential research, essential practice* (Proceedings of the 30th annual conference of the Mathematics Education Research Group of Australasia, Hobart, Vol. 2, pp. 834-842). Adelaide: MERGA.

Ya-amphan, D., & Bishop, A. (2004) Explaining Thai and Japanese student errors in solving equation problems: The role of the textbook. In I. Putt, R. Faragher, & M. McLean (Eds.), *Mathematics education for the third millennium: Towards 2010* (Proceedings of the 27th annual conference of the Mathematics Education Research Group of Australasia, Townsville, Vol. 2, pp. 597-604). Sydney: MERGA.

Zevenbergen, R., Atweh, B., Kanes, C., & Cooper, T. (1996). Social and cultural contexts in mathematics education. In B. Atweh, K. Owens, & P. Sullivan (Eds.), *Research in mathematics education in Australasia: 1992-1995* (pp. 11-40). Sydney: MERGA.

AFFILIATIONS

Wee Tiong Seah
Monash University, Australia

Bill Atweh
Curtin University of Technology, Australia

Philip C. Clarkson
Australian Catholic University, Australia

Nerida Ellerton
Illinois State University, USA

PETER GROOTENBOER, GREGOR LOMAS, & NAOMI INGRAM

THE AFFECTIVE DOMAIN AND MATHEMATICS EDUCATION

Australasian research in the affective domain is examined in this chapter. During the review period, the domain has continued to be dominated by a focus on beliefs, yet there was more research emerging in the areas of self-concept, identity, motivation, and engagement. There remains, however, a lack of theorising in aspects of the domain. While there has been an increase in mixed methods studies, there has been little change in methodological approaches since the last review. With some exceptions, only a limited range of methodological instruments has been used. Data are still primarily self-reported rather than observational, and this issue remains an aspect to be addressed in future research.

Key words: attitudes, beliefs, emotions, engagement, identity, methodological approaches, motivation

INTRODUCTION

This is the second chapter on affective issues to appear in MERGA reviews of research in mathematics education and as such reflects the ongoing importance of affective issues to the mathematics education research community. The first such chapter (Schuck & Grootenboer, 2004) noted a continuing move away from studies on attitudes to projects on beliefs and the consideration of a broader range of affective aspects. This broadening of areas of interest in exploring affect in mathematics education was reflective of worldwide trends. This is evidenced by significant recent publications such as: the seminal book edited by Leder, Pehkonen, and Törner (2002) on beliefs in mathematics education; a special issue of the Mathematics Education Research Journal (August, 2005) whose editorial gave a brief overview of the previous 10 years of MERGA related research in the affective domain; a chapter in the *Handbook of Research in the Psychology of Mathematics Education* by Leder and Forgasz (2006) that discussed, among other matters, developments in the use of different methodological approaches and the potential of new technologies; and a Special Issue of Educational Studies in Mathematics (October, 2006) where a series of papers employed six complimentary theoretical frameworks to examine affective issues with a particular emphasis on the place of emotion in the affective domain.

H. Forgasz et al. (eds.), Research in Mathematics Education in Australasia 2004-2007, 255-269.

The evolution of interest in the affective domain in mathematics education from considerations of attitudes and anxiety in the 1960s and 1970s, to a focus on the relationship between affect and cognition in problem solving, and gender, in the 1980s, and to McLeod's (1992) identification of the three affective concepts – belief, attitudes and emotions – and the later addition of a fourth – values (sometimes including ethics and morals), are discussed in Zan, Brown, Evans, and Hannula (2006). They state that arising from this history are two main trends in current research on affect: the development of theoretical/conceptual frameworks and of the development of new methodological instruments. In contrast to these trends, Leder and Grootenboer (2005) noted the overall lack of theorising and the limited range of methodological instruments used in Australasian research and presented a theoretical model of the relationships between the four concepts (see Figure 1) as a starting point in theorising the relationship between aspects of the affective domain.

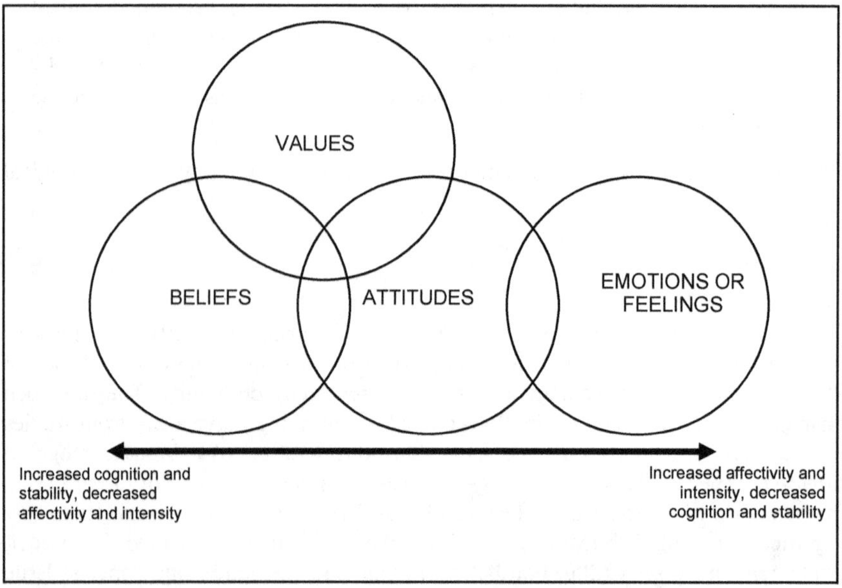

Figure 1. The affective domain (Leder & Grootenboer, 2005).

Parallelling the global situation reported by Zan et al. (2006), Leder and Grootenboer (2005) reported a predominance of belief studies, a diminishing number of studies on attitudes, a small number of studies on "emotions, feelings, attributions and motivation" (p. 3) and a few on values.

Given the relative paucity of work in Australasia on clarifying concepts and the development of theoretical frameworks, there has been little evolution of definitions of the four key aspects of affect considered in the last review chapter (Schuck & Grootenboer, 2004). While the positions taken there are used in this chapter, Figure 1 captures a sense of a more holistic view of the affect domain as a continuum (McLeod, 1992) and its inclusion of the cognitive while delineating the four generally acknowledged concepts within affect – beliefs , attitudes, emotions and values.

Beliefs are positions held by individuals that they feel to be true and their nature cannot be directly observed but must be inferred from actions. Although there are a variety of definitions of attitudes the common elements to these definitions are that attitudes are learnt and are evident in responses to a situation or object, and are seen as positive or negative. Emotions or feelings are described in terms of their transitory and unstable nature arising as an affective response to particular events/contexts whereas values are seen as criteria by which choices or assessments, in terms of desired/desirable outcomes or behaviours, are made (Schuck & Grootenboer, 2004). The first three of these will be discussed, as will be self-concept/identity and engagement as in the previous review chapter. Following this, certain methodological aspects will be considered.

The following sections consider Australasian mathematics education research on the affective domain, as outlined above, except for work on values which is addressed in the chapter by Seah et al. in this book.

TOPICS/FINDINGS

In the broad area of affect in mathematics education, research has focussed on a number of particular aspects, but about half of the studies have centred on beliefs. Other affective aspects that have been prevalent include attitudes, emotions, identity and engagement (and values).

Beliefs

The strong focus on beliefs is a continuing trend as was highlighted by Schuck and Grootenboer (2004) in the previous MERGA review. The studies published in the period 2004-2007 have again been mostly undertaken with teachers as their participants with many exploring the relationship between beliefs and practice. Two Australian writers, who have regularly addressed this area both on their own, and in conjunction with others, are Anderson and Beswick.

Anderson, Sullivan and White (2004, 2005) reported on her work with primary school teachers in New South Wales where she explored the relationship between the teachers' problem solving teaching approaches and their beliefs about the role of problem solving in mathematics. Significantly, Anderson et al. (2005) built on a range of earlier work to theorise and model the relationship between teachers' beliefs and practices for problem solving. After collecting data through a range of methods they developed and refined their model so that it included the significant influence of the social context. A particular feature of Anderson, Sullivan, and White's report was the authors' attention to theorising, setting this report apart from many other, often largely descriptive, reports on teachers' beliefs.

Beswick is another author who has reported extensively and for an extended period of time on beliefs (Beswick, 2005a, 2005b, 2007; Beswick, Watson, & Brown, 2006), focussing primarily on secondary teachers. Amongst other findings, she identified nine beliefs held by secondary teachers about the nature of mathematics, mathematics learning and the role of the teacher. Furthermore, Beswick (2005a) noted some associations between particular clusters of beliefs and the classroom environment, but also highlighted that these relationships were complex and contextually bound. Other studies on teachers' beliefs include: Barkatsas and Malone's (2005) exploration of the beliefs Greek mathematics teachers and the relationship with their practice; McDonough and Clarke's (2005) inquiry into changes in the practices and beliefs of primary teachers involved with a numeracy professional development program; Nisbet and Grimbeek's (2004) quantitative review of primary teachers' beliefs about compulsory numeracy testing; Perry, Way, Southwell, White, and Pattison's (2005) examination of the relationship between preservice teachers' beliefs and mathematical achievement; Prescott and Cavanagh's (2006) qualitative study into preservice secondary teachers beliefs about effective mathematics teaching; and, Scott's (2005) mixed method investigation and modelling of preservice primary teachers' beliefs and intentions about mathematics education.

Studies into students' beliefs about mathematics and mathematics learning have been less frequent than ones on teachers' beliefs, but nevertheless, there has been some work done in this area. A focussed area for this study has been the University of Waikato (New Zealand) where Taylor, Hawera, and Young-Loveridge (2005), Young-Loveridge and Taylor (2005), and Young-Loveridge, Taylor, Sharma, and Hawera (2006) have reported often on primary students' beliefs about mathematics learning, particularly in the context of the New Zealand Numeracy Project. Various reports showed that many students had clear expectations about the teachers' role in teaching them mathematics and deep-seated views about mathematics, and these influenced their learning.

In 2007 there were also reports from Anderson (2007), Grootenboer (2007), Ng and Stillman (2007), Pierce, Stacey, and Barkatsas (2007), and Sullivan and McDonough (2007) on the development of theory-based quantitative instruments to measure student's affective responses to mathematics. While at this stage the findings from the use of these instruments are limited, it is timely that some more robust and rigorous scales are developed.

Attitudes and emotions

In the period 2004 to 2007 there were only a few reports of research that focussed specifically on attitudes and emotions. Pierce and Stacey (2006) reported on a qualitative study that explored the improvement of students' attitudes to mathematics through association with simple pleasures from real world contexts. They found that real world contexts appeared to lead to improved student attitudes to mathematics, but they cautioned that this was due to superficial pleasures rather than substantive mathematics. These authors also worked with Barkatsas (Pierce, Stacey, & Barkatsas, 2007) to develop the *Mathematics and Technology Attitudes*

Scale (MTAS) that monitored students' attitudes to learning mathematics with technology. In this study they reported some small gender differences between the attitudes and confidence of the boys and girls related to learning mathematics with technology.

In terms of students' attitudes to learning a particular aspect of mathematics, Norton and Irvin (2007) reported on a classroom intervention, which was designed to develop students' positive perceptions of their ability to learn algebra. The six-week intervention was based around a selection of activities designed to enable students to make connections between materials, verbal language, and symbolic language. The study found that the students valued the instructional discourse much more than they did their normal mathematics learning experience and had an increased confidence in their capacity to understand algebra.

Another attitudinal study focussed on secondary mathematics teachers' attitudes to a range of alternative assessment methods (Watt, 2005). In short, Watt found that these teachers were generally satisfied with traditional tests as valid measures of student achievement and they did not favour alternative assessments as they were seen as too subjective.

Studies that centred particularly on emotional aspects of learning mathematics were rather scarce. Hawera (2004) reported on a tertiary course designed to address the feelings of some mathematically anxious preservice primary teachers. She found that by creating a safe learning environment and by building the course around social constructivist principles, the participants were able to go some way to addressing their mathematical anxiety. Frankcom's (2006) thesis examined mathematics anxiety and self efficacy in preservice primary teachers and found, as expected, that these were closely negatively correlated but that, unexpectedly, success with mastery experiences in their mathematics education courses did not reduce anxiety or increase self efficacy significantly. In Shannon's (2004) case study with one senior secondary school girl, she found that there were a range of factors that were often not necessarily mathematical, that caused the student to dislike and avoid participation in mathematics.

Other affective aspects

Other studies related to the affective domain emerging amongst the MERGA community focus on identity and engagement. Walshaw (2004a, 2004b) and Klein (2004) focussed on the construction of identity with preservice teachers. They both highlighted the complexity of identity development and the many discourses that influence the process. A symposium held at MERGA 29 in 2006 on *researching identity in mathematics education* provided a general overview (Grootenboer, Smith, & Lowrie, 2006), and then explored the topic from a range of perspectives including Bourdieuian (Zevenbergen, 2006), Wenger's social theory of learning (Smith, 2006), and feminist poststructuralism (Letts, 2006). The symposium was not grounded in empirical research, but rather it gave some theoretical underpinning to work that may emerge in this area. Grootenboer and Zevenbergen (2007) and Ingram (2007) have presented subsequent papers that further explore identity in empirical studies on mathematical learning and secondary students' affective responses to mathematics respectively.

Motivation and engagement have been explored by Sullivan, McDonough and their colleagues (Sullivan & McDonough, 2007; Sullivan, McDonough, & Turner Harrison, 2004; Sullivan, Tobias, & McDonough, 2006). Their large scale and continuing work has already produced a number of findings including the role peer pressure and classroom culture can play in inhibiting students' learning opportunities in mathematics. Cretchley (2005) examined engagement through lecture attendance in a tertiary context and found that there was a relationship between participation and mathematics performance.

Critique of the topics/findings

In the preceding section we have provided a brief and concise summary of much of the Australasian research that has been reported on the affective domain in mathematics education in the period 2004-2007. While each report appears to have individual merit, there are some themes and issues that are worth noting. This critique is not so much about individual reports or studies, but rather about the topics, findings and directions of the body of research in this area.

In the period 2004-2007 a significant proportion of the reports have focussed on beliefs; a trend that has continued since the last review of research in this area (Schuck & Grootenboer, 2004). There is a sense that some of the research has not so much built on the previous work but is largely more of the same. Indeed, many are largely descriptive studies of teachers' beliefs, and lack any significant theorising. It seems that in other parts of the world they are beginning to consider the theoretical foundations of research into beliefs and other affective aspects (e.g., Zan et al., 2006), and it is time for a greater theorising about the general findings that are, seemingly, being consistently reported in Australasia.

Many of the reports reviewed focus on a particular aspect of the affective domain and this has given some illumination on these dimensions in mathematics teaching and learning. However, due to the complex nature of the affective domain, and the difficulties in clearly defining its various components, it would be valuable to see affectivity explored in a more holistic manner. Similarly, there appears to be only a few studies that look at the relationship between affectivity and cognition and achievement (e.g., Watson, Beswick, Brown, & Callingham, 2007). That said, Walshaw and Cabral (2005) contend that there are a number of studies into affective and cognitive aspects of mathematics learning that are grounded in theories of discursive practice, embodiment, somatic markers, neuroscience, representation and situated practice, although these are largely undertaken outside of Australasia (see Zan et al., 2006). Now that there appears to be a greater understanding of the affective domain and its components, it is time for a more focussed research agenda that examines these relationships. The emerging focus on identity may well provide an avenue for such research.

METHODOLOGICAL ISSUES

In this section we have focussed on methodological issues because of their importance when researching the affective domain in mathematics education. As understanding has grown of the complex nature of mathematics education, multiple perspectives and new approaches to research methodologies and paradigms have become valued. This is particularly true for the affective domain, which has unique methodological issues and limitations because of its complexity as evidenced by Bragg's (2007) discussion of attempts to reconcile contradictory indicators arising from quantitative and qualitative data. Also significant are the ethical issues surrounding its highly personal nature and the high level of inference required. In their review of affective literature, Leder and Forgasz (2006) stated "affect can not be measured directly but needs to be inferred from the way an individual behaves or responds to specifically designed instruments, cues, or situations" (p. 404). This reliance on inference can limit affective research to "beliefs, attitudes and feelings that participants are willing and able to share in either a verbal or written form" (Schuck & Grootenboer, 2004, p. 12), creating a high proportion of self-reporting data collection methods (Leder & Forgasz, 2006).

Anthony (2004) in her review of trends in mathematics education, commented that grouping research papers by methodology was "fraught with difficulty" (p. 6) because of a multiplicity of descriptors used by researchers. This was reflected in the Australasian literature during the current review period. Philosophical assumptions and methodological frameworks were sometimes absent and only the methods and instruments used described. There are many reasons why this may be the case including the limited space available in many contexts, but, "metaphorically, unarticulated philosophical perspectives are like an unlighted room … one is likely to stumble over objects and to misunderstand the nature of the rooms within which one is stumbling around" (Maykut & Morehouse, 1994, p. 22).

In this research period, excluding a small number of non-empirical position and review papers, about a quarter of the studies were quantitative (26%) and there were about equal numbers of qualitative (38%) and mixed method studies (36%); the increase in mixed method studies reflecting general research trends (Creswell, 2003). Mixed method studies include a simultaneous or sequential collection of elements of qualitative and quantitative methods in their design (Creswell, 2003; Lankshear & Knobel, 2004). Research in this domain can perhaps be better seen as lying on a continuum between qualitative and quantitative. There are a number of studies in this research period that *tend* to be more qualitative. They are qualitative in nature but have varying levels of simple numerical analysis, which help clarify patterns or can be used for descriptive or illustrative purposes (Lundy Dobbert & Kurth-Schai, 1992). McDonough and Clarke (2005), for example, in investigating change in teachers' beliefs and practice as a result of professional development in the Early Numeracy Research Project in Victoria, used some quantification in comparing the entry and exit questionnaires by using tables with frequencies of themes and a simple t-test to show differences in beliefs about the nature of mathematics, the role of the teacher, and attitude and confidence in teaching mathematics. Barkatsas and Malone (2005) is an example of a mixed method design that tends towards the quantitative. They used a sequential explanatory strategy of inquiry when they initially investigated 465 Greek mathematics teachers' beliefs through a quantitative survey, and then explored the links between these beliefs and classroom practice through a case study of one secondary school mathematics teacher. The qualitative data from the teacher was integrated into the overall analysis to help explain and interpret the findings of the larger quantitative study.

The last four yearly review stated a need for more classroom-based research using observational methods in primary and secondary classrooms (Schuck & Grootenboer, 2004). While this review period indicates that a better balance in the affective literature has been achieved with more studies researching, in particular, secondary students, this research can not necessarily be considered classroom-based. A large proportion of the student-centred studies were done through interviews and surveys. The most frequent instruments were Likert-scale or open-ended questionnaires and semi-structured interviews. Other methods used were classroom observations (e.g., Seah, 2004), the collection of assessment information (Cretchley, 2005; Seah, 2004), weekly conferences (Caswell & Nisbet, 2005), journals (Hawera, 2004; Wilson & Thornton, 2005, 2006), task-related interviews or teaching conversations (Sullivan et al., 2006), and artefact collection such as student work samples, written reflections, and enrolment information (Brady, 2007; Parnell, 2005; Williams, 2006).

Smith (2006) explored the innovative methodology of self-study through narrative inquiry using as a context a preservice mathematics curriculum subject of which she was the teacher. She found narrative inquiry allowed her "identities as classroom teacher, teacher educator, learner, and researcher to come together" (p. 477) enabling her to monitor her own journey, become a co-learner and therefore, maximise generative learning. Later in the review period Bailey (2007) also employed the use of narrative inquiry to pursue and change previously unknown personal beliefs about learning mathematics. Anderson (2005) used a Personal Journey Graph as an instrument in a case study that followed one teacher through the implementation of the New Zealand Intermediate Numeracy Project. A narrative framework was used and data gathered about changing attitudes and beliefs, mathematical and pedagogical content knowledge, and knowledge of student learning in numeracy. The Personal Journey Graph was drawn to reflect the teacher's "opinion of her ability to implement the approaches consistent with the Numeracy Project in her mathematics classroom" (p. 98). The teacher was asked to draw a curve that represented her implementation ability over time as low, medium or high. She was given the opportunity to reflect on and make changes to her graph at the end of the following year. The graph showed the teacher's initial confidence waning and then rapidly increasing in the latter half of the year allowing a unique insight into the teacher's reflections of her journey. This instrument was also being used by Ingram (2007) to understand the affective development of secondary school students through exploring their mathematical identities.

Leder and Forgasz (2006) explored other methodological possibilities available now with technological advances such as mobile phones and the conferencing facilities of the internet. They outlined the *Experience Sampling Method (ESM)*, which allow self-reports of mental processes by using a signal as a trigger between five and seven times per day (Leder & Forgasz, 2006). Participants chart their activities, and their reactions to and beliefs about those activities allowing a less intrusive way of capturing rich and comprehensive data. Forgasz and Leder (2006) used ESM to collect contextualised information to explore the work patterns and factors associated with stress in experienced and novice secondary mathematics teachers. Over a one-week period, the daily lives of 14 teachers were tracked using mobile phone text-messaging as a signal and data collected on their activities and their reactions to those activities. Four of the teachers were then interviewed. The study found that teachers' work-related activities extended beyond work hours and, while teaching mathematics was more likely to be the cause of stress for novice teachers, experienced teachers experienced high engagement and pleasure in the task of teaching mathematics, and were more likely to find administration work a stressor.

The consistencies and inconsistencies between teachers' professed and attributed beliefs remains a major topic of study. However, there have been few advances made to research designs, data collection, and data analysis methods. Speer (2005) cautions that reported inconsistencies in this field of research may be actually the consequence of methodological issues, specifically a lack of *shared understanding* among researchers and teachers about what descriptive terms mean, and a consequence of a lack of coordination between data on beliefs and practices. Furthermore, Speer (2005) contends "that it is inappropriate to classify any data on beliefs as purely professed. All claims about teachers' beliefs are, to greater or lesser extents, attributed to teachers by researchers" (p. 362). Anderson, Sullivan, and White (2004, 2005) go some way to addressing these concerns. They investigated the problem-solving beliefs and practices of 162 primary teachers through a mixed method study using surveys, semi-structured interviews, which used the survey responses as a starting point, and classroom observations of two of the teachers. The authors recognised the possibilities of misinterpretation in the survey questions and provided specific examples for the teachers to illustrate the terminology.

Longitudinal studies are an important way of studying affect, allowing understanding about what mathematical experiences the students have throughout their education that led to the development of their affective views (Schuck & Grootenboer, 2004). There were few studies in the review period, which were explicitly called longitudinal. Lomas (2004) in his doctoral research examined the perceptions of a primary preservice teacher cohort over the three year period of their preservice teacher education degree. This tracked the level of alignment of a 'reform' agenda promoted by lecturers with its acceptance by the preservice teachers. The findings show, over the three-year period of the study, high levels of alignment with the pedagogical aspects of reform but significantly lower levels with emancipatory aspects and only minor incremental change in students' alignment in both aspects. Many of the longitudinal studies in the review period (2004–2007) were reported in parts. The reasons for this are perhaps because research is being published as it is being analysed particularly in longitudinal studies relating to doctoral research or because of the space limitations in journal and conference proceedings. This can lead to problems when the general and methodological background for a paper is not fully presented due to their inclusion in an earlier one so the paper does not completely stand-alone.

There is also a relative absence of action research or interventionist type research in this period. The wealth of descriptive studies over the last decade have provided a rich awareness of affective concerns in mathematics education, and it is now time to undertake further studies that seek to not only report on, but also move forward, on some of these concerns. As we continue to improve our knowledge about the affective domain, this seems an important strategy for researchers to adopt, literally putting knowledge into practice.

In the last review, Schuck and Grootenboer (2004) questioned the disproportionate number of studies that involved preservice and inservice teachers, and wondered if this was because these groups were more readily available to the current field of researchers, often preservice teacher educators. This finding appears to be different from Anthony's (2004) review, which found that students were the main focus, although it may be that preservice teachers were considered here to be students. Anthony (2004) also remarked that there had been a noticeable decrease in the proportion of papers that relate to the wider tertiary sector. In this review period there have been several studies in the affective literature published about tertiary students, in particular with a focus on technology (e.g., Berger & Cretchley, 2005). However, there are still gaps in studies on tertiary mathematics lecturers and a noticeable (now glaring) gap in any research in the early childhood sector. There is also no research that concerns vocational education participants, the community, or parents. This is consistent with the last review (Schuck & Grootenboer, 2004) and also reflects Leder and Forgasz's (2006) comment that there has been little examination of general societal beliefs about mathematics or the community's concern with affectivity.

CONCLUDING COMMENTS

In the 2004 review a recommendation for a greater focus on school students' affectivity was made. Over the last 4 years, although there has been an increase in the number research reports dealing with school student perspectives, with a predominance of secondary focussed studies, preservice teacher and classroom teacher studies continue to dominate. This continues the trend reported by Leder and Grootenboer (2005) and adds weight to their call for more work with primary children and in early childhood.

While links between beliefs and practice continue to be a major area of interest, few studies advance any theoretical considerations that might underpin causal relationships between the two. Implications in studies are rarely developed beyond a local frame of reference and indeed the testing or development of theoretical frameworks is largely absent. The need for theory development to underpin research is apparent elsewhere in the world and needs to be a feature of Australasian affect research in the near future.

Changes in methodological approaches have seen an increase in mixed method studies with roughly equal numbers of qualitative, quantitative and mixed method studies being represented. However, the actual extent of this change is difficult to determine due to the ongoing reporting of some larger studies in small segments, and with a variety of authors, which leads to multiple counts of some approaches. This is further compounded by apparent variation in what is understood by mixed methods; in some studies labelled mixed methods by the researchers, the reviewers held a different view. Alongside this trend toward mixed methods studies with multiple data sources has been the continued use of familiar data collection methods. Indeed, the use of new types of instruments has been relatively limited. The methods of data collection used remain predominantly that of self-reporting via survey and interview, with classroom observation being largely absent. This situation tends to restrict affect research to individual and group perceptions. If belief/practice investigations are to move to more fertile ground then it is important that observational data of teachers and students in classroom environments become an integral part of research projects.

REFERENCES

Anderson, J. (2005). I didn't know what I didn't know: A case study of growth in teacher knowledge within the Intermediate Numeracy Project. In P. Clarkson, A. Downton, D. Gronn, M. Horne, A. McDonough, R. Pierce, & A. Roche (Eds.), *Building connections: Research, theory and practice* (Proceedings of the 28th annual conference of the Mathematics Education Group of Australasia, Melbourne, Vol. 1, pp. 97-104). Sydney: MERGA.

Anderson, J., Sullivan, P., & White, P. (2004). The influence of perceived constraints on teachers' problem-solving beliefs and practices. In I. Putt, R. Faragher, & M. McLean (Eds.), *Mathematics education for the third millennium: Towards 2010* (Proceedings of the 27th annual conference of the Mathematics Education Research Group of Australasia, Townsville, Vol. 1, pp. 39-46). Sydney: MERGA.

Anderson, J., Sullivan, P., & White, P. (2005). Using a schematic model to represent influences on, and relationships between, teachers' problem-solving beliefs and practices. *Mathematics Education Research Journal, 17*(2), 9-38.

Anderson, V. (2007). An online survey to assess student anxiety and attitude response to six different mathematical problems. In J. Watson & K. Beswick (Eds.), *Mathematics: Essential research, essential practice* (Proceedings of the 30th annual conference of the Mathematics Education Research Group of Australasia, Hobart, Vol. 1, pp. 93-102). Adelaide: MERGA.

Anthony, G. (2004). MERGA: A community of practice. In I. Putt, R. Faragher, & M. McLean (Eds.), *Mathematics education for the third millennium: Towards 2010* (Proceedings of the 27th annual conference of the Mathematics Education Research Group of Australasia, Townsville, Vol. 1, pp. 2-15). Sydney: MERGA.

Bailey, J. (2007). Mathematical investigations: A primary teacher educator's narrative journey of professional awareness. In J. Watson & K. Beswick (Eds.), *Mathematics: Essential research, essential practice* (Proceedings of the 30th annual conference of the Mathematics Education Research Group of Australasia, Hobart, Vol. 1, pp. 103-112). Adelaide: MERGA.

Barkatsas, A., & Malone, J. (2005). A typology of mathematics teachers' beliefs about teaching and learning mathematics and instructional practices. *Mathematics Education Research Journal, 17*(2), 69-90.

Berger, M., & Cretchley, P. (2005). Technology and mathematics learning at university level: A South African perspective. *African Journal of Research in Mathematics, Science, and Technology Education, 9*(2), 97-108.

Beswick, K. (2005a). The beliefs/practice connection in broadly defined contexts. *Mathematics Education Research Journal, 17*(2), 39-68.

Beswick, K. (2005b). It depends on the students: Influencing teachers' beliefs about the ends and means of numeracy teaching. In P. Clarkson, A. Downton, D. Gronn, M. Horne, A. McDonough, R. Pierce, & A. Roche (Eds.), *Building connections: Research, theory and practice* (Proceedings of the 28th annual conference of the Mathematics Education Group of Australasia, Melbourne, Vol. 1, pp. 137-144). Sydney: MERGA.

Beswick, K. (2007). Teachers' beliefs that matter in secondary mathematics classrooms. *Educational Studies in Mathematics, 65*(1), 95-120.

Beswick, K., Watson, J., & Brown, L. (2006). Teachers' confidence and beliefs about the ends and means of numeracy teaching. In P. Grootenboer, R. Zevenbergen, & M. Chinnappan (Eds.), *Identities, cultures and learning spaces* (Proceedings of the 29th annual conference of the Mathematics Education Research Group of Australasia, Canberra, Vol. 1, pp. 68-83). Adelaide: MERGA.

Brady, K. (2007). Imagined classrooms: Prospective primary teachers visualise their ideal mathematics classroom. In J. Watson & K. Beswick (Eds.), *Mathematics: Essential research, essential practice* (Proceedings of the 30th annual conference of the Mathematics Education Research Group of Australasia, Hobart, Vol. 1, pp. 143-152). Adelaide: MERGA.

Bragg, L. (2007). Students' conflicting attitudes towards games as a vehicle for learning mathematics: A methodological dilemma. *Mathematics Education Research Journal, 19*(1), 29-44.

Caswell, R., & Nisbet, S. (2005). Enhancing mathematical understanding through self-assessment and self-regulation of learning: The value of meta-awareness. In P. Clarkson, A. Downton, D. Gronn, M. Horne, A. McDonough, R. Pierce, & A. Roche (Eds.), *Building connections: Research, theory and practice* (Proceedings of the 28th annual conference of the Mathematics Education Group of Australasia, Melbourne, Vol. 1, pp. 209-216). Sydney: MERGA.

Creswell, J. (2003). *Research design: Qualitative, quantitative, and mixed method approaches* (2nd ed.). California: Sage Publications.

Cretchley, P. (2005). Mathematics and dumping lectures: Another perspective on the shift towards learner pragmatism. In M. Bulmer, H. MacGillivray, & C. Varsavsky (Eds.), *Kingfisher Delta '05: Proceedings of the fifth Southern Hemisphere conference on undergraduate mathematics and statistics teaching and learning, Fraser Island, Qld.* (pp. 22-26). International Delta Steering Committee.

Forgasz, H., & Leder, G. (2006). Work patterns and stressors. *Australian Mathematics Teacher, 62*(3), 36-40.

Frankcom, G. (2006). *Maths anxiety in third year pre-service primary student teachers: Fact or fiction?* Unpublished Masters thesis, University of Auckland.

Grootenboer, P. J. (2007). Measuring students' affective responses to mathematics and mathematics education. In C. S. Lim, S. Fatimah, G. Munirah, S. Hajar, M. Y. Hashimah, W. L. Gan, & T. Y. Hwa (Eds.), *Meeting challenges of developing quality mathematics education culture: Proceedings of the 4th East Asia Regional Conference on Mathematics Education* (pp. 273-280). Penang, Malaysia: Universiti Sains Malaysia.

Grootenboer, P. J., Smith, T., & Lowrie, T. (2006). Researching identity in mathematics education: The lay of the land. In P. Grootenboer, R. Zevenbergen, & M. Chinnappan (Eds.), *Identities, cultures and learning spaces* (Proceedings of the 29th annual conference of the Mathematics Education Research Group of Australasia, Canberra, Vol. 2, pp. 612-615). Adelaide: MERGA.

Grootenboer, P. J., & Zevenbergen, R. (2007). Identity and mathematics: Towards a theory of agency in coming to learn mathematics. In J. Watson & K. Beswick (Eds.), *Mathematics: Essential research, essential practice* (Proceedings of the 30th annual conference of the Mathematics Education Research Group of Australasia, Hobart, Vol. 1, pp. 335-344). Adelaide: MERGA.

Hawera, N. (2004). Addressing the needs of mathematically anxious preservice primary teachers. In I. Putt, R. Faragher, & M. McLean (Eds.), *Mathematics education for the third millennium: Towards 2010* (Proceedings of the 27th annual conference of the Mathematics Education Research Group of Australasia, Townsville, Vol. 1, pp. 287-294). Sydney: MERGA.

Ingram, N. (2007). A story of a student fulfilling a role in the mathematics classroom. In J. Watson & K. Beswick (Eds.), *Mathematics: Essential research, essential practice* (Proceedings of the 30th annual conference of the Mathematics Education Research Group of Australasia, Hobart, Vol. 1, pp. 450-459). Adelaide: MERGA.

Klein, M. (2004). Preservice teacher education in mathematics: Thinking innovatively about innovation. In I. Putt, R. Faragher, & M. McLean (Eds.), *Mathematics education for the third millennium: Towards 2010* (Proceedings of the 27th annual conference of the Mathematics Education Research Group of Australasia, Townsville, Vol. 2, pp. 328-335). Sydney: MERGA.

Lankshear, C., & Knobel, M. (2004). *A handbook for teacher research: From design to implementation.* England: Open University Press.

Leder, G., & Forgasz, H. J. (2006). Affect and mathematics education. In A. Gutiérrez & P. Boero (Eds.), *Handbook of research on the psychology of mathematics education: Past, present and future* (pp. 403-427). Rotterdam, The Netherlands: Sense Publishers.

Leder, G., & Grootenboer, P. (2005). Affect and mathematics education. *Mathematics Education Research Journal, 17*(2), 1-8.

Leder, G., Pehkonen, E., & Törner, G. (2002). *Beliefs: A hidden variable in mathematics education?* Dordrecht, The Netherlands: Kluwer Academic.

Letts, W. (2006). Bringing feminist poststructuralism to bear on [mathematics] teacher education. In P. Grootenboer, R. Zevenbergen, & M. Chinnappan (Eds.), *Identities, cultures and learning spaces* (Proceedings of the 29th annual conference of the Mathematics Education Research Group of Australasia, Canberra, Vol. 2, pp. 623-627). Adelaide: MERGA.

Lomas, G. (2004). *The relationship between mathematics educators' beliefs and their teaching practices.* Unpublished Doctoral Thesis, Curtain University of Technology, Perth.

Lundy Dobbert, M., & Kurth-Schai, R. (1992). Systematic ethnography: Towards an evolutionary science of education and culture. In *The handbook of qualitative research in education.* California: Academic Press.

Maykut, P., & Morehouse, R. (1994). *Beginning qualitative research: A philosophic and practical guide.* London: The Falmer Press.

McDonough, A., & Clarke, B. (2005). Professional development as a catalyst for changes in beliefs and practice: Perspectives from the Early Numeracy Research Project. In P. Clarkson, A. Downton, D. Gronn, M. Horne, A. McDonough, R. Pierce, & A. Roche (Eds.), *Building connections: Research, theory and practice* (Proceedings of the 28th annual conference of the Mathematics Education Group of Australasia, Melbourne, Vol. 2, pp. 521-528). Sydney: MERGA.

McLeod, D. B. (1992). Research on affect in mathematics education: A reconceptualisation. In D. Grouws (Ed.), *Handbook of research on mathematics teaching and learning* (pp. 575-596). New York: NCTM and Macmillan.

Ng, K. E. D., & Stillman, G. (2007). Interdisciplinary learning: Development of mathematical confidence, value, and the interconnectedness of mathematics scales. In J. Watson & K. Beswick (Eds.), *Mathematics: Essential research, essential practice* (Proceedings of the 30th annual conference of the Mathematics Education Research Group of Australasia, Hobart, Vol. 2, pp. 533-542). Adelaide: MERGA.

Nisbet, S., & Grimbeek, P. (2004). Primary teachers' beliefs and practices with respect to compulsory numeracy testing. In I. Putt, R. Faragher, & M. McLean (Eds.), *Mathematics education for the third millennium: Towards 2010* (Proceedings of the 27th annual conference of the Mathematics Education Research Group of Australasia, Townsville, Vol. 2, pp. 406-413). Sydney: MERGA.

Norton, S., & Irvin, J. (2007). Developing positive attitudes towards algebra. In J. Watson & K. Beswick (Eds.), *Mathematics: Essential research, essential practice* (Proceedings of the 30th annual conference of the Mathematics Education Research Group of Australasia, Hobart, Vol. 2, pp. 561-570). Adelaide: MERGA.

Parnell, S. M. (2005). Do student reflections increase undergraduate mathematics learning? In M. Bulmer, H. MacGillivray, & C. Varsavsky (Eds.), *Kingfisher Delta '05: Proceedings of the fifth Southern Hemisphere conference on undergraduate mathematics and statistics teaching and learning, Fraser Island, Qld.* (pp. 95-102). International Delta Steering Committee.

Perry, B., Way, J., Southwell, B., White, A. L., & Pattison, J. (2005). Mathematical beliefs and achievement of pre-service primary teachers. In P. Clarkson, A. Downton, D. Gronn, M. Horne, A. McDonough, R. Pierce, & A. Roche (Eds.), *Building connections: Research, theory and practice* (Proceedings of the 28th annual conference of the Mathematics Education Group of Australasia, Melbourne, Vol. 2, pp. 625-632). Sydney: MERGA.

Pierce, R., & Stacey, K. (2006). Enhancing the image of mathematics by association with simple pleasures from real world contexts. *Zentralblatt fur Didaktik der Mathematik, 38*(3), 214-225.

Pierce, R., Stacey, K., & Barkatsas, A. (2007). A scale for monitoring students' attitudes to learning mathematics with technology. *Computers in Education, 48*(2), 285-300.

Prescott, A., & Cavanagh, M. (2006). An investigation of pre-service secondary mathematics teachers' beliefs as they begin their teacher training. In P. Grootenboer, R. Zevenbergen, & M. Chinnappan (Eds.), *Identities, cultures and learning spaces* (Proceedings of the 29th annual conference of the Mathematics Education Research Group of Australasia, Canberra, Vol. 2, pp. 424-431). Adelaide: MERGA.

Schuck, S., & Grootenboer, P. (2004). Affective issues in mathematics education. In B. Perry, C. Diezmann, & G. Anthony (Eds.), *Review of mathematics education in Australasia 2000-2003* (pp. 53-74). Flaxton, Qld: Post Pressed.

Scott, A. L. (2005). Preservice teachers' intentions to provide good examples and help children replicate them. In P. Clarkson, A. Downton, D. Gronn, M. Horne, A. McDonough, R. Pierce, & A. Roche (Eds.), *Building connections: Research, theory and practice* (Proceedings of the 28th annual conference of the Mathematics Education Group of Australasia, Melbourne, Vol. 2, pp. 656-663). Sydney: MERGA.

Seah, W. T. (2004). Shifting the lens of inquiry into the socialisation of mathematics teachers: Nature of value differences. In I. Putt, R. Faragher, & M. McLean (Eds.), *Mathematics education for the third millennium: Towards 2010* (Proceedings of the 27th annual conference of the Mathematics Education Research Group of Australasia, Townsville, Vol. 2, pp. 501-508). Sydney: MERGA.

Shannon, F. (2004). Classics counts over calculus: A case study. In I. Putt, R. Faragher, & M. McLean (Eds.), *Mathematics education for the third millennium: Towards 2010* (Proceedings of the 27th annual conference of the Mathematics Education Research Group of Australasia, Townsville, Vol. 2, pp. 509-516). Sydney: MERGA.

Smith, T. (2006). Self-study through narrative inquiry: Fostering identity in mathematics teacher education. In P. Grootenboer, R. Zevenbergen, & M. Chinnappan (Eds.), *Identities, cultures and learning spaces* (Proceedings of the 29th annual conference of the Mathematics Education Research Group of Australasia, Canberra, Vol. 2, pp. 471-478). Adelaide: MERGA.

Speer, N. M. (2005). Issues of methods and theory in the study of mathematics teachers' professed and attributed beliefs. *Educational Studies in Mathematics, 58*, 361-291.

Sullivan, P., & McDonough, A. (2007). Eliciting positive student motivation for learning mathematics. In J. Watson & K. Beswick (Eds.), *Mathematics: Essential research, essential practice* (Proceedings of the 30th annual conference of the Mathematics Education Research Group of Australasia, Hobart, Vol. 2, pp. 698-707). Adelaide: MERGA.

Sullivan, P., McDonough, A., & Turner Harrison, R. (2004). Students' perceptions of factors contributing to successful participation in mathematics. In M. Johnsen Høines & A. Berit Fugelstad (Eds.), *Proceedings of the 28th annual conference for the Psychology of Mathematics Education* (Vol. 3, pp. 289-296). Bergen, Norway: Bergen University College.

Sullivan, P., Tobias, S., & McDonough, A. (2006). Perhaps the decision of some students not to engage in learning mathematics in school is deliberate. *Educational Studies in Mathematics, 62*(1), 81-99.

Taylor, M., Hawera, N., & Young-Loveridge. (2005). Children's views of their teacher's role in helping them learn mathematics. In P. Clarkson, A. Downton, D. Gronn, M. Horne, A. McDonough, R. Pierce, & A. Roche (Eds.), *Building connections: Research, theory and practice* (Proceedings of the 28th annual conference of the Mathematics Education Group of Australasia, Melbourne, Vol. 2, pp. 728-734). Sydney: MERGA.

Walshaw, M. (2004a). Becoming knowledgeable in practice: The constitution of secondary teaching identity. In I. Putt, R. Faragher, & M. McLean (Eds.), *Mathematics education for the third millennium: Towards 2010* (Proceedings of the 27th annual conference of the Mathematics Education Research Group of Australasia, Townsville, Vol. 2, pp. 557-564). Sydney: MERGA.

Walshaw, M. (2004b). Pre-service mathematics teaching in the context of schools: An exploration into the constitution of identity. *Journal of Mathematics Teacher Education, 7*, 63-86.

Walshaw, M., & Cabral, T. (2005). Reviewing and thinking the affect/cognition relation. In H. L. Chick & J. L. Vincent (Eds.), *Proceedings of the 29th annual conference for the Psychology of Mathematics Education* (Vol. 4, pp. 297-303). Melbourne: PME.

Watson, J., Beswick, K., Brown, N., & Callingham, R. (2007). Student change associated with teachers' professional learning. In J. Watson & K. Beswick (Eds.), *Mathematics: Essential research, essential practice* (Proceedings of the 30th annual conference of the Mathematics Education Research Group of Australasia, Hobart, Vol. 2, pp. 785-795). Adelaide: MERGA.

Watt, H. M. G. (2005). Attitudes to the use of alternative assessment methods in mathematics: A study with secondary mathematics teachers in Sydney, Australia. *Educational Studies in Mathematics, 58*, 21-44.

Williams, G. (2006). Autonomous looking-in to support creative mathematical thinking: Capitilising on activity in Australian LPS classrooms. In D. Clarke, C. Kietel, & Y. Shimizu (Eds.), *Mathematics classrooms in twelve countries: The insider's perspective* (pp. 221-236). Rotterdam, The Netherlands: Sense.

Wilson, S., & Thornton, S. (2005). 'I am really not alone in this anxiety': Bibliotheraphy and pre-service primary teachers' self-image as mathematicians. In P. Clarkson, A. Downton, D. Gronn, M. Horne, A. McDonough, R. Pierce, & A. Roche (Eds.), *Building connections: Research, theory and practice* (Proceedings of the 28th annual conference of the Mathematics Education Group of Australasia, Melbourne, Vol. 2, pp. 791-798). Sydney: MERGA.

Wilson, S., & Thornton, S. (2006). To heal and enthuse: Developmental bibliotherapy and pre-service primary teachers' reflections on learning and teaching mathematics. In P. Grootenboer, R. Zevenbergen & M. Chinnappan (Eds.), *Identities, cultures and learning spaces* (Proceedings of the 29th annual conference of the Mathematics Education Research Group of Australasia, Canberra, Vol. 1, pp. 36-44). Adelaide: MERGA.

Young-Loveridge, J., & Taylor, M. (2005). Children's views about mathematics learning after participation in a numeracy initiative. *Research in Education, 74*, 83-90.

Young-Loveridge, J., Taylor, M., Sharma, S., & Hawera, N. (2006). Students' perspectives on the nature of mathematics. In P. Grootenboer, R. Zevenbergen, & M. Chinnappan (Eds.), *Identities, cultures and learning spaces* (Proceedings of the 29th annual conference of the Mathematics Education Research Group of Australasia, Canberra, Vol. 2, pp. 583-590). Adelaide: MERGA.

Zan, R., Brown, L., Evans, J., & Hannula, M. S. (2006). Affect in mathematics education: An introduction. *Educational Studies in Mathematics, 63*(2), 113-121.

Zevenbergen, R. (2006). Teacher identity from a Bourdieuian perspective. In P. Grootenboer, R. Zevenbergen, & M. Chinnappan (Eds.), *Identities, cultures and learning spaces* (Proceedings of the 29th annual conference of the Mathematics Education Research Group of Australasia, Canberra, Vol. 2, pp. 616-618). Adelaide: MERGA.

AFFILIATIONS

Peter Grootenboer
Griffith University, Australia

Gregor Lomas
University of Auckland, New Zealand

Naomi Ingram
University of Otago, New Zealand

COLLEEN VALE & HANNAH BARTHOLOMEW

GENDER AND MATHEMATICS: THEORETICAL FRAMEWORKS AND FINDINGS

The meaning of gender equity, the degree and nature of gender equity in mathematical outcomes and pedagogical practices, and the theoretical position of researchers of gender and mathematics are the concerns of the review of literature presented in this chapter. Findings generally reveal few significant gender differences in mean scores for achievement in Australia and New Zealand for the period under review, but gender differences favouring males in a range of affective factors, and in senior secondary participation, persist. Feminist and post-modern theories influenced some research into pedagogical practices, however most of the Australasian research conducted in the current period drew upon difference or deficit theory.

Key words: achievement, affect, deficit theory, equity, feminism, post-modern theory

INTRODUCTION

In this chapter we continue the series of reviews of Australasian research into gender and mathematics education (Barnes, 1988; Barnes & Horne, 1996; Forgasz, Leder, & Vale, 2000; Leder; 1984; Leder & Forgasz, 1992; Vale, Forgasz, & Horne, 2004). These reviews have provided a continuous chart of progress towards, and achievements and concerns with respect to, gender equity in mathematics teaching and learning. Changes since the early 1980s in gender differences in various mathematical learning outcomes including participation, achievement, and affect, and in pedagogies with respect to gender equity were mapped in these reviews. Theoretical frameworks informing and emerging from the research were also documented.

Following the period leading up to the turn of the century in which there was a trend toward narrowing and disappearing gender differences in achievement and affect (Forgasz et al., 2000), the authors of the most recent review argued that evidence of continuing and re-emerging gender differences warranted on-going attention and research into gender and mathematics education (Vale et al., 2004). They noted the anti-feminist political and social agenda at the turn of the century and called on researchers in the field to be clearer with respect to the theoretical frameworks and meanings of equity that they used.

In conducting the current review of the literature we were interested in exploring the impact of the social and political context on the purpose and objectives of research studies and the theories informing the research. We have used the

H. Forgasz et al. (eds.), Research in Mathematics Education in Australasia 2004-2007, 271-290.

meanings of equity discussed in the literature published during the period of the review to structure this chapter. In the first section we discuss the meaning of equity in relation to gender and the various feminist theories that have influenced the studies reviewed. In the second section papers concerned with equalities and inequalities in mathematical outcomes are reviewed. In this section research concerning affective and motivational factors and achievement, as well as research on the outcomes for indigenous students are reviewed. What the studies in this section have in common is that they primarily relate to processes taking place within individual students. In the final section we shift focus to studies that are more concerned with contextual factors that serve to advantage or disadvantage particular groups and hence investigate factors concerning equitable access to learning. This section includes classroom studies, studies of single-sex learning environments, and particular learning environments and processes. Although individual learners are still present in many of the studies reviewed in this section, it is the ways in which they interact with their environment that is given most weight. Some of the issues of interest or concern identified in the previous review by Vale et al. (2004) were evident in the recently published literature.

THEORETICAL FRAMEWORKS

Equality or equity

In a review of the literature Bishop and Forgasz (2007) discussed the meanings of equity in education and in so doing mapped the theoretical development of these ideas through research concerned with particular social groups, including gender. They reported that, in one sense, equity is defined as an outcome of education with access to learning a condition for achieving equity. This meaning of equity was associated with equality and the three aspects that Fennema (1995) originally identified with respect to gender: equal outcomes, equal opportunities and equal treatment. Bishop and Forgasz (2007) also presented a second meaning, where equity is considered a criterion for evaluating many aspects of education including outcomes, along with access, disposition, and the quality of teaching. In this sense equity is a quality involving fairness and justice and Bishop and Forgasz (2007) cited a number of studies to show that in this sense equity is not the same as equality. Anthony and Walshaw (2007) argued that equity was "relational" and situated:

> Equity is about interactions between contexts and people; it is not about equal outcomes and approaches. Neither is it about equal access to people and curriculum materials. Setting up equitable arrangements for learners requires different pedagogical strategies and paying attention to the different needs that result from different home environments, different mathematical identifications, and different perspectives. (p. 10)

Given the contested meaning of equity we need to acknowledge and discuss the theoretical perspectives that inform these definitions from a gender perspective and for the studies reviewed in this chapter.

Deficit theory, difference theory, identity theory and agency

The theories and paradigms of researchers of gender equity as described by Kaiser and Rogers (1995) and more recently by Jungwirth (2003) were used to critique studies in previous chapters in which Australasian research on gender and mathematics was reviewed (e.g., Forgasz et al., 2000; Vale et al., 2004). According to deficit theory, differences in educational outcomes occur because of inherent deficiencies or weaknesses in girls' experiences, knowledge, and skills. Liberal feminist researchers argued, however, that these deficiencies were due to socialisation and that by attending to these deficiencies through particular educational programs equality of outcomes could be achieved. Radical feminist researchers, on the other hand, embraced differences between the genders and argued that patriarchal structure denied women the opportunity to use these differences as strengths for learning and achieving in mathematics. Radical feminists argued for changes to curriculum and teaching approaches. Post-modern researchers were critical of difference theory because it essentialised learners. They argued that groups of learners were not homogenous, rather individuals have multiple and shifting identities and these are shaped by the context in which they are situated. Analysing the relationships and the power within these relationships in mathematics classroom explained differences in learning behaviours and outcomes and the identities formed by learners in these classrooms. According to this theory, paying attention to the relationships within the classroom, the different identities, and hence the different needs of students in the mathematics classroom, are central to equity. In their review of research of mathematics education, Anthony and Walshaw (2007) drew attention to pedagogies that meet these requirements and are effective for student learning. More recently critical theorists have questioned the notion of shifting identity and its implications for choice. The gendered social norms that underpin the choices that girls make, they have argued, deny girls agency, that is, the power to make a difference in their lives (Hirschmann, 2003).

Perhaps the most striking feature of the studies reviewed in this chapter is that the meaning of equity and the framework informing the study are rarely made explicit. We have tried to identify the assumptions made by researchers and the purpose of their research study. As Leder (2004) observed, liberal feminism, with an emphasis on helping females to assimilate, is perhaps still the dominant perspective in research on gender and mathematics education. However, as will be seen, particularly in the later sections of this review, a growing body of research is emerging which draws, often implicitly, on other paradigms. So, for example, while a number of researchers remained concerned with persuading girls of the value of studying mathematics to a high level (e.g., Watt, 2004), others were beginning to question the basic assumptions underlying such work, and instead investigated the valid decisions that girls made in relation to mathematics (e.g., Shannon, 2004). Such a move away from a deficit view of gender differences in mathematics is also evident in the increasing prevalence of studies in which the interplay of a range of factors in shaping the experiences of girls and boys in mathematics classrooms were explored. The influence of post-modern perspectives was evident in the reluctance of some researchers to treat gender as a straightforward and stable category for the purposes of analysis (e.g., Barnes, 2005).

Feminist theory applied more generally

A theoretical development worth noting in the context of this review is that feminist theory, in various guises, appears as a theoretical tool in work in which issues other than gender differences were explored. For example, drawing on feminist post-structuralist theory, Klein (2006) explored pre-service primary teachers' constructions of what constitutes an engaging pedagogy in mathematics. Atweh (2004) also drew on the ideas of feminist theorists when writing about international collaborations in mathematics education.

This "mainstreaming" of feminist theory is to be welcomed. Like the shift away from an emphasis on the deficiencies of girls and women in relation to mathematics, it provides evidence that the ghettoisation of the female experience is fading. "Gender issues" and "feminist theories" are increasingly regarded as being of relevance to many concerned with mathematics education.

STUDIES OF MATHEMATICAL OUTCOMES FOR LEARNERS

In this section studies in which gender equity in mathematics education was evaluated in terms of mathematical outcomes are reviewed. Studies reviewed in the previous four-year period, at the turn of the century, did not reveal consistent findings with respect to gender differences in achievement or attitudes; the findings varied according to mathematical knowledge and skills, the method of assessment and the affective variables that were being measured (Vale et al., 2004). Vale, Forgasz, and Horne (2004) also reported that gender differences in achievement and affect also varied according to the socio-economic status of students. In the period of the current review, gender comparisons in achievement, participation and affect were included in a broad range of studies at all levels of schooling.

Achievement

The Third International Mathematics and Science Study (TIMSS) of 2002-3 and the Program of International Student Assessment (PISA) of 2003 revealed gender equality in mean scores in mathematics and mathematics literacy of Australian and New Zealand primary and secondary males and females. That is, there were no significant gender differences in the mean scores of nine year-old or 13 year-old students in mathematics (Ministry of Education, NZ, 2004a, 2004b; Thomson & Fleming, 2004), or among 15 year-old students in mathematical literacy (Thomson, Cresswell, & De Bortolli, 2004). However, three years later, the PISA 2006 results revealed significant gender differences favouring 15 year-old boys in mathematical literacy in Australia and New Zealand (OECD, 2007). In Australia the gender difference was greater than the OECD average, and gender differences favouring males were also apparent in the results for each state/territory. However, the overall gender difference (14 points) was much smaller than the significant difference in scores between students from low and high socio-economic status groups (78 points) (Thomson & De Bortoli, 2008). The instability in findings with respect to gender differences in large-scale international studies of mathematics is, perhaps, not surprising since there are contradictory findings among researchers studying achievement for other age groups. Horne (2004) found that significant gender differences favouring males in average performance of students emerged during the first three years of primary schooling for number, although there were no differences in the other content domains of space and measurement. For senior secondary mathematics in Victoria, Cox, Leder and Forgasz (2004) reported that females received higher mean scores than males on the vast majority of comparisons made, although no inferential statistics accompanied these findings.

While the comparison of mean scores in these studies provides a snapshot of gender differences relevant for assessing the performance of education systems, they are not sufficient for confirming gender equity or inequity in mathematics education. It is also important to compare the distribution of results for males and females and the distribution of results for students of particular social groups since demonstrating particular mathematical proficiencies or performing at particular levels of achievement is what provides students with access to the power that achievement in mathematics affords.

The analysis of the distribution of scores reported in the studies cited above reveal evidence of generally persistent gender differences in achievement favouring males and that these discrepancies are of concern for particular social or cultural groups. Males were consistently more highly represented among the highest achievers in mathematics for studies involving primary, junior secondary and senior secondary students (Cox et al., 2004; Thomson et al., 2004; Thomson & Fleming, 2004). Almost twice as many 15 year-old Australian male as female students performed at the highest level in mathematical literacy (PISA) in 2003 and 2006 (Thomson et al., 2004; Thomson & De Bortoli, 2008). These findings were also apparent in some Australian states with large gender differences among the highest achievers favouring males in New South Wales (NSW), Victoria, and Tasmania. The only exception was in the Australian Capital Territory (ACT) where females were more highly represented among the top achievers. Males were also more highly represented among the top achievers in studies of gifted and talented mathematics students (Leder, 2006).

Mathematics content and cognitive domains

Comparisons of performance in particular mathematical domains and cognitive domains have also been reported. Items used in TIMSS 2002-3 were of three cognitive types: knowing facts, procedures and concepts; applying knowledge and understanding; and reasoning. There were no gender differences on these types of items for grade 5 students in New Zealand, though grade 9 girls recorded significantly higher scores on reasoning items than boys (Ministry of Education, 2006). Meaney (2005) included a gender comparison in the ways in which final year secondary students in New Zealand presented justifications for algebra items but results were not conclusive for this small pilot study.

Horne (2004) found that in the first four years of schooling in Victoria boys developed more efficient strategies for counting, addition and subtraction, and multiplication and division and had a stronger grasp of place value than girls, suggesting greater fluency with operations and number sense among boys than girls. The findings of Young-Loveridge and Taylor (2005) with respect to strategies in multi-digit computation among 9-11 year old primary students in New Zealand contradicted these findings to some extent. In their study girls were more likely to give an alternate strategy, suggesting more flexible thinking or fluency in number operations. Thomson and Fleming (2004) reported that, on average, 13-year old Australian boys performed better than girls on number and measurement items in the TIMSS 2002-3.

Nine year-old primary girls in Australia and New Zealand recorded higher mean scores than 9 year-old boys for the geometry items on TIMSS 2002-3 (Ministry of Education, 2004a; Thomson & Fleming, 2004) but 15 year-old Australian boys performed better than girls on space and shape items in the PISA 2003 study (Thomson et al., 2004). Lowrie and Diezmann (2005) found that boys in grade 4 performed better than girls on six types of graphical "languages", suggesting that boys had higher-order visual-spatial reasoning capacity than girls. On the surface these different findings appear contradictory since the same content domain was the concern in each of these studies. Teaching and learning practices at different year levels may explain these findings; so too may the differences in the cognitive demands of the mathematical tasks used in the studies.

Forgasz, Griffith, and Tan (2006) compared the achievement of males and females in the Victorian Certificate of Education (senior secondary mathematics). They were particularly interested in the effect of digital technology and analysed the results of students taking equivalent advanced mathematics subjects, one in which graphics calculators were required and the other in which students used CAS (computer algebra system) calculators. They found that higher proportions of male than female students were awarded the highest grades (A^+, A and B^+) in the CAS environment than in the graphics calculator environment.

Of interest in this review is the theoretical position and recommendations made by the researchers arising from findings such as those reported above. In the past, deficit theory was used to explain these findings; and this explanation of gender difference was still evident in some studies in which mathematical achievement for particular content and skills was compared. For example, Lowrie and Diezmann (2005) focussed on the deficit of girls' knowledge, suggesting that teachers need to develop girls' skills in using and interpreting graphic images involving shape, size and orientation (even though average gender differences were only significant for number line tasks in their study). Horne (2004) on the other hand asked how the teacher and teaching practices, in other words, the socialisation practices within the classroom may be contributing to the likelihood of girls being "rule compliant" and boys being "risk takers". Young-Loveridge and Taylor (2005) who also investigated children's views of mathematics found that "more boys than girls reported that effort put into a problem was more important than getting a correct solution" (p. 83). It is not clear whether this finding is consistent with gendered 'rule compliance' in mathematics classrooms but the apparent contradictory findings from these studies, occurring in different countries and education systems, show that teaching practices are relevant. Forgasz, Griffith, and Tan (2006) argued that teachers were not aware of possible gendered effects of using technology in senior secondary mathematics but it is not at all clear whether they were arguing against the use of technology, for teachers to intervene to improve girls' performance in CAS environments, or for teachers to change their teaching practices in some way.

Davis, Clarke, and van den Heuvel-Panhuizen (2005) tested difference theory as an explanation for gender differences in mathematics achievement among primary students. Based on the TIMSS 2002-3 results of children in the Netherlands that showed significant gender differences favouring males, test items were defined as "boys items" or "girls items". For the sample of grade 6 Australian children these items followed the predicted pattern of gender difference in correct responses but the difference was only significant for one "boys item". This finding shows that it is important not to essentialise girls and boys with respect to mathematics skills. The authors argued that the nature of the curriculum and teaching practices may explain the different results for Australian children. However in reflecting on this finding the authors argued that the performance of girls could be improved if they had the opportunity to rehearse "boys items", but they offered no suggestions regarding improving boys' performance on "girls items", perhaps because they thought that boys were performing well on the important mathematics items, that is, those requiring flexibility in number sense and operations.

Indigenous students

In Australia, indigenous secondary students scored significantly lower than non-indigenous students in the TIMSS 2002-3 mathematics and PISA 2003 mathematical literacy, with distributions skewed toward lower levels of achievement (Thomson et al., 2004; Thomson & De Bortoli, 2008; Thomson, McKelvie, & Murnane, 2006). Anthony and Walshaw (2007) noted similar differences between Maori and non-Maori students in New Zealand. In Australia 15 year-old female indigenous students performed particularly poorly on mathematical literacy in the 2003 PISA study (Thomson et al., 2006). Achievement above the highest international benchmark was rare for indigenous female students and higher proportions of indigenous female students (45%) than indigenous male students (29%) failed to reach the lowest international benchmark. The difference between indigenous and non-indigenous students is most striking for females since only 10% of non-indigenous females did not reach the lowest benchmark. There was no significant difference in the results for PISA 2003 and 2006 with respect to average differences in mean scores for indigenous and non-indigenous students. The differences remained vast (80 score points) and significant (Thomson & Bortoli, 2008). In 2006, the proportion of indigenous males and females performing at the lowest level (38% and 40% respectively) was closer than in the earlier survey. Perhaps the nature of the sampling was responsible for this apparent improvement for indigenous females and deterioration for indigenous males. Clearly though, participation and access to mathematical learning, or rather limited or restricted access is the only reasonable explanation for these overall findings.

Anthony and Walshaw (2007) argued for the consideration of socio-political realities that shape students' identities, and mathematical teaching practices that acknowledge diversity, respect cultural identity and provide an ethic of care, explicit instruction and respectful exchange of mathematical ideas.

Summary

These results indicate that deficit and difference theories are not helpful in explaining gender differences observed in mathematics achievement. The contradictory findings with respect to mathematical domains and skills suggest that teachers need to be more aware of the way in which their teaching approach and the demands of the documented curriculum might influence the achievement of the girls and boys in their classrooms. The deterioration in gender equity apparent in the re-emergence of gender differences in mean scores in mathematical literacy (PISA 2006) in both New Zealand and Australia, warrant closer attention and consideration by governments and mathematics curriculum writers.

Participation in mathematics

Forgasz (2006) reported on a study of enrolment in year 12 mathematics subjects of Australian students from 2000-2004. She focussed her analysis on the mathematics subjects in the different state jurisdictions in Australia that are the minimum pre-requisites for continued mathematics study in tertiary education. National and international concerns regarding declining enrolments in tertiary mathematics were the impetus for this study. Of equal interest was the gendered pattern of enrolments favouring males reported in earlier studies, and the desire to discern whether these patterns were still evident and whether or not they had improved or deteriorated. The findings for Australia overall showed that for the five-year period enrolments in these particular mathematics subjects had declined. There were more males than females enrolled in these subjects in each state, except the ACT. Declining enrolments by girls during this period contributed to the overall fall in enrolments in each of the states that recorded a decline in enrolments in this subject. These findings contradict the trend of closing the gender gap in participation observed during the 1990s (Forgasz et al., 2000). Forgasz (2006) claimed that the data "indicate that females are continuing to limit their career options" (p. 220) suggesting the need to amend the girls' choices; a more critical approach would be to investigate the curriculum and gendered social norms that are influencing girls.

Affect

The authors of the previous MERGA review noted that while research associated with participation and self-concept continued to report differences favouring males, findings from studies considering the gender-stereotyping of mathematics were more equivocal, with new instruments throwing up some surprising results. A further finding to emerge from the literature was that the increasing use of computers in mathematics classrooms was resulting in "a retrograde shift in attitudes towards mathematics being viewed, once again, as a male domain" (Vale et al., 2004, p. 86). In the current review period research findings followed similar patterns.

Building on earlier work, Forgasz, Leder, and Kloosterman (2004) continued to investigate 'mathematics as a male domain'. They suggested that some items from the original Fennema-Sherman Mathematics Attitude Scales, developed 30 years earlier, might be anachronistic and no longer reliably interpretable. Two new scales were developed and administered in Australia and the US, *Mathematics as a Gendered Domain* and *Who and Mathematics*. It was found that in both countries, most students regarded mathematics as a *neutral* domain. The items relating to enjoyment in mathematics showed the surprising result that among the Australian students, females were thought more likely than males to enjoy mathematics. This reverses trends that have been found previously, with boys thought more likely to enjoy mathematics.

However, as will be seen below, the affective responses of individuals tell a less up-beat story, and this eradication, or even reversal, of previous stereotypes relating to mathematics does not extend to the ways in which males and females respond to their own mathematics learning. It is also the case that students' achievements in mathematics have a bearing on their affective responses; unsurprisingly Banfield (2005) found that mathematically gifted adolescent boys recorded higher levels of self-concept, self-esteem, and mathematical self-efficacy than boys of average mathematical ability. The two large international studies during the current review period, TIMSS 2002-3 and PISA 2003, included items exploring affective issues that were analysed by gender. In the TIMSS study of 2002-3 Australian males scored significantly higher levels of self-confidence in mathematics learning than girls for both 9-year old and 13-year old students (Thomson & Fleming, 2004). Thomson and Fleming (2004) also found that self-confidence was more strongly correlated with achievement than was value of mathematics. Findings from the PISA study were similar. Fifteen-year old Australian males reported higher levels of enjoyment, interest, self-efficacy and self-concept, and held stronger beliefs on the value of mathematics than did females (Thomson et al., 2004). Females were more anxious than males but reported more favourable relations with teachers and a stronger sense of belonging. Using Hierarchical Linear Modelling, Thomson, Cresswell, and De Bortolli (2004) found that self-efficacy and self-concept had the strongest relationship with mathematical literacy among the affective variables measured in the PISA study.

Based on the findings of the PISA 2003 study, Thomson, Cresswell, and De Bortolli (2004) argued that the participation of females in tertiary mathematics and related disciplines could be increased if teachers attended to girls' attitudes:

> PISA suggests a reason for [lower participation of females], finding that there are much larger gender differences at age 15 in approaches to learning mathematics than performance itself. Females appear to be less engaged, more anxious, and less confident in mathematics than males. This finding suggests that approaches to reducing these gender differences need to start at

> an early age in order to increase females' engagement in mathematics and build their confidence in their mathematical abilities. (p. xvi)

Attitudes to digital technology in mathematics

Using different survey instruments and methods three different studies of junior secondary students in Victoria reported gender-related attitudes to the use of digital technology in mathematics favouring males. In a mixed methods study of students in two secondary mathematics classes (grades 8 and 9) in which computers were regularly used in mathematics lessons, Vale and Leder (2004) collected data through questionnaires, interviews and observations. They observed that, in general, girls and boys were positive about the use of computers but girls viewed the use of computers in mathematics less favourably than boys. While boys thought that computers provided pleasure and relevance in mathematics, girls were more concerned about whether computers aided their learning or enabled success in mathematics. Analysis of responses to the Attitudes to Computer-Based Mathematics Scale (ACBM), composed of eleven items, confirmed these findings. Boys' attitudes were more strongly correlated with the desire to perform well at computing than were girls' attitudes.

Pierce, Stacey, and Barkatsas (2007) and Barkatsas (2005) developed the Mathematics and Technology Attitudes Scale (MTAS). In contrast to the ACBM scale that referred only to computers, the items measuring attitudes to use of technology for learning mathematics (MT) in the MTAS scale can be adapted to refer to computers, graphics calculators or other digital technologies. The MTAS instrument was used to gather responses from 350 Victorian students in grades 8-10. The decision to use the graphics calculator version of the MT sub-scale was not explained by the researchers, although they did comment that the low scores on this subscale at one school might be explained by the lack of experience with using technology, and that the very high scores at another school might have been due to students' recent experience with graphics calculators. Significant differences favouring males were recorded for mathematics confidence, affective engagement, confidence in using technology, and attitude to the use of technology for learning mathematics, but not for behavioural engagement. They also found that attitude to learning with technology was positively correlated to confidence with technology for boys, and negatively correlated with mathematics confidence for girls.

Forgasz (2004) also reported significant gender differences in attitude to the use of computers in mathematics favouring males. Her study involved 1613 Victorian students from grades 7-10 and used a five-item scale (Attitudes to computers for learning mathematics: ACM). She also compared attitudes and ownership of computers according to other equity factors: socio-economic status, language, Aboriginality and location. Personal ownership was more likely for boys, high socio-economic status, non-English speaking background and metropolitan students. In common with the other two studies, Forgasz (2004) also found a strong and significant correlation between attitudes to computers for learning mathematics and attitudes to computers, but not with attitudes to mathematics. Multivariate analysis was not used in this study but may have illuminated the relative strength of the various factors included in the study. Within gender differences were commented on in each of these studies.

Theories linking affect and participation

In a series of papers exploring the interplay between a range of affective factors and students' participation in mathematics courses among students in NSW, Watt (2004, 2005a, 2005b, 2006; Watt, Eccles, & Durik, 2006) found persistent gender differences. Watt (2006) looked at the intended and actual participation of girls and boys in New South Wales in senior high school mathematics programmes, and the mathematics-relatedness of the students' intended careers. The results showed a "remarkably robust" and statistically significant tendency for boys to plan and take higher-level mathematics subjects than girls, and for boys to be more likely to plan mathematical careers than girls. These gender differences in the actions of girls and boys were mirrored by gender differences in their self-perceptions of mathematics talent and their expectations of success that also favoured boys. However, gender differences in mathematics subjects taken, and planned careers were not accounted for by differences in achievement (Watt, 2005b). Furthermore despite similar success in mathematics, boys rated their talent more highly than did girls. Although prior success in mathematics was found to influence subject choice, students who rated the intrinsic value of mathematics, and their self-perception higher, took higher-level mathematics subjects, even when prior success was controlled for. Furthermore, while boys who saw mathematics as moderately useful were likely to aspire to mathematics related careers, only girls who saw mathematics as highly useful were likely to do so. Watt (2006) argued that:

> Boys maintained higher intrinsic value for maths and higher maths related self-perceptions than girls throughout adolescence. Since the present study has identified the importance of these factors in maths participation choices, both for senior high and planned careers, girls' lower levels are particularly problematic. We need to understand how it is that boys come to be more interested and like maths more than girls; and also why girls perceive themselves as having less talent, even when they perform similarly. (p. 319)

She concluded that educators might focus on explicating the high utility value of mathematics to girls.

In a survey of year 11 students in Victoria conducted in 2001 and 2003, Forgasz (2005) investigated whether the role of digital technology, and computers in particular, was influencing students' decision to enrol in senior mathematics subjects that provided a pathway to tertiary mathematics in ways that may explain the gendered pattern of enrolments in senior secondary mathematics. Her findings were not conclusive. For one year only, a significantly higher proportion of males stated that the use of computers in mathematics had contributed to their decision to continue studying mathematics. Forgasz (2005) urged further investigation of the relationship between digital technology and mathematics participation.

Summary

These studies indicate that the beliefs, participation and feelings of individuals continue to show differences along gendered lines. Males are more likely to exhibit self-confidence, to regard mathematics as relevant and of intrinsic value and to be motivated by the use of technology in mathematics. These factors appear to be responsible for continuing lower participation rates in advanced mathematics among females, which cannot be explained by differences in achievement.

Significantly, however, these differences are not mirrored in students' more general perceptions of who can do mathematics; when they are not responding about themselves, they see few differences between the sexes, and even respond that girls may be more likely than boys to enjoy mathematics. This apparent inconsistency suggests that although females no longer believe that pursuing mathematics would be inappropriate for them by virtue of their gender, there remain factors that tend to leave them, as a group, less enthusiastic than males about mathematics and about using technology in mathematics. While the studies reviewed have focused on explanations located within individual girls, a radical or critical feminist interpretation of these findings would advocate change in approaches to teaching and curriculum.

STUDIES OF CONTEXT FOR MATHEMATICAL LEARNING

As noted above, equity can only be achieved when learners have access to mathematical learning. Researchers in the past have found that access to learning is concerned with more than enrolment and participation in mathematics learning, but that the curriculum, teaching practices and social norms of the classroom may exclude some students from learning and result in differential experiences and effects on boys and girls (Forgasz et al., 2000). In the previous four-year review Vale, Forgasz, and Horne (2004) reported that technology based classroom settings and general mathematics classrooms advantaged boys' learning, especially those identifying with hegemonic masculinities. Which boys and which girls benefited in mathematics classrooms depended on the mathematical discourse, the teaching methods, the use of technology, and the attitudes of teachers. During the period of the current review, Australasian researchers continued to explore the effects of different classroom structures and teaching approaches on girls' and boys' engagement in mathematical learning.

Single-sex classrooms

Single-sex classrooms were an intervention strategy originally implemented by liberal feminists to target the needs of girls in mathematics. However, at the turn of the century teachers and parents were advocating that these settings enabled teachers to tailor their teaching approaches to suit the needs of boys in single-sex and co-educational settings (Forgasz et al., 2000), and 'boy friendly' methods of teaching gained attention in the education community (Vale et al., 2004). Not surprisingly then, in the current period of review Australasian researchers investigated the case of single-sex classrooms as settings that may improve the performance of boys.

Interviews with the principal, teachers, and students from a primary school in which single-sex classes were introduced for grade 7 students to address the educational and social needs of boys, revealed that the gendered pedagogies that the teachers implemented reinforced rather than challenged hegemonic masculinities (Martino, Mills, & Lingard, 2005). Improvements in boys' emotional literacy were noted by the researchers but they also reported that the "hands on", competitive, activity based learning used by teachers in the boys' classrooms reinforced teachers' essentialised views about how boys learn, and denied boys in this program access to higher order learning. Mulholland, Hansen, and Kaminski (2004) reported the effects on mathematics and English achievement in their study of single-sex classes at grade 9 that were implemented in a co-educational secondary school to improve the under-achievement of males. In the short time frame for which data were collected, the English performance of boys and girls in the single-sex classes improved compared to the co-educational classrooms, but there was no improvement for girls or boys in advanced mathematics in the single-sex classes. Interviews with the teachers revealed that they had implemented gender-stereotyped teaching and learning strategies, that is, "girl-friendly" or "boy-friendly" approaches, as appropriate, in English and mathematics in the single-sex settings. These approaches appeared to have had some impact in English, but not in mathematics. The authors argued however that the program at this school allowed the students to elect to join the single-sex classes and that this may be an important source of empowerment for them. No data on affective factors were collected to support this view however.

Classroom processes

While many studies include an element of classroom-based research, a relatively smaller number can be said to be classroom studies in the sense that their focus is on classroom processes. Notable among those studies that do focus specifically on the classroom is Mary Barnes' exploration of collaboration in high school mathematics (Barnes, 2004). Using positioning theory, this work explored the range of positions taken up by students during mathematics lessons in which students were working in groups. By looking at commonalities in the ways students took up various positions, a number of common patterns of participation were identified. The study did not set out to explore gender differences in particular, and Barnes (2004) argued that this approach avoided the tendency to dichotomise that is evident in much work on gender and mathematics, and which can "create differences as artifacts of the research process" (p. 8). By introducing gender as a variable only in the final stage of the analysis, it was possible to explore the complexities of classroom interactions. Considerable variation was found in the positions taken up by both boys and girls in the class. Some patterns of participation, for example, "Attention-Avoiders" and "Responsive Intellectuals", were found among both male and female students. Furthermore, she found that even those girls who took up stereotypically feminine positions were not powerless (as they are often constructed as being). Rather, all students acted to promote their own ends in lessons.

In a study of two intermediate school classrooms, Darragh (2005) also looked at student interactions and peer learning, this time in the context of the New Zealand numeracy project. Her focus, however, was not those occasions when collaborative work was planned and organized, but the informal interactions between students who have been told ‘you can work with a partner if you like’. Like Barnes (2004), she did not set out to explore gender differences, but gender did emerge as an interesting variable during the course of the analysis, particularly in connection with the attitudes to their work of students in the most advanced mathematics groups. She found that while many of these students were critical of the work that they were given, boys were much more likely to actively reject the activities they had been set. In accounting for this difference, Darragh (2005) drew on Walkerdine’s (1989) work into the ways different behaviours are validated and reinforced in girls and in boys. Darragh’s (2005) work also offers insights into the ways in which fellow students join in ‘policing’ the ways in which their peers engage with mathematics.

A significant feature of recent studies of the learning environment was a trend towards considering the benefits for *all* students of “girl friendly pedagogies”. Thus Barnes (2004) argued that collaborative learning, traditionally seem as benefiting girls, can be of benefit to students of both sexes. Similarly, in a PhD thesis focusing on ‘second chance’ mathematics courses, Miller-Reilly (2006) used the literature on connected knowing and teaching (Gilligan, 1982; Belenky et al., 1986; Rossi Becker, 1995) to theorise the success of one-to-one teaching sessions with a mature male student who had a debilitating fear of mathematics. In both studies, approaches growing out of female responses to mathematics have been found to have a broader relevance.

A different finding with respect to the general usefulness of “girl friendly” pedagogies emerged from the discussion of the ‘classroom culture’ in mathematics lessons in two schools (Bartholomew, 2004). Bartholomew (2004) critiqued the ostensibly ‘girl friendly’ environment of mathematics lessons at a single sex girls school. She found that while teachers at the school experienced considerable success at creating a ‘safe’ environment ensuring fairness and minimizing differences between students, it came at the expense of giving students opportunities to take responsibility for their own learning.

Such perspectives illustrate the shift away from a focus on interventions that may enable girls to ‘catch up’ with boys, towards a greater recognition that the experiences of girls may illustrate more general issues and point the way towards improving the mathematical experiences of students more generally. This, too, marks a departure from the dichotomising of girls and boys, and can be regarded as part of a move away from a liberal feminist perspective towards radical or post-modern understandings of gender issues in mathematics.

Technology contexts – digital and otherwise

Pedagogies that support the learning of socially disadvantaged groups were the focus of a case study on the use of digital technologies in mathematics reported by Vale (2006). As was the situation in other studies reviewed, gender was not the main focus of this study, rather the nature of the setting, an all girls school, provided insights into the ways that mathematics can be connected to the personal and social interests of students and their mathematical learning needs. In this case, the students used digital tools in their mathematical investigation of phenomena for integrated curriculum and the teacher scaffolded students' understanding, drawing attention to key mathematical ideas by using the dynamic feedback afforded by digital technologies in a whole class investigation setting.

Norton (2005, 2006) found that the scaffolding of mathematical learning through guided discussion and modelling was also important for girls' engagement with, and attitudes to, mathematics. Norton (2006) used a design based methodology to investigate the use of technology integrated learning with two different classes, one a single-sex girls class. Norton took an essentialised and possibly deficit view of girls, believing that by integrating technology with mathematics girls may become more interested in mathematics-related careers. The focus was proportional reasoning and, in the first stage of the study, students constructed machines using lego-technics to solve particular design problems involving gear ratios. While the test results showed that girls improved their mathematical understanding of proportion, their attitude to mathematics was worse at the end of the first stage of the study. Interviews with the girls in the class revealed their dissatisfaction with the lack of modelling and the disconnection of the phenomenon in mathematics and technology lessons. In the second stage of the study, conducted in an all girls setting, the integration of mathematics and technology used multiple contexts and technologies. The teacher drew attention to the proportional phenomenon being modelled in each of the problems. Both affect and achievement improved in the second stage. In both of these studies, the teacher's skills in drawing attention to the features of the mathematics that matter for learning were critical to student engagement with mathematics.

Learner identity

In a number of studies, students' responses to mathematics in relation to their identities have been theorised. Identity is useful in that it provides a theoretical tool that can interrogate the interplay between the individual and the social. In a case study of a mathematically talented grade 13 female student in a New Zealand school, Shannon (2004) used a post-structuralist framework to theorise her rejection of mathematics. In an interview, the student, Jane, drew on common discourses of mathematics as irrelevant and "taught differently" from other subjects, and having no space for creativity. The ways in which these constructions of mathematics intersected with Jane's feminine identity were examined. Shannon concluded that for students like Jane, with many pathways open to them, it was unsurprising that Jane opted out of mathematics.

Rodd and Bartholomew (2006) interviewed a group of female students who had opted to pursue mathematics to a high level, in a study of female, final year mathematics undergraduates in a British University. They found that, although many of the women were highly successful students, they were also strangely 'invisible'. Many had a great deal of difficulty 'owning' their success, clearly becoming uncomfortable when discussing the matter, and preferring to tell a story of hard work and a good memory than of ability. The ways in which their narratives drew on common discourses of women and mathematics suggested that balancing success in mathematics with their feminine identities was problematic for these students. This finding was reinforced by observations of some of their lectures: for example, on one occasion when one of the female students correctly answered a lecturer's question, she endured a negative reaction from the group. The conclusion to Shannon's (2004) article was also pertinent in relation to this study:

> I see encouraging girls into advanced mathematics as a noble goal. However, 'fixing' mathematics so that girls are not alienated by its current pedagogical structure is, in my opinion, a much more important and critical goal to be addressed. Until girls *enjoy* mathematics, I believe that we will find it difficult to justify to them that it is socially relevant and a way of improving their quality of life. (Shannon, 2004, p. 515)

Summary

The salient lesson from these studies is that it is important not to dichotomise girls and boys or to think that all girls will be interested or disinterested in mathematics and particular contexts and processes within mathematics lessons.

CONCLUSION

A very wide range of studies is reviewed here, and it is hard to make generalisations about them, but there are a number of trends worth noting. Internationally, over the past decade, Australia and New Zealand have had a very good record with respect to gender equity in mathematics. However the most recent finding of gender differences favouring males in the 2006 PISA survey on mathematical literacy, and the persistent finding in a number of studies of gender inequalities in the distribution of achievement results, with boys more often among the highest achievers, put this reputation in jeopardy. Boys also continue to be more likely to enrol in higher-level mathematics subjects in senior secondary mathematics. An interesting finding to emerge from the affective studies was that while the old gender stereotypes around mathematics appear to have broken down, there remains a difference in the ways that male and female students respond to their own mathematical experiences. In a number of studies boys reported higher levels of enjoyment, interest and self-efficacy in mathematics than did girls, and boys more highly valued the use of technology in mathematics. A number of studies reviewed in this chapter confirm or support the argument that there is a positive relationship between these affective factors and enrolment in more advanced mathematics subjects in senior secondary mathematics.

Attention to gender equity in the achievement and affect of particular groups of girls and boys by researchers was not as prevalent in the period of the current review as was evident in previous reviews. However the studies show that indigenous students remain particularly disadvantaged in mathematics, and that respectful and engaging teaching approaches are needed for indigenous and Maori students, especially indigenous girls.

Studies of classroom settings, pedagogical practices, and students' mathematical identities revealed that teachers should not assume that particular gender-preferred or stereotyped interventions, teaching approaches, contexts and tools will necessarily improve the performance of, or be preferred by, all girls or boys in their classrooms. Collaborative approaches and teaching from a "care" perspective have broader appeal than their "girl-friendly" pedagogical origins. Yet further research is needed on suitable pedagogies for high achieving girls and on the curriculum and pedagogical practices in senior secondary mathematics to advance gender equity in mathematics.

In the majority of studies reviewed, there was no explicit theoretical framework. In terms of numbers of papers, the majority was located within an implicit liberal feminist framework. When differences were looked for, these differences were understood to be located within individuals (relatively independently of context), and recommendations centred on finding ways of eradicating the differences. There remained an assumption that when girls were opting out of mathematics, the solution lay in persuading them of the importance of continuing with the subject. Having said this, the influence of radical feminist or post-modern perspectives was increasingly in evidence, not just in those few studies in which these frameworks were explicitly adopted. A number of researchers were looking at the context and suggesting ways of widening the appeal of mathematics, even if they remained located within a perspective that essentialised the sexes and adopted a somewhat deficit view of women.

REFERENCES

Anthony, G. & Walshaw, M. (2007). *Effective pedagogy in mathematics/pāngarau: Best evidence synthesis iteration (BES).* Wellington, NZ: Ministry of Education.

Atweh, B. (2004). Towards a model of social justice in mathematics education and its application to critique of international collaborations. In I. Putt, R. Faragher & M. McLean (Eds.), *Mathematics education for the third millennium: Towards 2010* (Proceedings of the 27th annual conference of MERGA, Townsville, Vol. 1, pp. 47-54). Pymble, NSW: MERGA.

Banfield, T. (2005). Ability grouping for mathematically gifted adolescent boys. *International Education Journal, 6*(2), 141-149.

Barkatsas, A. (2005). A new scale for monitoring students' attitudes to learning mathematics with technology (MTAS). In P. Clarkson, A. Downton, D. Gronn, M. Horne, A. McDonough, R. Pierce & A. Roche (Eds.), *Building connections: Theory, research and practice* (Proceedings of the 28th annual conference of MERGA, Melbourne, Vol. 1, pp. 129-136). Sydney: MERGA.

Barnes, M. (1988). Research on gender and mathematics: An annotated bibliography of Australian research, 1984-1987. In D. Blane & G. C. Leder (Eds.), *Mathematics education in Australia: A selection of recent research* (pp. 22-28). Melbourne: MERGA.

Barnes, M. (2005). Exploring how power is enacted in small groups. In H. Chick & J. Vincent (Eds.), *Proceedings of the 29th conference of the International Group for the Psychology of Mathematics Education* (Vol. 2, pp. 137–144). Melbourne: University of Melbourne.

Barnes, M. & Horne, M. (1996). Gender and mathematics. In B. Atweh, K. Owens & P. Sullivan (Eds.), *Research in mathematics education in Australasia 1992-1995* (pp. 51-87). Sydney: MERGA.

Bartholomew, H. (2004). Equity and empowerment in mathematics: Some tensions from the secondary classroom. In I. Putt, R. Faragher & M. McLean (Eds.), *Mathematics education for the third millennium: Towards 2010* (Proceedings of the 27th annual conference of MERGA, Townsville, Vol. 1, pp. 71-78). Pymble, NSW: MERGA.

Belenky, M. F., Clinchy, B. M., Goldberger, N. R. & Tarule, J. M. (1986). *Women's ways of knowing: The development of self, voice and mind.* New York: Basic Books.

Bishop, A. J. & Forgasz, H. J. (2007). Issues in access and equity in mathematics education. In F. Lester (Ed.), *Second handbook of research in mathematics teaching and learning* (Vol. 2, pp. 1145-1167). Reston, VA: NCTM.

Cox P. J., Leder, G. C., & Forgasz, H. J. (2004). Victorian Certificate of Education: Mathematics, science and gender. *Australian Journal of Education, 48*(1), 27-46.

Darragh, L. (2005). *Students' peer learning choices in mathematics education.* Unpublished masters dissertation. University of Auckland, Auckland, New Zealand.

Davis, L., Clarke, D. & van den Heuvel-Panhuizen, M. (2005). How unusual is the gender specificity of mathematical test item types reported for Dutch primary school students? In P. Clarkson, A. Downton, D. Gronn, M. Horne, A. McDonough, R. Pierce & A. Roche (Eds.), *Building connections: Theory, research and practice* (Proceedings of the 28th annual conference of MERGA, Melbourne, Vol. 1, pp. 281-288). Sydney: MERGA.

Fennema, E. (1995). Mathematics, gender and research. In B. Grevholm & G. Hanna (Eds.), *Gender and mathematics education. An ICMI study in Stiftsgarden, Akersberg, Hoor, Sweden, 1993* (pp. 21-38). Lund: Lund University Press.

Forgasz, H. J. (2004). Equity and computers for mathematics learning: Access and attitudes. In M. J. Johnsen Høines & A. B. Fuglestad (Eds.), *Proceedings of the 28th conference of the International Group for the Psychology of Mathematics Education* (Vol. 2, pp. 399–406). Bergen, Norway: Bergen University College.

Forgasz, H. J. (2005). Why study grade 11 mathematics: What have computers got to do with it? In M. Goos, C. Kanes, & R. Brown (Eds.), *Mathematics education and society. Proceedings of the 4th International Mathematics Education and Society conference* (pp. 166-175). Gold Coast, Qld: Centre for Learning Research, Griffith University.

Forgasz, H. J. (2006). Australian year 12 "intermediate" level mathematics enrolments 2000-2004: Trends and patterns. In P. Grootenboer, R. Zevenbergen, & M. Chinnappan (Eds.), *Identities, cultures and learning spaces* (Proceedings of the 29th annual conference of MERGA, Canberra, Vol. 1, pp. 211-220). Adelaide: MERGA.

Forgasz, H. J., Griffith, S., & Tan, H. (2006). Gender, equity, teachers, students and technology use in secondary mathematics classrooms. In C. Hoyles, J. Lagrange, L. H. Son, & N. Sinclair (Eds.), *Proceedings of the seventeenth ICMI study: Technology revisited, Hanoi, 3-8 December 2006.* Hanoi: ICMI. [Available on CD-ROM.]

Forgasz, H. J., Leder, G. C., & Kloosterman, P. (2004). New perspectives on the gender stereotyping of mathematics. *Mathematical Thinking and Learning, 6*(4), 389-420.

Forgasz, H., Leder, G.. & Vale, C. (2000). Gender and mathematics: Changing perspectives. In K. Owens & J. Mousley (Eds.), *Mathematics education research in Australasia: 1996-1999* (pp. 305 – 340). Turramurra, NSW: MERGA.

Gilligan, C. (1982). *In a different voice.* Cambridge, MA: Harvard University Press.

Horne, M. (2004). Early gender differences. In M. J. Johnsen Høines & A. B. Fuglestad (Eds.), *Proceedings of the 28th conference of the International Group for the Psychology of Mathematics Education* (Vol. 3, pp. 65-72). Bergen, Norway: Bergen University College.

Hirschmann, N. (2003). *The subject of liberty: Toward a feminist theory of freedom.* Princeton, NJ: Princeton University Press.

Kaiser, G., & Rogers, P. (1995). Introduction: Equity in mathematics education. In P. Rogers & G. Kaiser (Eds.), *Equity in mathematics education: Influences of feminism and culture* (pp. 1-10). London: Falmer Press.

Klein, M. (2006). Engaging pedagogies in early mathematics education: Fostering autonomy or the cruellest regulation? *Proceedings of the 2006 annual conference of the Australian Association of Research in Education, Adelaide.* Retrieved 1 November, 2007 from: http://ww.aare.edu.au/06pap/code06.htm#K

Jungwirth, H. (2003). What is a gender-sensitive mathematics classroom? In L. Burton (Ed.), *Which way social justice in mathematics education?* (pp. 3-26). Westport, CT: Praeger.

Leder, G. C. (1984). Girls and mathematics: An annotated bibliography of Australian research. In J. Briggs (Ed.), *Summary of research in mathematics education in Australia* (pp. 63-81). Brisbane: MERGA.

Leder, G. C. (2004). Gender differences among gifted students: Contemporary views. *High Ability Studies, 15*(1), 103-108.

Leder, G. C. (2006). Catering for individual differences: Lessons learnt from the Australian Mathematics Competition. In P. Grootenboer, R. Zevenbergen & M. Chinnappan (Eds.), *Identities, cultures and learning spaces* (Proceedings of the 29th annual conference of MERGA, Canberra, Vol. 2, pp. 336-343). Adelaide: MERGA.

Leder, G. C., & Forgasz, H. J. (1992). Gender: A critical variable in mathematics education. In B. Atweh & J. Watson (Eds.), *Research in mathematics education in Australasia 1988-1991* (pp. 67-95). Brisbane: MERGA.

Lowrie, T. & Diezmann, C. (2005). Fourth-grade students' performance on graphical languages in mathematics. In H. L. Chick & J. L. Vincent (Eds.), *Proceedings of the 30th annual conference of the International Group for the Psychology of Mathematics Education* (Vol. 3, pp. 265–272). Melbourne: University of Melbourne.

Martino, W., Mills, M. & Lingard, B. (2005). Interrogating single-sex classes as a strategy for addressing boys' educational and social needs. *Oxford Review of Education, 31*(2), 237-254.

Meaney, T. (2005). The use of algebra in senior high school students' justifications. In P. Clarkson, A. Downton, D. Gronn, M. Horne, A. McDonough, R. Pierce, & A. Roche (Eds.), *Building connections: Theory, research and practice* (Proceedings of the 28th annual conference of MERGA, Melbourne, Vol. 2, pp. 537-544). Sydney: MERGA.

Miller-Reilly, B. (2006). *Affective change in adult students in second chance mathematics courses: Three different teaching approaches.* Unpublished doctoral dissertation. University of Auckland, Auckland, New Zealand.

Ministry of Education. (2004a). *Mathematics and science achievement in New Zealand. First results from the Trends in International Mathematics and Science Study (TIMSS) 2002-2003 for year 9 students.* Wellington: Ministry of Education.

Ministry of Education. (2004b). *Mathematics and science achievement in New Zealand. First results from the Trends in International Mathematics and Science Study (TIMSS) 2002-2003 for year 5 students.* Wellington: Ministry of Education.

Ministry of Education. (2006). *Achievements in mathematics cognitive domains: Trends in International Mathematics and Science Study (TIMSS) 2002-2003.* Wellington: Ministry of Education.

Mulholland, J., Hansen, P., & Kaminski, E. (2004). Do single-gender classrooms in coeducational settings address boys' underachievement? An Australian study. *Educational Studies, 30*(1), 19-32.

Norton, S. (2005). Mathematics and the construction of feminine gender identity. In P. Clarkson, A. Downton, D. Gronn, M. Horne, A. McDonough, R. Pierce, & A. Roche (Eds.), *Building connections: Research, theory and practice* (Proceedings of the 28th annual conference of MERGA, Melbourne, Vol. 2, pp. 585-592). Melbourne: MERGA.

Norton, S. (2006). Pedagogies for the engagement of girls in the learning of proportional reasoning through technology practice. *Mathematics Education Research Journal, 18*(3), 69-99.

OECD. (2007). *PISA 2006 science competencies for tomorrow's world* (Vol. 2). New York: Program for International Student Assessment, OECD.

Pierce, R., Stacey, K. & Barkatsas, A. (2007). A scale for monitoring students' attitudes to learning mathematics with technology. *Computers and Education, 48*, 285-300.

Rodd, M., & Bartholomew, H. (2006). Invisible and special: Young women's experiences as undergraduate mathematics students. *Gender and Education, 18*(1), 35-50.

Rossi Becker, J. (1995). Women's ways of knowing in mathematics. In P. Rogers & G. Kaiser (Eds.), *Equity in mathematics education: Influences of feminism and culture* (pp. 163-174). London: The Falmer Press.

Shannon, F. (2004). Classics counts over calculus: A case study. In I. Putt, R. Faragher, & M. McLean (Eds.), *Mathematics education for the third millennium: Towards 2010* (Proceedings of the 27th annual conference of MERGA, Townsville, Vol. 2, pp. 509-516). Pymble, NSW: MERGA.

Thomson, S., Cresswell, J., & De Bortolli, L. (2004). *Facing the future: A focus on mathematical literacy among Australian 15-year-old students in PISA 2003*. Melbourne: ACER.

Thomson, S., & De Bortolli, L. (2008). *Exploring scientific literacy: How Australia measures up (The PISA 2006 survey of students' scientific, reading and mathematical literacy skills)*. Melbourne: ACER.

Thomson, S., & Fleming, N. (2004). *Summing it up: Mathematics achievement in Australian schools in TIMSS 200.* (TIMSS Australia Monograph No 6). Melbourne: ACER.

Thomson, S., McKelvie, P., & Murnane, H. (2006). *Achievement of Australia's early secondary indigenous students: Findings from TIMSS 2003* (TIMSS Australia Monograph No 10). Melbourne: ACER.

Vale, C., & Leder, G. (2004). Student views of computer based mathematics in the middle years: Does gender make a difference? *Educational Studies in Mathematics, 56*(3), 287-312.

Vale, C. (2006). Equity and technology: A case study. In P. Grootenboer, R. Zevenbergen, & M. Chinnappan (Eds.), *Identities, cultures and learning spaces* (Proceedings of the 29th annual conference of MERGA, Canberra, Vol. 2, pp. 512-518). Adelaide: MERGA.

Vale, C., Forgasz, H., & Horne, M. (2004). Gender and mathematics. In B. Perry, G. Anthony, & C. Diezmann (Eds.), *Research in mathematics education in Australasia: 2000-2003* (pp. 75-100). Flaxton, Qld: Post Pressed.

Walkerdine, V. (1989). *Counting girls out*. London: Virago.

Watt, H. M. G. (2004). Development of adolescents' self-perceptions, values, and task perceptions according to gender and domain in 7th- through 11th-grade Australian students. *Child Development, 75*(5), 1556–1574.

Watt, H. M. G. (2005a). Explaining gendered math enrolments for NSW Australian secondary students. *New Directions for Child and Adolescent Development, 110*, 15-29.

Watt, H.M.G. (2005b). Exploring adolescent motivations for pursuing maths-related careers. *Australian Journal of Educational and Developmental Psychology, 5*, 107-116.

Watt, H. M. G. (2006). The role of motivation in gendered educational and occupational trajectories related to maths. *Educational Research and Evaluation, 12*(4), 305-322.

Watt, H. M. G., Eccles, J. S., & Durik, A. M. (2006). The leaky mathematics pipeline for girls: A motivational analysis of high school enrolments in Australia and the USA. *Equal Opportunities International, 25*(8), 642-659.

Young-Loveridge, J., & Taylor, M. (2005). Children's views about mathematics learning after participation in a numeracy initiative. *Research in Education, 74*, 83-90.

AFFILIATIONS

Colleen Vale
School of Education,
Victoria University, Australia

Hannah Bartholomew
Department of Mathematics,
University of Auckland, New Zealand

MERRILYN GOOS, TRACEY SMITH, & STEVE THORNTON

RESEARCH ON THE PRE-SERVICE EDUCATION OF TEACHERS OF MATHEMATICS

This review is organised around four conceptual themes reflecting current research perspectives on mathematics teacher education: (1) pre-service teacher learning as growth of mathematical and pedagogical content knowledge (learning as *knowing*); (2) the role of beliefs and affects in pre-service teacher learning (learning as *believing* and *feeling*); (3) pre-service teacher learning as a sociocultural activity (learning as *becoming*); (4) teacher educators researching and reflecting on their own practice (learning as *inquiring*). Recommendations for advancing research in this field address the need for larger scale multidimensional studies, integration and synthesis of existing studies, and for teacher educators to develop a more critical stance in researching their own practices.

Key words: affect, beliefs, reflection on practice, sociocultural perspectives, teacher educators, teacher knowledge

INTRODUCTION

This chapter provides a critical review of Australasian research, published or presented in the period from 2004 to 2007, on the pre-service education of teachers of mathematics at all levels of schooling. To ensure that the chapter is both comprehensive in coverage and conceptual in focus we organise our review around current research perspectives on mathematics teacher education. Lerman (2001) identifies two main theoretical perspectives within this research: one perspective grounded in theories of teacher beliefs and knowledge, and another that conceptualises learning to teach as increasing participation in sociocultural activity. Studies of pre-service teachers' knowledge and beliefs also figure prominently in Llinares and Krainer's (2006) review of thirty years of teacher education research presented at Psychology of Mathematics Education (PME) conferences. Their review introduces an additional perspective by considering research on the learning and development of mathematics teacher educators. These international reviews suggest the following framework for our analysis of Australasian research:

- Pre-service teacher learning as growth of mathematical and pedagogical content knowledge (learning as *knowing*);
- The role of beliefs and affects in pre-service teacher learning (learning as *believing* and *feeling*);

H. Forgasz et al. (eds.), Research in Mathematics Education in Australasia 2004-2007, 291-311.

- Pre-service teacher learning as a sociocultural activity (learning as *becoming*);
- Teacher educators researching and reflecting on their own practice (learning as *inquiring*).

The conclusion to the chapter aligns our synthesis and critique of studies with the issues identified by Adler, Ball, Krainer, Lin, and Novotna (2005) in their analysis of the state of research on mathematics teacher education, and uses these issues to frame a set of recommendations for advancing research in this field.

We begin with some background on the structure of pre-service teacher education programs in Australia and New Zealand.

PRE-SERVICE TEACHER EDUCATION PROGRAMS IN AUSTRALASIA

There is no single system of teacher education in Australia because education is largely the responsibility of the individual states and territories. This has led to many differences across the nation in the way school education is organised and teachers are prepared. Nevertheless, it is possible to make some general statements about pre-service teacher education.

Pre-service teacher education programs usually require four years of full-time university study and include at least one year of professional education studies. This requirement may be fulfilled via either a four year Bachelor of Education degree or a three year undergraduate degree followed by a one year Graduate Diploma in Education.

The content of teacher education programs includes:

- professional studies in education, teaching and learning, and curriculum;
- discipline studies to provide the necessary subject matter knowledge (may occur in the initial undergraduate degree or the four year BEd);
- field experience (practicum).

The preparation of secondary teachers is distinguished from that of primary teachers by the requirement to specialise in one or two curriculum areas. Whereas in primary schools each class group is taught almost all subjects by only one teacher (English, mathematics, science, social studies etc; although there is some use of specialist teachers in subjects such as music, languages, and physical education), secondary school classes are organised around subjects taught by teachers who are meant to have specialised in those areas. Some institutions also offer a Bachelor of Education specialising in early childhood that enables graduates to teach in the early years and primary schools.

Initial teacher education programs in New Zealand are similar in design to those in Australia. However, some undergraduate teacher education programs for primary school are available as a three year option.

PRE-SERVICE TEACHER LEARNING AS GROWTH OF MATHEMATICAL AND PEDAGOGICAL CONTENT KNOWLEDGE

In their review of mathematics teacher education research, Llinares and Krainer (2006) note that both subject matter knowledge and pedagogical content knowledge (Shulman, 1986, 1987) are assumed to be required for effective teaching. Previous research has identified apparent weaknesses in primary pre-service teachers' mathematical knowledge, and recent Australasian studies continue to investigate this issue. Southwell and Penglase (2005) tested 78 primary pre-service teachers who were taking an extra mathematics course to address gaps in their knowledge. As well as finding difficulties with decimals and place value concepts (e.g., *Calculate* 14.83×0.06; 26.9% of students gave correct response), calculation with fractions (e.g., *Find* $\frac{4}{5}+\frac{2}{3}$; 57.7% correct), and conversion between metric units (e.g., *Convert* 6.23 *km to m*; 75.6% correct), the study also noted that mathematics-specific language was a major issue for these pre-service teachers. These findings should be interpreted with caution, because by enrolling in the extra mathematics course the participants had already identified themselves as lacking adequate mathematical knowledge and skills, and the test merely confirmed this. However, other studies involving a more representative sample of pre-service teachers have produced similar results.

Mays (2005) used a test made up of five mental computation items and thirty items from the 1999 TIMSS (Trends in International Mathematics and Science Study) test to assess the mathematical knowledge of 159 primary pre-service teachers taking their first university course. She found that only 10% of students demonstrated mastery by obtaining at least 80% on the initial test. Question types that caused the most difficulty included mental computation, calculation with fractions, decimals and place value, ratios, conversion between metric units, use of algebraic rules to extend patterns, and probability. Appropriate use of problem solving strategies was another area of weakness identified by the test (e.g., misreading the question, not checking work, making arithmetic errors). Analysis of common errors (e.g., adding numerators and denominators when adding fractions, writing $n \times n \times n$ as $3n$ instead of n^3) assisted the teacher education staff in designing workshops to address students' mathematical misconceptions. Afamasaga-Fuata'i, Meyer, and Falo (2007) administered the same test to 140 Samoan primary pre-service teachers and used Rasch analysis to identify their main misconceptions. Although these researchers did not directly compare their findings with those obtained by Mays, it appears that the same question types were handled poorly by both Australian and Samoan students. Other studies by Sharma (2005) and Zevenbergen (2005) reported comparable problems with pre-service primary teachers' understanding of statistical variation and measurement concepts.

Collectively, the studies suggest that the mathematical knowledge of many pre-service primary teachers tends to be fragmented with a focus on procedures. Zevenbergen (2005) found that even when presented with a course that promoted approaches such as estimation and drawing a diagram, the pre-service teachers' reliance on algorithmic approaches proved resistant to change. This resistance was heightened by practicum experiences that also promoted algorithmic thinking, creating a disconnection between on-campus learning and experiences in schools that has been the subject of investigation in many other studies of pre-service teacher education (see studies discussed in sections 3 and 4 of this chapter).

Complementing the descriptive research summarised above are studies that offer strategies for improving pre-service teachers' mathematical knowledge. Ryan and McCrae (2005/2006) developed and trialled a test for first year teacher education students covering the strands of Number, Measurement, Space and Shape, Chance and Data, Algebra, and Reasoning and Proof, and a Rasch analysis of the results produced individual profiles that identified patterns of errors. The researchers suggest these profiles could be used by teacher educators to provide diagnostic feedback on misconceptions and errors, and by pre-service teachers as the basis for examining their own areas of misunderstanding. Zevenbergen (2004) reported on a more deliberate intervention designed to improve primary pre-service teachers' mathematical knowledge through the use of study groups. Although there was some initial resistance, positive learning outcomes included improved knowledge of specific content, changes in intended teaching strategies used during the practicum, and a developing sense of power in teaching something to someone else. The study groups had cognitive, social and affective benefits.

Despite continuing interest in researching the subject matter knowledge of pre-service teachers, few studies explicitly attend to the methodological and conceptual challenges in developing valid measures of this type of knowledge. From a methodological perspective, can we be sure that errors on written tests arise from lack of understanding rather than insufficient warning or time to prepare for the tests? Do correct answers indicate procedural rather than conceptual knowledge? More importantly, how does pre-service teachers' subject matter knowledge contribute to student achievement? The work of Ball and colleagues (e.g., Ball, 1990; Ball & Bass, 2000; Ball, Lubienski, & Mewborn, 2001) has reframed this question by suggesting that

> … at least in mathematics, *how* teachers hold knowledge may matter more than *how much* knowledge they hold. In other words, teaching quality might not relate so much to performance on standard tests of mathematics achievement as it does to whether teachers' knowledge is procedural or conceptual, whether it is connected to big ideas or isolated into small bits, or whether it is compressed or conceptually unpacked (Hill & Ball, 2004, p. 332, original emphasis).

Ball argues that research should be concerned with what is required to know *mathematics for teaching*. Rather than a single focus on mathematical knowledge, studies that focus on *pedagogical* content knowledge are concerned with the most useful ways of representing and formulating mathematics that make it comprehensible to others (Shulman, 1986).

In pre-service teacher education, prospective primary school teachers are the main target for research on pedagogical content knowledge. Beswick (2005) analysed responses to an examination question at the end of a mathematics curriculum course, which asked pre-service teachers to indicate agreement or otherwise with a statement about the feasibility of helping children to achieve relational understanding in mathematics. Her analysis identified eight categories of misunderstanding about instrumental versus relational understanding (Skemp,

1978) that were displayed by about 30% of the pre-service teachers. Beswick concluded that research is needed to investigate the effectiveness of pre-service courses constrained by non-ideal circumstances (large classes, limited face to face contact time), and to increase integration of on-campus coursework and school experience for pre-service teachers. Verenikina and Chinnapan (2006) investigated primary pre-service teachers' understanding of scaffolding, an important pedagogical strategy for improving students' learning. Analysis of survey responses revealed that while participants frequently mentioned aspects of scaffolding such as the active positioning of the child, the existence of joint activity, and the segmentation of tasks into smaller components, they showed little understanding of the tool-mediated nature of teacher-child interaction or of links between the concepts of scaffolding and the Zone of Proximal Development. Each of these studies illustrates the challenges for teacher educators in developing pre-service teachers' understanding of abstract pedagogical concepts within the constraints of a university course.

Most studies of pedagogical content knowledge reviewed for this chapter aimed to bring about change rather than simply report on the extent of pre-service teachers' knowledge, and these studies illustrate a range of theoretical approaches. Chambers and Stacey (2006) describe two multimedia resources they developed to help pre-service teachers examine children's thinking and project themselves into their future role as teachers. The advantage of such resources over 'live' observation of classrooms is that they capture complex behaviours that can be reviewed on demand. The resources highlighted the value of research to inform teaching in pre-service teacher education. McDougall, McRobbie, and Nason (2004) used a different type of computer-based resource, a Knowledge Forum environment, to link ten pre-service teachers with two experts to collaboratively develop lesson plans. Analysis of archived contributions to the electronic discussion showed high levels of participation and found evidence of social development of knowledge that the researchers claimed was enhanced by the Computer Supported Collaborative Learning (CSCL) environment.

Other researchers investigated the development of pre-service teachers' pedagogical content knowledge in and through professional practice, such as during practicum placements. Vale and Davies (2005) describe a project that constructed partnerships between primary pre-service teachers and Year 5/6 teachers. They looked for evidence of knowledge growth and reflection in a range of data sources: reflective journals, videotape records, and evaluation questionnaires completed by the pre-service teachers, university staff, school teachers, and children. The partnerships challenged the pre-service teachers' ideas about what children were capable of doing and led to surprising insights into mathematical ideas. Case studies of two pre-service teachers illustrated interesting contrasts in the outcomes. The more confident pre-service teacher experienced contradictory thoughts in rejecting teaching by telling (e.g., when her students discovered facts about multiples) while still recommending that the students be grouped by skill level. On the other hand, the pre-service teacher who was less confident gave authority to students' sense making and showed openness in her own learning. The partnerships produced insightful discussions incorporating practical wisdom and brought about generative practices. A similar approach was adopted by Bobis (2007), who reports on a program in which teams of pre-service teachers work with groups of students in an alternating program of university study and practical experience. The situated learning, co-teaching, and embedded assessment incorporated in the program empowered pre-service primary teachers to effectively translate theory into practice. During the experience, the pre-service teachers improved their pedagogical content knowledge, began to ask higher order questions, and oriented their activities increasingly towards significant mathematics.

Frid and Sparrow (2004) used portfolios to support reflective practice in two cohorts of primary pre-service teachers. Four aspects of the portfolio process emerged as integral to engendering reflective practice: development of a personal mathematics teaching philosophy, submission of draft portfolio items for feedback, lecturer and peer mentoring, and formal assessment interviews. The effect of interaction was powerful, especially for weak pre-service teachers. Portfolios were recommended as tools for addressing lack of confidence and pedagogical knowledge.

While each of these studies contributes to an already substantial research base on pre-service teacher knowledge, few studies seem to build on an existing body of evidence or to accumulate evidence of knowledge growth over time. Almost all of the studies reviewed here involved prospective primary school teachers. Although pre-service teachers who choose to specialise in secondary mathematics may have more opportunities than primary pre-service teachers to build strong subject matter knowledge, the development of pedagogical content knowledge is surely relevant to mathematics teachers (and teacher educators) at all levels of schooling and deserves to be a focus for research in secondary mathematics teacher education. Investigating pedagogical content knowledge is methodologically challenging because the most authentic, yet unpredictable, demonstration of this type of knowledge occurs during real-life classroom teaching situations and is thus different for each teacher and each mathematical topic. However, Chick and colleagues' research with practising teachers showed that a questionnaire and interview methodology was effective for revealing significant aspects of their pedagogical content knowledge in relation to a range of mathematical topics (Chick & Baker, 2005; Chick, Baker, Pham, & Cheng, 2006), and this is an approach that may be useful in future research with pre-service teachers.

THE ROLE OF BELIEFS AND AFFECTS IN PRE-SERVICE TEACHER LEARNING

Research on teachers' beliefs about mathematics, mathematics learning and teaching, and their attitudes towards mathematics often investigates relationships between beliefs, affects, mathematical knowledge, and teachers' classroom practices. There is generally acknowledgment that pre-service teachers' beliefs and attitudes can influence their approaches to teaching. It is also accepted that pre-service teachers may enter the university program with mathematical beliefs that are inconsistent with current models of curriculum and pedagogy and resistant to change. Prescott and Cavanagh (2006) interviewed sixteen secondary mathematics pre-service teachers immediately before starting their university program or in the first week of classes in order to discover their beliefs about the qualities of a good teacher and good teaching. The interviews asked about memories of their own experiences as a school student, their beliefs about the teaching of mathematics, and their reasons for wanting to become a secondary mathematics teacher. The researchers found that the beliefs of these pre-service teachers – all of whom had been high achievers in school and university mathematics – were based largely on their personal experience of learning mathematics at school and memories of their favourite teachers. For them, a good mathematics lesson involved listening to the teacher's careful explanation of well chosen examples and completion of many graded exercises to practise their skills, while good teachers gave clear explanations, made mathematics relevant by using real life examples, and managed the classroom so students could concentrate on learning. As these pre-service teachers had not yet begun their university coursework or had any practicum experiences it is not surprising that their beliefs were largely influenced by interpretations of how their own mathematics teachers had taught them.

Scott (2005a, 2005b, 2006) compared the beliefs of two groups of pre-service teachers who were in the beginning and final stages of their university program. She surveyed 163 commencing and 186 graduating primary pre-service teachers and interviewed a smaller sample from each cohort with the aim of determining the teaching strategies they intended to use and the experiences that influenced development of their mathematical beliefs. The theoretical framework for the study identified three types of beliefs and the corresponding sources of influence on their formation: descriptive beliefs derived from personal experiences as learners (cf. Prescott & Cavanagh, 2006); inferential beliefs arising from recent observations of teaching contexts; and informational beliefs formed from pre-service coursework experiences or discussions with friends and family members. Scott found that inferential beliefs seemed to be most significant for these pre-service teachers, so that the experience of observing other teachers during recent practicum placements became more influential than personal learning experience from school days long past or decontextualised university coursework. Scott's comparison of pre-service teachers' intentions in literacy and numeracy also showed that they had a more developed understanding of pedagogy in literacy than in numeracy.

Perry, Way, Southwell, White, and Pattison (2005) used a survey to investigate the beliefs of primary pre-service teachers, and they also tested mathematical knowledge and looked for relationships between beliefs and knowledge. The beliefs espoused by the pre-service teachers were strongly oriented towards constructivist approaches (although there were some contradictory responses), but their performance on the mathematics achievement test was poor. Some statistically significant (but small) correlations were found relating beliefs and achievement. For example, pre-service teachers who believed that mathematics is about searching for pattern and mathematics learning is enhanced by challenge in a supportive environment were more likely to perform well in the test, while those who thought that mathematics problems should be quickly solvable and that right answers are more important than the solution process had lower achievement scores. The research of Prescott and Cavanagh (2006), Scott (2005a, 2005b, 2006), and Perry et al. (2005) is suggestive of a persistent cycle that links school experiences in learning mathematics to the formation of mathematical beliefs, which may be reinforced by the teacher's practicum and initial professional experiences. In an analysis of pre-service primary teachers' imagined classrooms, Brady (2007) found similar evidence of the impact of school experiences on potential practice. While the pre-service teachers in the study recognised the value of both direct teaching, and teaching as facilitation, and described a range of potential strategies to cater for diversity of students, they saw mathematics as a useful collection of skills rather than a unified body of knowledge.

Lack of confidence with subject matter knowledge has been proposed as one possible source of mathematics anxiety, and several recent Australasian studies have investigated, or attempted to remediate, this potential relationship. Southwell, White, Way, and Perry (2005) gave a mathematics attitude survey and an achievement test to primary pre-service teachers enrolled in their first university mathematics course. Factor analysis of the attitude survey responses produced two factors, insecurity and confidence, but only the former was (negatively) correlated with achievement scores. The attitude scale was then divided into two sections, *attitude to teaching mathematics* (expressed by items such as "Generally, I feel secure about the idea of teaching mathematics") and *attitude to myself as a mathematics student* (expressed by items such as "Mathematics makes me feel inadequate"). Factor analysis of both subscales again determined two factors, insecurity and confidence. As for the total attitude scale, the insecurity factor from both subscales was found to be negatively correlated with achievement, but subscale relationships between the confidence factor and achievement were more complex. Confidence was positively correlated with achievement for the *attitude to teaching mathematics* subscale, but unrelated to achievement for the *attitude to myself as a mathematics student* subscale. The researchers suggested that confidence in one's own mathematical ability may be a less convincing predictor of mathematical achievement than confidence in teaching mathematics.

Several studies have attempted to change primary pre-service teachers' attitudes by improving their subject matter knowledge. For example, Hawera (2004, 2006) reports on a university course that aims to address the needs of mathematically anxious Maori pre-service teachers. This study links mathematics anxiety to lack of conceptual understanding, previous school experiences and teaching methods, a belief that the teacher has a purely transmissive role, and a belief that mathematical ability is innate. The course taught by Hawera is based on principles of social constructivism (relevance, interaction, use of manipulatives) and humanistic learning theory (providing a safe learning environment). Some modest but positive changes were observed in the pre-service teachers' views about mathematics, views about how mathematics is learned and taught, conceptual understanding of mathematics, and attitudes and confidence. Uusimaki and Kidman (2004) took a similar approach in working with sixteen self-identified mathematically anxious primary pre-service teachers. They offered workshops designed to develop mathematical knowledge, and administered an online survey before and after the intervention to analyse its effectiveness. Analysis of the survey results showed an increase in positive feelings (comfort, confidence) about mathematics; however it is difficult to judge the grounds for the success of the workshops as few details of their content or the teaching approach are provided.

Wilson and Thornton (2005a, 2005b; Wilson, 2007) took an innovative approach to dealing with the mathematical anxiety and negative self-image of primary pre-service teachers by introducing them to bibliotherapy – a method for overcoming "emotional turmoil related to a real-life problem by reading literature on the topic" (2005a, p. 791) – as a form of reflection that allows for identification with the problem while maintaining distance from it. Pre-service teachers read about children's mathematical anxiety and the reasons why this anxiety arises, and kept reflective journals recording responses to the readings. The bibliotherapy approach helped the pre-service teachers change their self-image as learners of mathematics and reassess their capacity to teach mathematics. This research differs from other types of intervention – such as those described above – that identify the source of pre-service teachers' anxiety as lack of confidence in their own subject matter knowledge and then attempt to improve this knowledge.

Even if university coursework is successful in changing pre-service teachers' attitudes towards mathematics, these studies do not always take into account the potential impact of the school-based practicum. Grootenboer's (2005/2006) research is particularly useful in this regard because it investigates the attitudes of prospective primary teachers at three points during the first year of their pre-service program: before the start of the program, after their first mathematics education course, and after their school-based practicum. Analysis of questionnaire responses showed that the students developed more positive attitudes following their coursework, but that these gains were partially reversed after the practicum. Interviews with the pre-service teachers revealed that some enjoyed positive practicum experiences that affirmed the affective change arising during the course. But for around half of the participants, practicum experiences involving streamed classes and teacher and textbook centred approaches conflicted with the type of mathematics teaching approaches promoted by the university course and served to reinforce some of the negative attitudes they reported on entering the course.

Recent Australasian research on mathematical beliefs and attitudes has focused largely on primary pre-service teachers, possibly because it seems that these are the teachers who suffer most from mathematical anxiety and negative attitudes brought about by their own lack of confidence in their mathematical abilities. Mathematical beliefs are certainly an issue for secondary mathematics teachers as well, but most research in this area has been carried out with practising teachers rather than pre-service teachers (e.g., Anderson, White, & Sullivan, 2004). Although the relatively short duration of many secondary pre-service programs (e.g., one year postgraduate diploma) can make it difficult to research changes in beliefs, there is potential for this issue, and other issues related to learning as *becoming*, to be investigated in longitudinal studies that span the transition from pre-service to beginning teaching.

PRE-SERVICE TEACHER LEARNING AS A SOCIOCULTURAL ACTIVITY

Australasian research in this category includes studies that have applied cultural metaphors and sociocultural and post-structuralist theories to the learning of pre-service teachers. A common theme across these studies is the formation and transformation of pre-service teacher identities.

Anderson, Averill, Eaton, and Smith (2005) describe their attempts to develop an inclusive learning community by using traditional Maori cultural activities as an overarching theme for content and pedagogical knowledge development in a pre-service course comprising non-Maori students. The goals were for the pre-service teachers to identify the mathematics in such activities and to understand implications of the Treaty of Waitangi for mathematics pedagogy and curriculum design. The pre-service teachers were found to be more confident with the former than the latter aspect of the course, which led the researchers to conclude that more work needs to be done to connect Maori cultural contexts with the Treaty.

Several Australian and New Zealand researchers have adapted sociocultural theories of learning to investigate identity formation in pre-service teachers, mainly at the secondary school level. The interest in secondary rather than primary pre-service teachers in this category of research (cf. the sections on knowledge, beliefs and affects) might indicate expectations that secondary teachers form a stronger identity as specialist teachers of *mathematics*. One line of inquiry pursued by Goos (2005a, 2005b, 2006) elaborates on the inter-relationships between factors that influence the process of learning to teach, using Valsiner's (1997) concepts of the Zones of Proximal Development (ZPD), Free Movement (ZFM) and Promoted Action (ZPA). In a teacher education context, the ZPD is influenced by the pre-service teacher's knowledge and beliefs. The ZFM represents environmental constraints within the school context, such as student characteristics, curriculum and assessment requirements, and availability of teaching resources, while the ZPA symbolises the efforts of a teacher educator or more experienced colleague to promote particular teaching skills or approaches. Goos (2006) applied this framework to suggest that conditions necessary to promote and sustain professional growth in beginning teachers reside in the interplay between individual teachers' knowledge and beliefs, the school or curricular environment, and available sources of assistance such as mentoring or professional development.

Researchers working with sociocultural perspectives have also drawn on the work of Lave and Wenger (Lave, 1988, 1997; Lave & Wenger, 1991; Wenger, 1998) and the concept of a community of practice to investigate identity formation in pre-service and beginning teachers. Goos and Bennison (2004, 2005, 2006, 2008) theorised their own practice in terms of establishing and maintaining an online community of practice focused on becoming a teacher of secondary school mathematics. Emergence of the online community was associated with the teacher educators' role in facilitating professional dialogue amongst the pre-service teachers, the voluntary and unstructured nature of participation, initial face to face interaction during university classes that created familiarity and trust, and the convenience of communicating via email rather than logging on to a website. The researchers concluded that the emergent design of the community contributed to its sustainability in allowing the pre-service and beginning teachers to define their own professional goals and values.

Smith (2004), working with primary pre-service teachers, used Wenger's social learning theory to describe a process of becoming for teacher educators and pre-service teachers that involves narrative practices. She claims that the use of narrative practices such as the writing of case stories and oral poster presentations led to the development of a community of inquiry that fostered pre-service teachers' identities as inquirers. Similar studies involving narrative practices have been undertaken with primary pre-service teachers using mathematical investigations and journals (Bailey, 2007) and autobiographical writing to address mathematical anxiety (McNaught, 2007). All of these studies reported the transformative potential of the writing process for all participants in terms of capturing changes in beliefs, identifying factors that influence pedagogical practice, and sharing these new understandings more publicly so that they can become sources of learning for others.

Walshaw (2004a) theorises teacher identity in terms of how identities are produced by meanings and myths about teaching. She justifies the use of Lave's social practice theory by arguing that learning to teach is a distinctly social activity. Her study involving secondary pre-service teachers problematised the simplicity of learning to teach and highlighted the complex social relationships that constitute the formation of a teaching identity. From a similar perspective, Cavanagh and Prescott (2007) identified the competing demands that secondary pre-service teachers from two university settings encountered on their practicum placements as they were socialised into the school community of practice. These researchers identified that the major influences for pre-service teachers' classroom practice were the pedagogy of the supervising teacher, the academic ability of pupils, and concerns about classroom management. Hudson (2007) developed a survey instrument – Mentoring for Effective Mathematics Teaching (MEMT) – to benchmark 147 pre-service teachers' perceptions of effective mentoring that may help to address the issue of competing demands during practicum placements. His study identified perceived gaps in mentoring practices used by pre-service primary teachers' mentors (supervising teachers) during practicum experiences. By identifying gaps in mentors' personal attributes, system requirements, pedagogical knowledge, modelling, and feedback, Hudson suggests that well-constructed mentoring programs can be designed and implemented using these data to address these perceived issues and increase the quality of school-based practicum experiences and university-community collaboration.

A symposium presented at the MERGA conference in Canberra in 2006 also made use of sociocultural theories to argue that the concept of identity provided a deeper understanding of what impacts on teaching and learning in mathematics education. The participants presented an analysis of a scenario of a pre-service teacher from various sociocultural perspectives that included Bourdieu's notions of field, habitus, and cultural capital (Zevenbergen, 2006), Wenger's social theory of learning (Smith, 2006a), and feminist post-structuralism (Letts, 2006).

Collectively, the papers in the symposium continued to problematise the notion that learning to teach mathematics can be understood from a single perspective.

Klein (2004a, 2004b) and Walshaw (2004b) make use of post-structuralist theories to foreground concepts of identity and agency in learning to teach, and to confront and interrupt relations of power and the way in which identities are constituted. Klein (2004a) critiques a policy document and explores the positioning of pre-service teachers as more than teachers with skills and knowledge. She argues that teacher educators need to create the space for pre-service teachers to invent themselves as innovative teachers/learners through their own modelling of innovative practices in their university programs. Klein's (2004b) analysis makes visible previously unseen aspects of pedagogical practice in mathematics teacher education, and emphasises the importance of creating a discursive space that unsettles "taken for granted" notions of learners as rational and autonomous agents. Walshaw (2004b) uses Foucault to analyse the "making" of a primary school teacher during a practicum experience by asking: "What does the development of a teaching identity look like?" The paper problematises taking a teacher's identity for granted, and offers a more complex and layered notion of learning to teach that views the pre-service teacher's identity as being constituted through experience. Teaching identity is thus complex and multiple, fractured and fragmented.

Researchers who view learning to teach as a sociocultural activity – as *becoming* – look at "individual teachers through their social settings" (Lerman, 2001, p. 44) to describe connections between knowledge, beliefs, attitudes, and practices. From this perspective, learning to teach is about developing an identity as a person in the profession rather than changing internal knowledge and belief structures. Socioculturally oriented research in teacher education generates theories and frameworks that are related to broader theories of learning and opens up more critical perspectives on teacher education practice. There is potential here for developing a stronger and more coherent theoretical foundation for research in teacher education.

TEACHER EDUCATORS RESEARCHING AND REFLECTING ON THEIR OWN PRACTICE

Research in pre-service teacher education is conducted mostly by teacher educators working with their own students. Some of the studies previously described in this chapter implicitly involve the authors analysing their own practice as teacher educators; however, some who research in this area have begun to write explicitly about issues and dilemmas of teacher education practice, or to report on efforts to systematically evaluate or transform their own practice. This is a relatively new development that is gaining attention in international reviews of teacher education research (e.g., Llinares & Krainer, 2006).

The few Australasian studies carried out in this area typically focus on the teacher educator's own pedagogical practice and reflect their distinctive theoretical positions. For example, Klein (2006) used a post-structuralist lens to investigate her own beliefs and practices in the context of a new one-year postgraduate teacher education program that was designed so graduates could teach across Years 1-10. The program differed from the four year undergraduate degree previously offered in that methods courses for each curriculum area were no longer taught separately, but were condensed into transdisciplinary studies, thus resulting in reduced teaching time for mathematics curriculum. As Klein pondered the question of "what to leave out" of the course (p. 328), the new teacher education environment prompted her to re-examine the epistemological and ontological assumptions that informed her teaching. These assumptions were expressed through the three intersecting elements of her practice: building mathematical and pedagogical content knowledge; addressing the power imbalances in teacher-student social relationships; and pursuing an inquiry approach to mathematics teaching and learning. She argued that pre-service teachers' "ways-of-being" (p. 335) as teachers of mathematics are produced by their previous and current experiences as learners in schools and teacher education courses, and that the task of teacher educators is to make it possible for pre-service teachers to recognise and question the "taken for granted" aspects of these experiences in the process of forming professional identities. Thus the question is not what to leave out of the course, but how to adapt teacher education practice in order to maintain a commitment to producing inquiry-oriented teachers of mathematics.

Lomas (2005) positioned his investigation of teacher educator pedagogies within the discourse of constructivist classroom learning environments. He identified three approaches to constructivism: radical constructivism, which focuses on cognitive development; social constructivism, which is concerned with social and cultural aspects of classrooms; and critical constructivism, where the goal is transformation of educational theory and practice. The pre-service program at his university espouses a critical constructivist perspective; hence the aims of his study were to determine the extent of alignment between (a) the university program goals and teacher educators' practice (as perceived by pre-service teachers) and (b) pre-service teachers' perceptions of the ideal and actual learning environments. Two surveys were used to gather data. Items on the *Questionnaire on Teacher Interactions* (QTI) reflected a social constructivist perspective; pre-service teacher responses revealed strong alignment between this perspective and teacher educator practice, and between their perceptions of the actual and ideal learning environments. Items on the *Classroom Learning Environment Survey* (CLES) reflected a critical constructivist perspective; the results here showed only weak alignment between this perspective and teacher educator practice, and between perceptions of the actual and ideal learning environments. Lomas concluded that teacher educator practice aligned with social constructivism rather than the critical perspective espoused in the program goals, but noted that pre-service teachers were uncomfortable with the critical perspective, even when it was weakly enacted. These findings show that the theory-practice gaps identified in studies of pre-service teachers also exist for teacher educators.

Grootenboer (2006) reflects on his own research in the field of affects and beliefs to propose a set of moral and ethical dilemmas for primary teacher educators working to change the attitudes and beliefs of pre-service teachers. He reveals his own discomfort with the unanticipated role he has taken on as counsellor and therapist to pre-service teachers and contrasts this with his former role as a mathematics teacher, where he assumed his job was simply to present the mathematics to the students he taught. (Grootenboer's reference to therapy as a "dilemma" provides an interesting counterpoint to the work of Wilson and Thornton (2005a), whose intentional use of bibliotherapy suggests they accepted a counselling role as part of their practice as teacher educators.) Grootenboer also challenges teacher educators to consider the ethical question of whose beliefs are "correct", and whether they really have the right to change pre-service teachers' beliefs to meet their own ideological goals. He asks about the consequences for pre-service teachers who might not display "correct" beliefs in their assessment work, or those who feel obliged to espouse certain beliefs in order to satisfy the lecturer and course assessment requirements. A third dilemma he discusses relates to the need for realistic expectations about changing pre-service teacher beliefs, since few research studies have been able to bring about lasting change.

Brandenburg (2004) describes her self-study approach to researching and transforming her practice as a teacher educator. The incentive for change came from her dissatisfaction with the mismatch between her commitment to constructivist teaching approaches and the transmissive practices she realised she had gradually adopted. She developed a roundtable strategy as an alternative to tutorials organised around prescribed readings in order to encourage pre-service teachers to raise their own issues and make sense of their experiences. This had the effect of granting them a "voice" and placing the teacher educator in the new role of co-learner.

Smith (2006b) explicitly describes a methodology of self-study through narrative inquiry as a way of critically examining her pedagogical practice. As was the case with Brandenburg's (2004) study, and the study undertaken by Bailey (2007) cited earlier, Smith noted that this approach paved the way for establishing "an inquiry landscape for co-learning" (p. 473). In particular, making explicit the narrative practice of "noticing, naming and reframing" (p. 473) enhanced teacher learning and teacher identity formation. However, she sounded a note of caution in discussing the moral dimensions of narrative research, since this methodology, although intended to be empowering for pre-service teachers, asks them to publicly share their innermost thoughts, vulnerabilities and uncertainties. Smith reminded teacher education researchers that we need to constantly question and monitor the ethical implications of research we conduct with our own students.

Teacher educators who research their own practice are concerned about critically reflecting on, and improving, their practice. It could be argued that the researchers in this section are "walking the talk" in terms of learning as inquiring and thus, by strategically inquiring into their own pedagogical practices and theorising their work as teacher educators, they make their practices publicly available to the professional community for deliberation, critique and judgement. However, there is a need for more explicit attention to be given to ethical considerations of this type of research, and, indeed, for any research where the participants are the teacher educator's own students. Although there is a strong sense of altruism underpinning the work of teacher-educator-researchers, evident, for example, when they write about "empowering" pre-service teachers or granting them a "voice", the unequal power relationship between the researcher and the researched deserves to be acknowledged and its implications for the validity of such research carefully considered.

CONCLUSION: IMPLICATIONS FOR FUTURE RESEARCH

Adler, Ball, Krainer, Lin, and Novotna (2005) conducted a survey of published research on mathematics teacher education, focusing on the major journals in the field (*Journal of Mathematics Teacher Education, Educational Studies in Mathematics, Journal for Research in Mathematics Education*) as well as international conferences such as PME. Their three main findings were:

- Most research involved small scale qualitative studies.
- Most research is conducted by teacher educators with their own students.
- There is a proliferation of theoretical constructs and frameworks.

One recommendation arising from these findings involved expanding the field through larger scale studies, including longitudinal, cross-national, and policy studies. The challenges here are related to the scale of funding and collaboration needed for such studies, together with strengthening of the quantitative methodology skills that would be required. The field might also be expanded via integration and synthesis of the multitude of published research reports. A second recommendation is to develop a more critical stance in researching our own teacher education practices and programs, by creating "scepticism" and "distance". This might imply subjecting our practice to the scrutiny of colleagues in other institutions and countries. Finally, there is a need to generate theories, constructs, and frameworks that relate to broader theories of learning.

Achieving synthesis and critique, as well as distance and scepticism, requires different forms of *collaboration.* Qualitative case studies might be undertaken via collaboration between researchers across institutions within a country or cross-nationally, while international comparative research could involve cross-national collaboration on large scale studies or collaboration between qualitative and quantitative researchers. A second issue is that these collaborations pre-suppose *research questions*, and we also need to ask how our syntheses can help us to formulate new questions and also to build theories. Some goals of such syntheses might be:

- To build a larger data set – *synthesising across cases* creates a more convincing body of evidence;
- To formulate research questions using an inductive approach – *synthesising across research questions* by starting with localised questions in specific contexts and identifying the broader question of which these are instances;
- To build theories that subsume the plethora of existing concepts and frameworks – *synthesising across theories* in order to link to bigger, more general theories such as constructivist and sociocultural theories of learning;

- To achieve a critical stance in research – *synthesising across researchers* through collaboration with colleagues who, through their critique of our practices, can imbue our research with the distance, scepticism, and ethical awareness that is so difficult to achieve when we only research our own practices with our own pre-service teachers.

Researching pre-service teacher education is a complex process. The analytical framework adopted in this chapter, where pre-service teacher education learning has been described as *knowing, believing, feeling, doing, becoming and inquiring*, may offer possibilities for imagining future research that encompasses all of these dimensions. Such a multidimensional perspective would simultaneously consider the intellectual, social, cultural and emotional dimensions of learning to teach mathematics, and learning to teach teachers of mathematics, that many of the researchers cited in this chapter have suggested are inseparable.

REFERENCES

Adler, J., Ball, D., Krainer, K., Lin, F-L., & Novotna, J. (2005). Reflections on an emerging field: Researching mathematics teacher education. *Educational Studies in Mathematics, 60*, 359-381.

Afamasaga-Fuata'I, K., Meyer, P., & Falo, N. (2007). Primary student teachers' diagnosed mathematical competence in semester one of their studies. In J. Watson & K. Beswick (Eds.), *Mathematics: Essential research, essential practice* (Proceedings of the 30th annual conference of the Mathematics Education Research Group of Australasia, Hobart, Vol. 1, pp. 83-92). Adelaide: MERGA.

Anderson, D., Averill, R., Eaton, H., & Smith, D. (2005). Use of a cultural metaphor in pre-service mathematics teacher education. In P. Clarkson, A. Downton, D. Gronn, M. Horne, A. McDonough, R. Pierce, & A. Roche (Eds.), *Building connections: Theory, research and practice* (Proceedings of the 28th annual conference of the Mathematics Education Research Group of Australasia, Melbourne, Vol. 1, pp. 81-88). Sydney: MERGA.

Anderson, J., White, P., & Sullivan, P. (2004). Using a schematic model to represent influences on, and relationships between, teachers' problem-solving beliefs and practices. *Mathematics Education Research Journal, 17*(2), 9-38.

Bailey, J. (2007). Mathematical investigations: A primary teacher educator's narrative journey of professional awareness. In J. Watson & K. Beswick (Eds.), *Mathematics: Essential research, essential practice* (Proceedings of the 30th annual conference of the Mathematics Education Research Group of Australasia, Hobart, Vol. 1, pp. 103-112). Adelaide: MERGA.

Ball. D. L. (1990). The mathematical understandings that prospective teachers bring to teacher education. *Elementary School Journal, 90*(4), 449-466.

Ball, D. L., & Bass, H. (2000). Interweaving content and pedagogy in teaching and learning to teach: Knowing and doing mathematics. In J. Boaler (Ed.), *Multiple perspectives on the teaching and learning of mathematics* (pp. 83-104). Westport, CT: Ablex.

Ball, D. L., Lubienski, S., & Mewborn, D. (2001). Research on teaching mathematics: The unsolved problem of teachers' mathematical knowledge. In V. Richardson (Ed.), *Handbook of research on teaching* (4th ed.) (pp. 433-456). New York: Macmillan.

Beswick, K. (2005). Pre-service teachers' understandings of relational and instrumental understanding. In H. L. Chick & J. L. Vincent (Eds.), *Proceedings of the 29th annual conference for the Psychology of Mathematics Education* (Vol. 1, pp. 161-168). Melbourne: University of Melbourne.

Bobis J. (2007). Empowered to teach: A practice-based model of teacher education. In J. Watson & K. Beswick (Eds.), *Mathematics: Essential research, essential practice* (Proceedings of the 30th annual conference of the Mathematics Education Research Group of Australasia, Hobart, Vol. 1, pp. 103-112). Adelaide: MERGA.

Brady, K. (2007). Imagined classrooms: Prospective primary teachers visualise their ideal mathematics classroom. In J. Watson & K. Beswick (Eds.), *Mathematics: Essential research, essential practice* (Proceedings of the 30th annual conference of the Mathematics Education Research Group of Australasia, Hobart, Vol. 1, pp. 143-152). Adelaide: MERGA.

Brandenburg, R. (2004). Roundtable reflections: (Re) defining the role of teacher educator and the preservice teacher as 'co-learners'. *Australian Journal of Education, 48*(2), 168-181.

Cavanagh, M., & Prescott, A. (2007). Professional experience in learning to teach secondary mathematics: Incorporating preservice teachers into a community of practice. In J. Watson & K. Beswick (Eds.), *Mathematics: Essential research, essential practice* (Proceedings of the 30th annual conference of the Mathematics Education Research Group of Australasia, Hobart, Vol. 1, pp. 182-191). Adelaide: MERGA.

Chambers, D. P., & Stacey, K. (2006). Developing and using multimedia effectively for undergraduate students. *Australian Journal of Educational Technology, 21*(2), 211-221.

Chick, H., & Baker, M. (2005). Investigating teachers' responses to student misconceptions. In H. L. Chick & J. L. Vincent (Eds.), *Proceedings of the 29th annual conference for the Psychology of Mathematics Education* (Vol. 2, pp. 249-256). Melbourne: University of Melbourne.

Chick, H., Baker, M., Pham, T., & Cheng, H. (2006). Aspects of teachers' pedagogical content knowledge for decimals. In J. Novotná, H. Moraová, M. Krátká, & N. Stehliková (Eds.), *Proceedings of the 30th annual conference for the Psychology of Mathematics Education* (Vol. 2, pp. 297-304). Prague: Charles University.

Frid, S., & Sparrow, L. (2004). Supporting reflection as pre-service primary teachers develop a mathematics teaching portfolio. In I. Putt, R. Faragher, & M. McLean (Eds.), *Mathematics education for the third millennium: Towards 2010* (Proceedings of the 27th annual conference of the Mathematics Education Research Group of Australasia, Townsville, Vol. 1, pp. 239-246). Sydney: MERGA.

Goos, M. (2005a). A sociocultural analysis of learning to teach. In H. L. Chick & J. L. Vincent (Eds.), *Proceedings of the 29th annual conference for the Psychology of Mathematics Education* (Vol. 3, pp. 49-56). Melbourne: University of Melbourne.

Goos, M. (2005b). A sociocultural analysis of the development of pre-service and beginning teachers' pedagogical identities as users of technology. *Journal of Mathematics Teacher Education, 8*, 35-59.

Goos, M. (2006). Licence to thrill or live and let die? *Principal Matters, 68*, 6-8.

Goos, M., & Bennison, A. (2004). Emergence of a pre-service community of practice. In I. Putt, R. Faragher, & M. McLean (Eds.), *Mathematics education for the third millennium: Towards 2010* (Proceedings of the 27th annual conference of the Mathematics Education Research Group of Australasia, Townsville, Vol. 1, pp. 271-278). Sydney: MERGA.

Goos, M., & Bennison, A. (2005). The role of online discussion in building a community of practice for beginning teachers of secondary mathematics. In P. Clarkson, A. Downton, D. Gronn, M. Horne, A. McDonough, R. Pierce, & A. Roche (Eds.), *Building connections: Theory, research and practice* (Proceedings of the 28th annual conference of the Mathematics Education Research Group of Australasia, Melbourne, Vol. 1, pp. 385-392). Sydney: MERGA.

Goos, M., & Bennison, A. (2006). An online community of practice for pre-service and beginning teachers of secondary mathematics. In J. Novotná, H. Moraová, M. Krátká, & N. Stehliková (Eds.), *Proceedings of the 30th annual conference for the Psychology of Mathematics Education* (Vol. 3, pp. 209-216). Prague: Charles University.

Goos, M., & Bennison, A. (2008). Developing a communal identity as beginning teachers of mathematics: Emergence of an online community of practice. *Journal of Mathematics Teacher Education*.

Grootenboer, P. (2006). Mathematics educators: Identity, beliefs, roles and ethical dilemmas. In P. Grootenboer, R. Zevenbergen, & M. Chinnappan (Eds.), *Identities, cultures and learning spaces*

(Proceedings of the 29th annual conference of the Mathematics Education Research Group of Australasia, Canberra, Vol. 1, pp. 270-277). Adelaide: MERGA.

Grootenboer, P. (2005/2006). The impact of the school-based practicum on pre-srvice teachers' affective development in mathematics. *Mathematics Teacher Education and Development, 7*, 18-32.

Hawera, N. (2004). Addressing the needs of mathematically anxious preservice primary teachers. In I. Putt, R. Faragher, & M. McLean (Eds.), *Mathematics education for the third millennium: Towards 2010* (Proceedings of the 27th annual conference of the Mathematics Education Research Group of Australasia, Townsville, Vol. 1, pp. 287-294). Sydney: MERGA.

Hawera, N. (2006). Maori preservice teachers' responses to mathematics investigations. In P. Grootenboer, R. Zevenbergen, & M. Chinnappan (Eds.), *Identities, cultures and learning spaces* (Proceedings of the 29th annual conference of the Mathematics Education Research Group of Australasia, Canberra, Vol. 1, pp. 286-292). Adelaide: MERGA.

Hill, H. C., & Ball, D. L. (2004). Learning mathematics for teaching: Results from California's mathematics professional development institutes. *Journal for Research in Mathematics Education, 35*(5), 330-351.

Hudson, P. (2007). Benchmarking preservice teachers' perceptions of their mentoring for developing mathematics teaching practices. In J. Watson & K. Beswick (Eds.), *Mathematics: Essential research, essential practice* (Proceedings of the 30th annual conference of the Mathematics Education Research Group of Australasia, Hobart, Vol. 1, pp. 412-420). Adelaide: MERGA.

Klein, M. (2004a). Preservice teacher education in mathematics: Thinking innovatively about innovation. In I. Putt, R. Faragher, & M. McLean (Eds.), *Mathematics education for the third millennium: Towards 2010* (Proceedings of the 27th annual conference of the Mathematics Education Research Group of Australasia, Townsville, Vol. 2, pp. 271-278). Sydney: MERGA.

Klein, M. (2004b). The premise and promise of inquiry based mathematics in pre-service teacher education: A poststructuralist analysis. *Asia-Pacific Journal of Teacher Education, 32*(1), 35-47.

Klein, M. (2006). What to leave out when pre-service mathematics education goes from four years to one year: A poststructural account. In P. Grootenboer, R. Zevenbergen, & M. Chinnappan (Eds.), *Identities, cultures and learning spaces* (Proceedings of the 29th annual conference of the Mathematics Education Research Group of Australasia, Canberra, Vol. 2, pp. 328-335). Adelaide: MERGA.

Lave, J. (1988). *Cognition in practice: Mind, mathematics and culture in everyday life*. Cambridge, UK: Cambridge University Press.

Lave, J. (1997). The culture of acquisition and the practice of understanding. In D. Kirshner & J. A. Whitson (Eds.), *Situated cognition: Social, semiotic, and psychological perspectives* (pp. 17-35). Mahwah, NJ: Lawrence Erlbaum.

Lave, J., & Wenger, E. (1991). *Situated learning: Legitimate peripheral participation*. New York: Cambridge University Press.

Lerman, S. (2001). A review of research perspectives on mathematics teacher education. In F-L. Lin & T. J. Cooney (Eds.), *Making sense of mathematics teacher education* (pp. 33-52). Dordrecht, The Netherlands: Kluwer.

Letts, W. (2006). Bringing feminist poststructuralism to bear on [mathematics] teacher education. In P. Grootenboer, R. Zevenbergen, & M. Chinnappan (Eds.), *Identities, cultures and learning spaces* (Proceedings of the 29th annual conference of the Mathematics Education Research Group of Australasia, Canberra, Vol. 2, pp. 623-627). Adelaide: MERGA.

Llinares, S., & Krainer, K. (2006). Mathematics (student) teachers and teacher educators as learners. In A. Gutièrrez & P. Boero (Eds.), *Handbook of research on the psychology of mathematics education* (pp. 429-459). Rotterdam: Sense Publishers.

Lomas, G. (2005). How primary pre-service teachers perceive mathematics teacher educators' practice: A case study. In P. Clarkson, A. Downton, D. Gronn, M. Horne, A. McDonough, R. Pierce, & A. Roche (Eds.), *Building connections: Theory, research and practice* (Proceedings of the 28th annual conference of the Mathematics Education Research Group of Australasia, Melbourne, Vol. 2, pp. 491-498). Sydney: MERGA.

Mays, H. (2005). Mathematical knowledge of some entrants to a pre-service education course. In M. Coupland, J. Anderson, & T. Spencer (Eds.), *Making mathematics vital* (Proceedings of the 20th biennial conference of the Australian Association of Mathematics Teachers, pp.186-193). Adelaide: AAMT.

McDougall, M. J., McRobbie, C. J., & Nason, R. (2004). Growth of teacher knowledge: The promise of CSCL, communities of practice. *Proceedings of the 2004 annual conference of the Australian Association for Research in Education, Melbourne*. Retrieved December 13, 2007 from http://www.aare.edu.au/04pap/nas04549.pdf

McNaught, K. (2007). The power of writing for all pre-service mathematics teachers. In J. Watson & K. Beswick (Eds.), *Mathematics: Essential research, essential practice* (Proceedings of the 30th annual conference of the Mathematics Education Research Group of Australasia, Hobart, Vol. 2, pp. 473-482). Adelaide: MERGA.

Perry, B., Way, J., Southwell, B., White, A., & Pattison, J. (2005). Mathematical beliefs and achievements of pre-service primary teachers. In P. Clarkson, A. Downton, D. Gronn, M. Horne, A. McDonough, R. Pierce, & A. Roche (Eds.), *Building connections: Theory, research and practice* (Proceedings of the 28th annual conference of the Mathematics Education Research Group of Australasia, Melbourne, Vol. 2, pp. 625-632). Sydney: MERGA.

Prescott, A., & Cavanagh, M. (2006). An investigation of pre-service secondary mathematics teachers' beliefs as they begin their teacher training. In P. Grootenboer, R. Zevenbergen, & M. Chinnappan (Eds.), Identities, cultures and learning spaces *(Proceedings of the 29th annual conference of the Mathematics Education Research Group of Australasia*, Canberra, Vol, 2, pp. 424-431). Adelaide: MERGA.

Ryan, J., & McCrae, B. (2005/2006). Assessing pre-service teachers' mathematics subject matter knowledge. *Mathematics Teacher Education and Development*, 7, 72-89.

Scott, A. L. (2005a). Pre-service teachers' experiences and the influences on their intentions for teaching primary school mathematics. *Mathematics Education Research Journal*, *17*(3), 62-90.

Scott, A. L. (2005b). Pre-service teachers' intentions to provide good examples and help children replicate them. In P. Clarkson, A. Downton, D. Gronn, M. Horne, A. McDonough, R. Pierce, & A. Roche (Eds.), *Building connections: Theory, research and practice* (Proceedings of the 28th annual conference of the Mathematics Education Research Group of Australasia, Melbourne, Vol. 2, pp. 656-663). Sydney: MERGA.

Scott, A. L. (2006). Implications of preservice teachers' intentions to use particular learning tasks. In P. Grootenboer, R. Zevenbergen, & M. Chinnappan (Eds.), *Identities, cultures and learning spaces* (Proceedings of the 29th annual conference of the Mathematics Education Research Group of Australasia, Canberra, Vol. 2, pp. 463-470). Adelaide: MERGA.

Sharma, S. (2005). Exploring pre-service teachers' reasoning about variability: Implications for research. In P. Clarkson, A. Downton, D. Gronn, M. Horne, A. McDonough, R. Pierce, & A. Roche (Eds.), *Building connections: Theory, research and practice* (Proceedings of the 28th annual conference of the Mathematics Education Research Group of Australasia, Melbourne, Vol. 2, pp. 743-750). Sydney: MERGA.

Shulman, L. S. (1986). Those who understand: Knowledge growth in teaching. *Educational Researcher*, *15*(2), 4-14.

Shulman, L. S. (1987). Knowledge and teaching: Foundations of the new reform. *Harvard Educational Review*, *57*(1), 1-22.

Skemp, R. R. (1978). Relational understanding and instrumental understanding. *Arithmetic Teacher*, *26*(3), 9-15.

Smith, T. (2004). Learning in a community of inquirers: Developing an inquiry stance. In I. Putt, R. Faragher, & M. McLean (Eds.), *Mathematics education for the third millenium: Towards 2010* (Proceedings of the 27th annual conference of the Mathematics Education Research Group of Australasia, Townsville, Vol. 2, pp. 525-532). Sydney: MERGA.

Smith, T. (2006a). Becoming a teacher of mathematics: Wenger's social theory of learning perspective. In P. Grootenboer, R. Zevenbergen, & M. Chinnappan (Eds.), *Identities, cultures and learning spaces* (Proceedings of the 29th annual conference of the Mathematics Education Research Group of Australasia, Canberra, Vol. 2, pp. 619-622). Adelaide: MERGA.

Smith, T. (2006b). Self-study through narrative inquiry: Fostering identity in mathematics teacher education. In P. Grootenboer, R. Zevenbergen, & M. Chinnappan (Eds.), *Identities, cultures and learning spaces* (Proceedings of the 29th annual conference of the Mathematics Education Research Group of Australasia, Canberra, Vol. 2, pp. 471-478). Adelaide: MERGA.

Southwell, B., White, A., Way, J., & Perry, B. (2005). Attitudes versus achievement in pre-service mathematics teacher education. *Proceedings of the 2005 annual conference of the Australian Association for Research in Education, Sydney.* Retrieved April 22, 2007 from http://www.aare.edu.au/05pap/sou05705.pdf

Southwell, B., & Penglase, M. (2005). Mathematical knowledge of pre-service primary teachers. In H. Chick & J. L. Vincent (Eds.), *Proceedings of the 29th annual conference for the Psychology of Mathematics Education* (Vol. 4, pp. 209-216). Melbourne: University of Melbourne.

Uusimaki, L., & Kidman, G. (2004). Reducing maths-anxiety: Results from an online anxiety survey. *Proceedings of the annual conference of the Australian Association for Research in Education, Melbourne.* Retrieved December 11, 2007 from: http://www.aare.edu.au/04pap/kid04997.pdf

Vale, C., & Davies, A. (2005). Glimpses of generative practice: Constructing pre-service teachers' learning in partnership. In P. Clarkson, A. Downton, D. Gronn, M. Horne, A. McDonough, R. Pierce, & A. Roche (Eds.), *Building connections: Theory, research and practice* (Proceedings of the 28th annual conference of the Mathematics Education Research Group of Australasia, Melbourne, Vol. 2, pp. 743-750). Sydney: MERGA.

Valsiner, J. (1997). *Culture and the development of children's action: A theory of human development* (2nd ed.). New York: John Wiley & Sons.

Verenikina, I., & Chinnapan, M. (2006). Scaffolding numeracy: Pre-service teachers' perspective. In P. Grootenboer, R. Zevenbergen, & M. Chinnappan (Eds.), *Identities, cultures and learning spaces* (Proceedings of the 29th annual conference of the Mathematics Education Research Group of Australasia, Canberra, Vol. 2, pp. 519-526). Adelaide: MERGA.

Walshaw, M. (2004a). Becoming knowledgeable in practice: The constitution of secondary teaching identity. In I. Putt, R. Faragher, & M. McLean (Eds.), *Mathematics education for the third millennium: Towards 2010* (Proceedings of the 27th annual conference of the Mathematics Education Research Group of Australasia, Townsville, Vol. 2, pp. 557-564). Sydney: MERGA.

Walshaw, M. (2004b). Pre-service mathematics teaching in the context of schools: An exploration into the constitution of identity. *Journal of Mathematics Teacher Education, 7*(1), 63-86.

Wenger, E. (1998). *Communities of practice: Learning, meaning, and identity.* Cambridge, MA: Cambridge University Press.

Wilson, S., & Thornton, S. (2005a). "I am really not alone in this anxiety": Bibliotherapy and pre-service primary teachers' self image as mathematicians. In P. Clarkson, A. Downton, D. Gronn, M. Horne, A. McDonough, R. Pierce, & A. Roche (Eds.), *Building connections: Theory, research and practice* (Proceedings of the 28th annual conference of the Mathematics Education Research Group of Australasia, Melbourne, Vol. 2, pp. 791-798). Sydney: MERGA.

Wilson, S., & Thornton, S. (2005b). The year in which my love of maths changed: Pre-service primary teachers' self-image as mathematicians. In M. Coupland, J. Anderson, & T. Spencer (Eds.), *Making mathematics vital* (Proceedings of the 20th biennial conference of the Australian Association of Mathematics Teachers, pp. 268-274). Adelaide: AAMT.

Wilson, S. (2007). My struggle with maths may not have been a lonely one: Bibliotherapy in a teacher education number theory unit. In J. Watson & K. Beswick (Eds.), *Mathematics: Essential research, essential practice* (Proceedings of the 30th annual conference of the Mathematics Education Research Group of Australasia, Hobart, Vol. 2, pp. 815-823). Adelaide: MERGA.

Zevenbergen, R. (2004). Study groups as a tool for enhancing preservice students' content knowledge. *Mathematics Teacher Education and Development, 6*, 4-22.

Zevenbergen, R. (2005). Primary preservice teachers' understandings of volume: The impact of course and practicum experiences. *Mathematics Education Research Journal, 17*(1), 3-23.

Zevenbergen, R. (2006). Teacher identity from a Bourdieuian perspective. In P. Grootenboer, R. Zevenbergen, & M. Chinnappan (Eds.), *Identities, cultures and learning spaces* (Proceedings of the 29th annual conference of the Mathematics Education Research Group of Australasia, Canberra, Vol. 2, pp. 616-618). Adelaide: MERGA.

AFFILIATIONS

Merrilyn Goos
The University of Queensland, Australia

Tracey Smith
Charles Sturt University, Australia

Steve Thornton
University of Canberra, Australia

JUDY ANDERSON, JANETTE BOBIS, & JENNI WAY

TEACHERS AS LEARNERS: BUILDING KNOWLEDGE IN AND THROUGH THE PRACTICE OF TEACHING MATHEMATICS

In the past ten years, research surrounding teacher professional learning and development in mathematics education has increased rapidly in both volume and focus. An examination of recent studies emanating from large-scale professional development projects has enabled the identification of key features associated with effective professional development. Similarly, trends from a large number of smaller-scale studies have emerged and helped raise important questions for future research into teacher professional development and learning. The review shows that there is still a great deal to learn about what makes professional development more effective. Recommendations for future research are presented.

Key words: professional development, professional learning, teacher education, teacher knowledge, teacher learning

INTRODUCTION

The current context of political and educational reform agendas places a major focus on the effectiveness of teacher learning and development. A government inquiry in teacher education acknowledged the ever-changing demands on teachers and recommended that teachers “commit to on-going professional learning” and that such learning be formalised as a requirement for registration (Commonwealth of Australia, 2007, p. 97). This recommendation is supported by professional associations (e.g., Australian Association of Mathematics Teachers [AAMT] Standards, 2006) and realised by the growing number of teacher registration institutions incorporating on-going professional learning as part of their requirements for accreditation (e.g., New South Wales [NSW] Institute for Teachers Standards, 2005). Such a political and educational context makes this a timely review.

Recent professional and curriculum documents take care to distinguish between ‘professional development’ and the ‘professional learning’ of teachers (e.g., NSW Department of Education and Training, 2004). In this chapter, ‘professional development’ refers to planned, focused activities and formal programs that teachers undertake to extend their professional learning. ‘Professional learning’, on the other hand, refers to the individual growth of a teacher’s expertise, which may be achieved through participation in professional development experiences, but may also be achieved through experience in classrooms, professional reading,

H. Forgasz et al. (eds.), Research in Mathematics Education in Australasia 2004-2007, 313-335.

postgraduate study and the like. While professional documents seem to focus on the professional *learning* of teachers, researchers often use the two terms interchangeably and with little or no clarification of their meanings. We believe the distinction between the two terms is important since 'professional learning' is more inclusive and recognises many activities teachers engage in during the course of their careers.

In the first section we review recent studies emanating from large-scale professional development projects. Large-scale was considered to include those studies that drew upon systemic or state-wide projects even when they may have only focused on a smaller subset of participants. An examination of this recent and rapidly expanding body of research enables the identification of key features associated with effective professional development. The uniqueness of such projects is exemplified in their sustained impact on the professional learning of teachers, on curriculum reform and ultimately on improved student outcomes. In the second section, the diverse and plentiful smaller-scale studies are reviewed. Acknowledging the importance of context, the studies in this section are explored in relation to growth in teacher learning within classrooms, within the school context, beyond the school context, or in transition. Finally, we discuss the emerging themes and conclude with some recommendations for future research.

LARGE-SCALE, SYSTEMIC OR STATE-WIDE PROJECTS

Literature on large-scale, systemic or state-wide professional development projects is dominated by research arising from three successful numeracy programs – the *Count Me In Too* project in New South Wales (CMIT), the *Early Numeracy Research Project* in Victoria (ENRP) and the *Numeracy Development Projects* in New Zealand (NDP). An overview of each program, including the essential components and their theoretical bases, key outcomes and the implications for on-going large-scale professional development is presented by Bobis, Clarke, Clarke, Thomas, Wright, Young-Loveridge, and Gould (2005). While many of the studies referred to in the article were conducted before the scope of this review, juxtaposing the three projects enabled the authors to highlight important commonalities, and to acknowledge each project's uniqueness. A significant feature of each of these projects has been their abilities to influence large numbers of teachers and students while still obtaining in-depth information about teacher growth and student learning. Information generated from each of these projects has had wide-sweeping implications for curriculum reform and professional development in New Zealand and Australia. For instance, three common features of these programs, described by Bobis et al., include:

1. a research-based framework of children's mathematical learning;
2. the use of an individual assessment interview; and
3. intensive whole-school focused professional development.

These features or 'ingredients' are regularly incorporated into other professional development projects.

'Ingredients' for effective professional development

Rather than supporting the existence of a standard recipe for effective professional development, studies exploring the impact of various aspects of large-scale programs support the "notion of rich ingredients" (Clarke, Cheeseman, Gervasoni, Gronn, Horne, McDonough, Montgomery, Roche, Sullivan, Clarke, & Rowley. 2002, p. 53). It is noted across a range of studies that professional development programs – particularly those that are large-scale in nature – have adopted or adapted such ingredients to suit their individual needs. For instance, Clarke, Stephens, Lewis, and Downton (2005) report on the effectiveness of the numeracy program being implemented in Victorian Catholic Schools – *Success in Numeracy Education* (SINE). SINE utilises a whole school approach and has incorporated the three main features characterising CMIT, ENRP and NDP as previously described.

Leadership and coordination at the individual school level have emerged as significant ingredients for the effective implementation of professional development programs. Cheeseman and Clarke (2005, 2006) examined exactly what the role of the school-based coordinator in the ENRP entailed and how such leadership roles evolved over time. Other researchers have focused on the effectiveness of such coordinators and the challenges they must overcome to ensure programs are maintained (Bobis, 2007; Higgins, 2005). Together, these studies point to the importance of appointing school-based coordinators who not only possess personal organisational and leadership qualities, but who are supported by school executive and external expert advisors, are provided with time and resources, and work in a whole-school collaborative environment.

Carroll (2005) surveyed 180 primary teachers from 13 schools to explore the ingredients contributing to effective professional development. Among those factors identified by teachers, the influence of 'significant others' – particularly colleagues with whom they maintained a professional relationship "over a period of time" (p. 207) – were perceived to have had greatest impact on their learning. Carroll concluded that this factor was linked to teachers' perceptions of professional development as a social activity rather than as a solitary one. This finding may have significant implications for professional bodies placing an increased focus on the professional *learning* of teachers. Given that such learning can originate from personal reflections on a teacher's own classroom experiences, professional reading and similar solitary pursuits, it is possible that many teachers may not view these activities as valid forms of professional development, possibly due to difficulties associated with being able to document what is learned. It is also possible that many teachers may not possess the skills or knowledge needed to pursue such solitary experiences to the degree necessary for significant professional learning to occur. These issues are in need of further exploration.

A widely debated issue was addressed by White, Mitchelmore, Branca, and Maxon (2004) when they examined two types of professional development – one focused on improving the pedagogy of mathematics teachers (CMIT teaching angles component) and the other focused on enriching the mathematical content knowledge of teachers (San Diego State University summer institute courses).

They compared the impact on teachers of undertaking both types of professional development with the aim of identifying desirable attributes for such programs. They concluded that three components were crucial to the effectiveness of the programs they examined: teacher pedagogy, mathematics content knowledge and the acknowledgment of teachers' professional learning through a formal accreditation body such as a university or institution for teacher registration. White et al. recommended that professional development programs adopt an appropriate balance of these three elements.

A number of studies have not only highlighted ingredients in professional development deemed most effective, they have also played a valuable role in identifying those most likely to inhibit change, growth and improvement. Kensington-Miller (2004) identified the lack of established supportive structures in low socio-economic school environments as a major inhibitor of pedagogical change. Other studies have reported on the difficulties that occur within different mentoring relationships (Kensington-Miller, 2005) and problems associated with lack of time, resources and the increased workload of teachers (Bobis, 2004).

Effectiveness of systemic and large-scale programs: Measuring outcomes

A general agenda for evaluating the impact of teacher professional development programs is suggested by Guskey (2002). He recommends evaluations be organised to report on five critical levels of information: teachers' reactions; teachers' learning; organisational support and change; teachers' use of new knowledge and skills; and student learning outcomes. Guskey claimed that the latter levels in this scale were more difficult to measure. It is noted that within the period 2004 to 2007, studies reporting the effects of large-scale professional development initiatives on student learning outcomes are less numerous. Understandably, the impact on student outcomes can only be measured once a program is well established. Additionally, the vast array of intervening variables does not allow simple causal inferences to be made between teacher professional development and improved student performances. For instance, White and Clements (2005) used Guskey's (2002) critical levels to guide their investigation, analysis and report on the *Active Mathematics in Classrooms* (AMIC) project in Brunei. This large-scale professional development project aimed at assisting upper-primary teachers to modify their teaching towards more student-centred practices. Using a combination of survey, interview and classroom observation, White and Clements reported positive outcomes in the first four critical levels – teachers' reactions and learning, organisational changes and teaching practices. However, they conceded that "detailed quantitative analyses" (p. 157) of student performance were not possible given the time needed for teachers' new knowledge and practices to have a significant impact on student performance levels.

Similarly, Tait-McCutcheon, and Sherley (2006) used Guskey's five critical levels of professional development evaluation to develop a rubric which 66 teachers involved in the NDP used to self-assess the impact of eight workshops on their professional development. While the researchers used results of 1621 students aged 5-13 on the NDP diagnostic interview to support teachers' self-reported improvements of student outcomes, they too conceded the difficulty of providing evidence of the fifth level.

Irwin and Britt (2005) compared the numeracy outcomes of students whose teachers were involved in the NDP with those who were not involved. They report that students involved in NDP employed algebraic reasoning to solve arithmetical problems more effectively. In a further investigation of student performance, Irwin and Irwin (2005) applied Rasch modelling to reveal differences and raise concerns about the progress of NDP students from various socio-economic areas over the course of a year. Their study revealed that while students from low socio-economic backgrounds made some progress, they generally did not attain the advanced cognitive strategies for solving additive, multiplicative, and proportional problems that students from high socio-economic areas attained. As suggested by Irwin and Irwin, this finding highlights the challenge in such research in that one of the main aims of the NDP is to improve the performance of students from disadvantaged backgrounds.

Several studies (e.g., Anderson, 2005a; Cheeseman, 2006) explore the impact of NDP and ENRP on teachers' beliefs, knowledge and practices. Julie Anderson (2005) used a case study to monitor change in these areas over a three-year period. Her conclusions are consistent with those of Cheeseman (2006) indicating the need for extended periods of exposure to professional development and on-going support for teaching practices to change. McDonough and Clarke (2005) report on the changes to teachers' beliefs and practices as an outcome of their involvement in the ENRP. The most commonly cited changes to teaching practices included the use of more open-ended tasks, the use of more probing questions, increased expectations of children's abilities, the use of more 'hands-on' activities and an increased emphasis on reflection and sharing of strategies.

Bobis et al. (2005) provided data drawn from various studies of systematic research surrounding each of the three large-scale programs – CMIT, ENRP and NDP. Taken together, the body of evidence was used by the researchers to confirm improvements at each of Guskey's five critical levels. In each case, large numbers of participants – students and teachers – were involved and confirm the overall impact on numeracy outcomes of these programs. For instance, in CMIT, the results of clinical interviews with 10 075 students aged between 4 years 5 months and 7 years 11 months, indicated that this group of students attained more advanced cognitive strategies for solving simple addition and subtraction problems than normally expected for this population. Similarly, clinical interviews conducted over a two-year period showed that students in ENRP Trial schools significantly outperformed students in Reference schools "at every grade level and in every mathematical domain" (p. 40).

Sustainability and expansion

With both CMIT and the NDP operating for around a decade, studies involving these two large-scale programs have moved beyond those levels suggested by

Guskey. There has emerged a need to assess a program's facility for sustainability and expansion (Knight, 2005). Consequently, the past four years have seen a rapid increase in the number of studies concerned with these two aspects of longer-running programs. Pritchard and McDiarmid (2006) employed action research to discover ways in which primary schools involved in the NDP acted to sustain numeracy practices acquired since their professional development. They also explored the factors that contributed to or inhibited professional learning within schools. Their findings were confirmed by Bobis (2007) when she examined the effectiveness of CMITs expansion to higher grades and its ability to be sustained via a school-based facilitator. In both cases, sustainability was shown to be dependent on a range of elements working together and included the need for a school plan that articulated shared goals and established firm structures for the on-going success of professional development, and the provision of 'time' for these shared goals to be realised. Above all else, appointing a school-based coordinator with suitable expertise and leadership qualities, whose responsibility it was to maintain the momentum of professional development in the school, was considered crucial for sustainability. Longer-term, some major questions remain: What happens in these schools if or when the main driving-force leaves? What structures or mechanisms can be put in place to cater for a facilitator's succession?

Identifying teachers' professional development needs

A number of large-scale studies have focused on identifying the professional development needs of teachers. Such studies provide important guidelines for the providers of professional development and can highlight priority areas. For instance, a component of a national survey reported by Panizzon and Pegg (2007), identified the areas of greatest need for 547 secondary teachers in rural, regional and metropolitan schools as being professional development opportunities for teaching higher-order thinking, classroom management, organization and alternative teaching methods, and release from face-to-face teaching for in-school collaborative activities.

Goos and Bennison (2004), as part of a state-wide survey of secondary mathematics teachers to explore teachers' practices and beliefs related to their use of technology, found that 33% of respondents (from a total of 485) expressed a need for professional development that specifically addressed 'how' to effectively integrate technology into their instructional practices.

In their survey of 121 primary teachers from 56 schools across Queensland, Nisbet and Grimbeek (2004) and Nisbet (2005) explored teachers' needs and responses to professional development. They provide evidence confirming that professional development of mathematics teachers has greater power to transform teacher practices than actual teaching experience. They also argue that their findings further affirm the important role of professional development for improving national numeracy outcomes. Overall conclusions drawn from these two studies provide vital evidence of the causal link between professional development and student achievement in mathematics.

NON-SYSTEMIC SMALL-SCALE PROFESSIONAL LEARNING

Investigations into the development of teacher knowledge using teaching experiments, self-study and investigations of small-group learning communities are evidenced in the research into teacher professional development in Australasia. Focused investigations of a small group of teachers reflect the complexity of teacher learning which is "to a large extent influenced by personal, social, organisational, cultural, and political factors" (Llinares & Krainer, 2006, p. 429). In order to explore such complexities, many studies reported here involve small numbers of teachers (usually fewer than ten), which has enabled the collection of rich, detailed data using multiple sources. The focus for change has included developing content knowledge and pedagogical content knowledge, or implementing new teaching methods or new technologies.

The structure of this section recognises the important role played by different contexts in which teacher professional learning takes place. Some studies have been conducted at the school level as teachers pursue self-study, develop communities of practice within the classroom, or work collaboratively with other staff members to develop their knowledge and understanding of particular aspects of practice. Other research explores the impact of professional development offered beyond the school context as teachers attend workshops or courses with teachers from other schools. The final group of studies reviewed in this section explores teachers in transition from teaching in other cultural contexts to Australian contexts, from not using to using new technologies, and from pre-service education to teaching placements in schools. All of the studies reviewed in this section support the notion that quality sustained professional development opportunities support teachers' professional learning in significant ways. An overarching characteristic of the studies, is that most rely solely on self-report data as evidence of teacher change. The small number of studies that incorporate observation data to validate self-reported claims seem to do so on few occasions and only involve one or two teachers. Possible reasons for these potential shortcomings are suggested in the discussion.

Building learning communities within classrooms: Negotiating new roles for teachers and students

Several studies report change in classrooms for individual or small groups of teachers as a result of implementing new approaches to teaching and learning which have collectively been referred to as reform-oriented teaching approaches (Anderson & Bobis, 2005). Typically these practices promote the use of investigations in mathematics with a focus on a range of processes supporting student engagement and learning such as reasoning, communicating, connecting and reflecting. The vehicles for promoting the use of these processes by students include open-ended questions, model-eliciting tasks, and unfamiliar problems.

Individual teachers were the focus for research in some studies where the teacher's role changed as students gradually took more responsibility for their learning and the teacher focused on listening and learning from students. Smith (2006a, 2006b) used narrative inquiry to support self-reflection as she implemented new approaches for teaching, learning and assessment in her Year 6 classroom. By using open-ended tasks and encouraging discussion and reflection on learning, Smith began to 'listen' more and 'tell' less. Smith acknowledges that post-graduate study and her role as a school consultant provided the impetus for change in her practice. Hunter (2005) observed the transformation of classroom communication norms for one teacher and her students as the promotion of a range of processes such as exploring, representing, reasoning and generalising, enabled the development of a community of practice and enhanced student engagement. A collaborative partnership between the teacher and researcher supported the change from teacher as sole authority to facilitator of learning.

Working with groups of teachers has led to similar results as researchers scaffold and support teacher learning of reform-oriented approaches. English and Doerr (2004) describe the powerful approach of 'learning-in-practice' where the mathematical knowledge of one middle grade teacher was challenged as she implemented modelling tasks in data analysis. In a subsequent paper, Doerr and English (2006) describe changes in mathematical knowledge as well as the adopted role in classrooms as two teachers used novel modelling tasks to support students' mathematical reasoning. In addition, the students engaged in 'interpretative listening' to evaluate solutions; thus reinforcing the notion that the development of a community of practice in classrooms requires a renegotiation of roles between teachers *and* students. Brown and Renshaw (2006) note the support of an expert 'critical friend', familiar with the scaffolding strategies of 'collective argumentation'. Their paper reports the outcomes of one teacher's implementation of collective argumentation to alter the learning environment for her Year 7 students. The change in practices highlighted the "previously hidden levels of student interest, involvement and communicative skill" (p. 105). Makar (2007) documented the influence of support mechanisms, or *connection levers*, in assisting upper primary teachers over the course of a year to develop confidence in teaching mathematics through inquiry.

Appropriately sequenced learning tasks and lessons have supported teacher learning in studies by Sullivan and his colleagues (Sullivan, Mousley, & Zevenbergen, 2004; Sullivan, Mousley, & Zevenbergen, 2006). Included in the research is the development of a model for planning and teaching which incorporates mathematical community norms encouraging students to make decisions and contribute to class discussions and lesson reviews (Sullivan et al., 2006). Teachers were supported in their learning as they used variations of open-ended tasks including enabling prompts and extending prompts (Sullivan et al., 2004).

Knowledge growth within the school context: Developing partnerships

The involvement of a mathematics educator or consultant in a teacher professional learning project with a school or cluster of schools, may eventuate in several ways:

initiation by the educator as a researcher, invitation by schools, invitation by systems, or appointment by a funding body as a ‘critical friend’. Whatever the genesis of the interaction, the mathematics educator’s role can take many forms, ranging from advisor to evaluator or a driving force behind the design and implementation of the project. Regardless of the participation role, all of the teacher learning projects discussed in this section have been reported from the perspective of the ‘expert’. While it is obvious that each case of teacher professional learning is unique and complex, there are several strong common elements to the project findings in recent years, all centred on relationships – relationships between participants, relationships between teacher practice and student learning, and relationships between teacher change and the school environment.

Several studies highlight relationships between external expert and teacher as an approach to promote professional learning in school contexts, including Hunter (2005) who worked with one teacher and Callingham (2005) and Rogers (2006, 2007) who worked in whole-school contexts. Callingham supported teachers from Years 3 to 6 in a small rural school in NSW to develop strategies promoting mental computation. Rogers (2006) describes the role of ‘mathematician in residence’ at a ten-teacher rural school with input sessions, lesson observations, teaching of ‘model’ lessons, team teaching, teacher partners and reflection/discussion sessions. All projects reported growth of teacher confidence, greater flexibility of teaching approaches, increase in students’ willingness to discuss thinking and more frequent professional discussion, particularly in the whole-school contexts. However, the work of Pritchard and Bonne (2007) with four teachers developing their questioning skills, highlights the value of minimising the gap between ‘expert’ and teacher by using methodologies that place the teacher in the genuine role of a researcher and serve to ‘problematise’ the teaching process rather than simplify it.

A number of studies report the impact of the strength of the relationships between teachers on the success of projects. A study of three teachers from the same school who worked closely together to plan and program learning experiences for their early years students reported beliefs and practices supporting reform-oriented approaches, and identified time and collegial support as key elements to change (Bobis & Anderson, 2006). McPhail (2004) discovered that success of a program with varying levels of consultancy support (as measured by both teachers and student learning) was most strongly influenced by the relationships within the school-based teams and the strength of commitment to improvement. In a comparison between the implementation of the Lesson Study model in an Australian school and two Malaysian schools, White (2006) emphasised the critical influence of the relationships within a learning community including teachers, school executive and external ‘experts’ on the outcomes of professional learning. However, care must be taken when interpreting these findings and the benefits of close teacher relationships. The success of even established models of professional development, such as ‘Lesson Study’, remains dependent on the unique characteristics of each situation. Also focusing on the impact of teacher relationships in the context of teachers in remote areas, Cooper,

Baturo, and Warren (2005) reported that the lack of productive professional relationships between inexperienced non-Indigenous teachers and experienced Indigenous teacher aides prohibited the exchange of professional learning about mathematics and about authentic learning contexts for Indigenous students.

Some studies have focussed on relationships between teacher practices and attitudes and student learning outcomes. Cooper, Baturo, and Grant (2006) found that examining student-learning outcomes allowed evaluation of the effectiveness of various instructional interactions and collaborations between researchers and teachers in bringing about changes in teacher-practice. The presence of all three types of pedagogies: technical (practical tips), domain (specific to the mathematics topic), and generic (generally applicable across mathematics topics) in planning, practice and reflection positively influenced student learning. An additional finding was the apparent relationship between the nature (mathematical structure) of a mathematics topic and the action of the pedagogical levels. Pegg and Panizzon (2004) focussed on teachers applying a theoretical framework for assessment of student cognitive development (SOLO taxonomy) as a catalyst for reflection and change in teaching practice. Beswick (2005) reported on the changes in attitudes and practice of teachers engaged in a professional development program aimed at improving the numeracy education of students with learning difficulties. A shift away from believing a skills-based curriculum to be more appropriate than the development of deep conceptual understanding was achieved.

Few studies in the last four years have explored constraints on teacher learning. However, one study used auto-ethnography to explore the impact of an "exam-driven school" on the teacher's practice (Hagan, 2005). Hagan's story reveals deep reflection on a year of teaching, raising many issues about the impact of context on practice as well as the influence of earlier learning environments.

Knowledge growth beyond the school context

Several studies have explored professional development beyond the school, thus bringing together teachers from different contexts to learn. While Hawley and Valli (1999) promote professional development which is connected to and derived from teachers' work with students, we argue that this does not necessarily suggest that effective professional learning can only occur in school contexts. The following studies provide evidence that bringing together teachers from a range of schools to learn about a common aspect of teaching and learning leads to important shared experiences and knowledge development.

Viewing teachers in action is a powerful tool for reflection and shared understandings about 'good' practice, as is the use of video recordings of one's own practice. In an investigation of effective numeracy teachers, Siemon, Virgona, Lasso, Parsons, and Cathcart (2004) and Siemon and Virgona (2003) used a one-way window for teachers from two clusters of three primary schools to observe colleagues teaching a small group of children. Participants were required to comment on what they noticed, thus becoming 'codifiers of practice in real time' with analyses of data revealing twelve scaffolding practices. Project teachers reported the 'behind-the-screen' activity as a highly significant outcome of the project since a common language of practice was developed and shared by teachers. Davies and Walker (2005, p. 273) examined the process of "learning to notice" in relation to the "constructed, enacted and negotiated knowledge" in the classrooms of eight teachers from three primary schools. Regular meetings facilitated knowledge growth in a community of learners as the teachers explored rich tasks and viewed videos of their practice. Change in 'noticing' was evident as teachers moved from comments about children's behaviour and procedural knowledge to deeper, conceptual knowledge. Teachers' growth in their understanding of the importance of the mathematical concepts was accompanied by a change in pedagogy. This was frequently accompanied by better questioning practices and longer wait time.

Grootenboer and Zevenbergen (2007) describe the process used in a professional development program with a group of primary school teachers who were encouraged to 'work like mathematicians' by solving problems in collaborative groups. The features for working like a mathematician (cognitively/affectively, socially/culturally, and mathematically) are presented as critical components to overcome the shallow teaching syndrome dominating many classrooms as well as to support the learning of teachers.

Curriculum reform as a context for professional development

New curriculum documents in several states in Australia have embraced reform-oriented approaches through process strands such as Working Mathematically (e.g., Board of Studies NSW, 2002) and have yielded several studies into curriculum implementation. Anderson and Moore (2005) evaluated an extended professional development course addressing key changes as well as the importance of the working mathematically processes from the NSW syllabus. In reflective writing from the 28 participants, data revealed increased understanding of the changes but there was little evidence of deeper reflection at the level of beliefs, even when teachers were prompted in interviews. In an independent study, Cavanagh (2006) conducted a survey of 193 secondary teachers in NSW to determine their engagement with the Working Mathematically strand of the new syllabus. Teachers' responses were classified into three groups – the dissenters, aspirants and supporters. The supporters were enthusiastic about the changes and reported changing practice as a result of the new curriculum. Cavanagh (2006) proposed that professional development might assist the aspirants but is not likely to be successful with the dissenters. As no observations were conducted in either of these studies, the extent of change in classrooms cannot be determined.

One aspect of working mathematically is problem solving. Judy Anderson's (2005) survey of 130 primary and 12 secondary mathematics teachers from NSW revealed commonly held beliefs about problem solving as an 'end' rather than as a 'means' to learning mathematics. However, teachers identified the need for opportunities to plan and learn together in collaborative environments in order to improve their knowledge and confidence in implementing problem solving. A new curriculum in Singapore promotes problem solving in mathematics. The changes to three teachers' problem-solving teaching approaches were investigated in a study of Singaporean classrooms by Ho and Hedberg (2005). An intervention involving professional development was provided with pre- and post-interviews and tests for students as well as lesson observations, which were coded for various aspects of problem-solving approaches. After the intervention, teachers spent more time on reflection, used group work for problem-solving questions, and used fewer standard textbook type questions. The short time frame for this study enabled the promotion of teaching *about* problem solving with a focus on Polya's phases of problem solving. While this approach was successful, a longer-term aim could be to explore the implementation of teaching *through* problem solving.

In the Tasmanian context, Watson and her colleagues (Brown, Watson, Beswick, & Fitzallen, 2006; Watson, Beswick, & Brown, 2006) investigated the implementation of a new curriculum framework in middle years classrooms as teachers sought to improve students' numeracy outcomes. Professional development was planned in response to teachers' identified needs including mathematical content, planning from the new framework, analysis of student work and case studies in schools. After the first year of intervention the measurements of change in student performance and attitude were inconclusive, with the direction of causal relationships difficult to establish, although some improvement in the achievement of students in the transition from primary to secondary was detected (Watson, Beswick, Brown, & Callingham, 2007). From the same context, a study of effective numeracy teachers led to the development of a reflective tool, which focuses on the important component of teacher questioning (Muir, 2006). An observation schedule was used to document one teacher's questioning and responding as well as her use of connections and examples. By discussing the outcome of the observations the teacher became more aware of the need to employ better questioning approaches. The study did not investigate whether this awareness led to change in her practice.

In the Queensland context, the introduction of a new curriculum has led to a focus on teachers' mathematical content knowledge. Warren, Cooper, and Lamb (2006) used the Vygotskian-based zones of learning theory of Valsiner (1997) to study the professional development of teachers taking part in an early algebra project. Warren et al. used the "illusory zone" to explain the teacher's apparent openness to the new approach but in reality she was struggling for survival since she failed to notice students' difficulties, or respond to misconceptions. They argue that the professional development did not take into account the teacher's 'zone of proximal development' (ZPD). In a related study, Warren (2006) applied the same framework to explore the interactions between the 'expert' and a group of six teachers across three schools, and the interactions among the teaching partners in each school. Two teachers' and their students' conceptual understanding of division, were investigated by Lamb and Booker (2004). Pen and paper tests and interviews were used to ascertain the level of conceptual understanding. Professional development in using games and concrete materials supported knowledge growth of both teachers and their students. In another study, Baturo

(2004) identified that good assessment instruments, and appropriately sequenced tasks supported the development of one experienced teacher's subject matter knowledge of fractions. Data from tests, observations and interviews were compared to a carefully matched control classroom revealing significantly improved learning outcomes in the experimental class. Goos, Dole, and Makar (2007) used action research cycles with 4 pairs of teachers from schools in Queensland to support the integration of working mathematically from the new curriculum into teachers' practices. It is apparent from this review, that curriculum reform is an important catalyst for research about teacher professional learning, its impact on teaching practices and on student learning. It is also clear that these studies are generally small scale in nature, are conducted over short timeframes and often rely on self-reported changes to teaching practices rather than actual observations of classroom teaching.

Teachers in transition

Some teachers experience considerable discomfort when taking on new roles in schools, when changing schools, and when changing countries to teach in quite different cultural environments. Also, transition from university pre-service education to the school setting can be overwhelming for beginning teachers, as can the implementation of new technologies. Several studies have explored these aspects of professional learning.

Teachers in transition from a familiar home culture to an Australian school context were the focus of a study by Seah (2005). Using values as the socio-cultural variable, he explored the professional practice of immigrant secondary mathematics teachers. Successful socialisation required amalgamation and appropriation to combine values from both cultures, establishing equilibrium rather than enculturation into the host culture. An outcome of the investigation was a strong recommendation that professional development needs to allow teachers in transition from different cultures to explore and negotiate differences so that dissonance is reduced.

In a collection of papers using zone theory (Valsiner, 1997) to track beginning teacher's learning and use of new technologies, Goos (2005, 2006) proposes a theory of teacher learning and development with three 'zones of influence' including teacher knowledge and beliefs representing the potential for development; professional context which defines the teaching actions allowed; and the sources of assistance available for promoting specific teaching actions. She represents the zones as three overlapping circles and argues that professional learning is most effective when teachers are challenged to reflect on beliefs and practices within a supportive learning environment.

Using new technologies to support the teaching and learning of mathematics has been the focus of several Australasian studies. Teacher renewal through professional development in the use of a range of technologies was explored by Yates (2005). Teachers working in pairs supported each other as they planned, implemented and evaluated a range of action research projects in their schools. A project tracking the implementation of a computer algebra system (CAS) for the first time in a senior mathematics course in Victoria, notes that teachers' development of understanding for the use of CAS was supported by materials and training. Ball and Stacey (2005) report four teachers' collaborative efforts to determine acceptable solutions to problems when a CAS is used, particularly in high-stakes examinations. The change in practice of one teacher's use of CAS as his experience and knowledge developed was investigated by Ball and Stacey (2006). They noted a change in CAS use from functional to pedagogical as the teacher began to explore CAS outputs to problems with his senior students.

The introduction of graphic calculators has also led to change in practice. Seven experienced teachers from two schools participated in professional development workshops before implementing graphics calculators into their practice (Thomas & Hong, 2005). Questionnaires, interviews and observations revealed different uses of the technology, with one group of teachers having an instrumental use while another, through their greater confidence, were able to focus more on the mathematics than the use of the technology. Fitzallen's (2005) case study of four teachers demonstrates that "the acquisition of ICT skills does not necessarily translate to transformative pedagogical practices" (p. 353). Apparently this was because there was insufficient support for the teachers to understand how the increased use of technology related to mathematical thinking and the enhancement of learning for students. Williams (2006) found a relationship between teachers' levels of confidence in using spreadsheets, their perceptions of the usefulness of spreadsheets and their ability to utilise spreadsheet activities that promote higher order thinking. With the increasing speed at which technology is absorbed into normal classroom teaching, a substantial and rigorous body of research surrounding how teachers learn to use it in the classroom and its impact on their pedagogy is sadly falling behind. In this sense, when it comes to technology as a stimulus for research about teacher learning – teachers will, it seems, always be in a state of 'transition'. Such a situation lends itself to the need for some 'bigger picture' research about the professional learning of teachers in a technologically-transient classroom.

DISCUSSION AND CONCLUSION

The fact that this is the first time an entire chapter has been dedicated solely to the learning and development of practicing teachers in the history of MERGA research reviews is testimony to the increased significance of this topic over the past decade. This rise in prominence is evidenced by the increasing number of publications represented in this chapter and is perceived to be a result of many factors. Foremost among these factors is the occurrence of the three large-scale professional development projects – CMIT, ENRP and NDP. These projects have not only been founded on a firm research-base, but have been accompanied by a systematic program of research to study their impact. Similarly, the Australian Government Quality Teaching Program (AGQTP) is responsible for generating a vast number of smaller-scale research studies. Importantly, each project has been funded by an Australian state, federally or the New Zealand national government and is a major indicator of the high political status the professional development of teachers, particularly those teaching mathematics, is now attributed. The proliferation of research reports and the dissemination resulting from these studies has been aided by the often mandated, liaison between government agencies, individual schools and an academic partner.

Unlike sections in previous MERGA reviews of pre-service teacher research, we have chosen to differentiate between professional *development* and professional *learning*. We believe this to be an important distinction that is still emerging in the professional community (e.g., New South Wales Institute of Teachers, 2005), but is still to be realised in the research literature. While we consider only a few researchers, such as Carroll (2005) and Smith (2006b), to have started scratching the surface of the much 'bigger picture' of teacher professional learning, there is still much to learn about this longer-term journey and the role that professional development plays in this journey.

What have we learned from this expanding body of research concerning teacher learning and development? Within the broader context of professional learning, research has shown that involvement in quality professional development can have more influence on teachers than actual teaching experience (Nisbet & Grimbeek, 2004). We also know that professional learning opportunities exist through a wide range of vehicles, from teacher involvement in system-led or school-initiated professional development programs, to individual programs of learning involving postgraduate study, professional reading, action research and self-study. Foremost in each of these are the importance of reflective practice (e.g., Siemon et al., 2004) and the building of communities of practice, learners and researchers (e.g., Brown & Renshaw, 2006).

While the volume of smaller-scale studies is a prominent feature of this review, this characteristic is in fact a significant point to be raised in the research community and has been noted by others concerned with the development in this field (Adler, Ball, Krainer, Lin, & Novotna, 2005). At first glance, the diversity of so many small-scale studies raises questions in our minds about their long-term impact and contribution to the advancement of knowledge in this field. However, the process of review has caused significant themes to emerge and helped raise important questions for future research. For instance, emerging from the many smaller-scale studies focussed on teacher professional learning is the importance of 'noticing', both in terms of children's learning and teacher pedagogy (e.g., Davies & Walker, 2005).

It was noted that the smaller-scale studies tended to rely on self-report data and that few incorporated significant amounts of observation data to help validate the self-reported findings. It is felt that this phenomena is due to the labour-intensive and high cost involved when studies incorporate classroom observation. A challenge to researchers in the future is to explore other ways in which observation data may be collected and analysed or explore other methods of validating self-reported data of teachers. Also in terms of future research, longitudinal studies that clearly focus on the professional *learning* of teachers are needed so as to understand the impact of, say, postgraduate study, teacher registration requirements or different forms of professional development.

Research on professional development programs indicates the significance of a number of rich ingredients. Most notable is the focus on children's mathematical development via research-based frameworks and individual diagnostic interviews. Other important ingredients include the need for school-based coordinators who possess leadership qualities and receive support from both school executive and external expert advisors (e.g., Cheeseman & Clarke, 2005, 2006). Insufficient time is perhaps the single most often-cited factor considered an inhibitor to professional development (e.g., Bobis, 2004) and, hence, an important ingredient to be considered by providers.

The large-scale projects have been well placed to provide more detailed evidence of improvements in teacher knowledge and confidence. However, few studies have taken the opportunity for long-term systematic data collection of this nature. The profiling instrument used by Watson, Beswick, Caney, and Skalicky (2005/2006) to measure change in various types of teacher knowledge should prove to be highly useful for this purpose in the future. Larger-scale projects have also been better placed to report their impact on student numeracy outcomes. However, it has been revealed that straight-forward causal links between teacher professional development and improved numeracy performances are difficult to ascertain given the large number of intervening variables. Hence, a 'big' question remains mostly unanswered: How can we be sure that professional development makes a difference to our students' learning outcomes in mathematics?

This review has shown how research surrounding teacher professional learning and development in mathematics has rapidly increased in both volume and focus in the past ten years. However, we still have a great deal to learn about what makes it more effective. In particular, we still need large-scale and long-terms studies to monitor the impact of teachers' practice in relation to student outcomes. Within the current context of political and educational reforms, with teacher registration requirements being stipulated by state and national bodies, there is also a need to study the long-term impact on teachers' professional learning.

REFERENCES

Adler, J., Ball, D., Krainer, K., Lin, F. L., & Novotna, J. (2005). Reflections on an emerging field: Researching mathematics teacher education. *Educational Studies in Mathematics, 60*, 359-381.

Anderson, Judy (2005). Implementing problem solving in mathematics classrooms: What support do teachers want? In P. Clarkson, A. Downton, D. Gronn, M. Horne, A. McDonough, R. Pierce, & A. Roche (Eds.), *Building connections: Theory, research and practice* (Proceedings of the 28th annual conference of the Mathematics Education Research Group of Australasia, Melbourne, pp. 89-96). Sydney: MERGA.

Anderson, Julie. (2005). I didn't know what I didn't know: A case study of growth in teacher knowledge within the Intermediate Numeracy Project. In P. Clarkson, A. Downton, D. Gronn, M. Horne, A. McDonough, R. Pierce, & A. Roche (Eds.), *Building connections: Theory, research and practice* (Proceedings of the 28th annual conference of the Mathematics Education Research Group of Australasia, Melbourne, pp. 97-104). Sydney: MERGA.

Anderson, J. A., & Bobis, J. (2005). Reform-oriented teaching practices: A survey of primary school teachers. In H. L. Chick & J. L. Vincent (Eds.), *Proceedings of the 29th annual conference for the Psychology of Mathematics Education* (Vol. 2, pp. 65-72). Melbourne: University of Melbourne.

Anderson, J. A., & Moore, M. (2005). *Evaluating the professional learning of secondary mathematics teachers: Reflecting on their reflections!* Refereed paper presented at the annual conference of the Australian Association of Research in Education, Sydney. Retrieved February 23, 2008 from: http://www.aare.edu.au/05pap/and05154.pdf

Australian Association of Mathematics Teachers (AAMT). (2006). *Standards for excellence in teaching mathematics in Australian schools*. Adelaide: AAMT.

Ball, L., & Stacey, K. (2005). Good CAS written records: Insight from teachers. In H. L. Chick & J. L. Vincent (Eds.), *Proceedings of the 29th annual conference for the Psychology of Mathematics Education* (Vol. 2, pp. 113-120). Melbourne: PME.

Ball, L., & Stacey, K. (2006). Coming to appreciate the pedagogical uses of CAS. In J. Novotná, H. Moraová, M. Krátká, & N. Stehliková (Eds.), *Proceedings of the 30th annual conference for the Psychology of Mathematics Education* (Vol. 1, pp. 105-112). Prague: Charles University.

Baturo, A. R. (2004). Empowering Andrea to help year 5 students construct fraction understanding. In M. J. Johnsen Høines & A. B. Fuglestad (Eds.), *Proceedings of the 28th annual conference for the Psychology of Mathematics Education* (Vol. 2, pp. 95-102). Bergen, Norway: Bergen University College.

Beswick, K. (2005). It depends on the students: Influencing teachers' beliefs about the ends and means of numeracy teaching. In P. Clarkson, A. Downton, D. Gronn, M. Horne, A. McDonough, R. Pierce, & A. Roche (Eds.), *Building connections: Theory, research and practice* (Proceedings of the 28th annual conference of the Mathematics Education Research Group of Australasia, Melbourne, pp. 137-144). Sydney: MERGA.

Board of Studies (BOS) NSW. (2002). *Mathematics K-6 syllabus*. Sydney: BOS NSW.

Bobis, J. (2004). Time, resources, information overload and classroom management: Issues surrounding professional development. In I. Putt, R. Faragher, & M. McLean (Eds.), *Mathematics education for the third millennium: Towards 2010* (Proceedings of the 27th annual conference of the Mathematics Education Group of Australasia, Townsville, Vol. 1, pp. 103-110). Sydney: MERGA.

Bobis, J. (2007). Facilitating change as part of a large-scale numeracy program. In J. Novotná & H. Moraova (Eds.), *Approaches to teaching mathematics at the elementary level* (Proceedings of the International Symposium of Elementary Mathematics Teaching, pp. 76-84). Prague: Charles University, Faculty of Education.

Bobis, J., & Anderson, J. A. (2006). Reform-oriented teaching practices and the influence of school context. In P. Grootenboer, R. Zevenbergen, & M. Chinnappan (Eds.), *Identities, cultures and learning spaces* (Proceedings of the 29th annual conference of the Mathematics Education Research Group of Australasia, Canberra, pp. 92-98). Sydney: MERGA.

Bobis, J., Clarke, B., Clarke, D., Thomas, G., Wright, R., Young-Loveridge, J., Gould, P. (2005). Supporting teachers in the development of young children's mathematical thinking: Three large scale cases. *Mathematics Education Research Journal, 16* (3), 27-57.

Brown, N., Watson, J., Beswick, K., & Fitzallen, N. (2006). *Numeracy in a reform-based learning environment.* Refereed paper presented at the annual conference of the Australian Association of Research in Education, Adelaide. Retrieved February 23, 2008 from: http://www.aare.edu.au/06pap/bro06369.pdf

Brown, R., & Renshaw, P. (2006). Transforming practice: Using collective argumentation to bring about teacher change in a Year 7 mathematics classroom. In P. Grootenboer, R. Zevenbergen, & M. Chinnappan (Eds.), *Identities, cultures and learning spaces* (Proceedings of the 29th annual conference of the Mathematics Education Research Group of Australasia, Canberra, pp. 99-106). Adelaide: MERGA.

Callingham, R. (2005). A whole-school approach to developing mental computation strategies. In H. L. Chick & J. L. Vincent (Eds.), *Proceedings of the 29th annual conference for the Psychology of Mathematics Education* (Vol. 2, pp. 201-208). Melbourne: University of Melbourne.

Carroll, J. (2005). Developing effective teachers of mathematics: Factors contributing to development in mathematics education for primary school teachers. In P. Clarkson, A. Downton, D. Gronn, M. Horne, A. McDonough, R. Pierce, & A. Roche (Eds.), *Building connections: Theory, research and practice* (Proceedings of the 28th annual conference of the Mathematics Education Research Group of Australasia, Melbourne, pp. 201-216). Sydney: MERGA.

Cavanagh, M. (2006). Implementing a reform-oriented mathematics syllabus: A survey of secondary teachers. In J. Novotná, H. Moraová, M. Krátká, & N. Stehliková (Eds.), *Proceedings of the 30th annual conference for the Psychology of Mathematics Education* (Vol. 2, pp. 273-280). Prague: Charles University.

Cheeseman, J., & Clarke, D. (2005). Early numeracy coordinators in Victorian primary schools: Components of the role, highlights and challenges. In P. Clarkson, A. Downton, D. Gronn, M. Horne, A. McDonough, R. Pierce, & A. Roche (Eds.), *Building connections: Theory, research and practice* (Proceedings of the 28th annual conference of the Mathematics Education Research Group of Australasia, Melbourne, pp. 225-232). Sydney: MERGA.

Cheeseman, J., & Clarke, D. (2006). Examining the changed role of numeracy coordinators. In P. Grootenboer, R. Zevenbergen, & M. Chinnappan (Eds.), *Identities, cultures and learning spaces* (Proceedings of the 29th annual conference of the Mathematics Education Research Group of Australasia, Canberra, pp. 103-110). Adelaide: MERGA.

Cheeseman, L. (2006). The numeracy journey: How long does it take to get on board? In P. Grootenboer, R. Zevenbergen, & M. Chinnappan (Eds.), *Identities, cultures and learning spaces* (Proceedings of the 29th annual conference of the Mathematics Education Research Group of Australasia, Canberra, pp. 131-138). Adelaide: MERGA.

Clarke, D. M., Cheeseman, J., Gervasoni, A., Gronn, D., Horne, M., McDonough, A. Montgomery, P.,

Roche, A., Sullivan, P., Clarke, B. & Rowley, G. (2002). *Early numeracy research project final report, February 2002*. Melbourne: Mathematics Teaching and Learning Centre, Australian Catholic University.

Clarke, D., Stephens, M., Lewis, G., & Downton, A. (2005). The evaluation of the Success in Numeracy Education Program. In P. Clarkson, A. Downton, D. Gronn, M. Horne, A McDonough, R, Pierce, & A. Roche (Eds.), *Building connections: Theory, research and practice* (Proceedings of the 28th annual conference of the Mathematics Education Research Group of Australasia, Melbourne, pp. 257-264). Sydney: MERGA.

Commonwealth of Australia (2007). *Top of the Class: Report on the inquiry into teacher education*. Canberra: Commonwealth of Australia.

Cooper, T. J., Baturo, A. R., & Grant, E. J. (2006). Collaboration with teachers to improve mathematics learning: Pedagogy at three levels. In J. Novotná, H. Moraová, M. Krátká, & N. Stehliková (Eds.), *Proceedings of the 30th annual conference for the Psychology of Mathematics Education* (Vol. 2, pp. 361-368). Prague: Charles University.

Cooper, T. J., Baturo, A. R., & Warren, E. (2005). Indigenous and non-Indigenous teaching relationships in three mathematics classrooms in remote Queensland. In H. L. Chick & J. L. Vincent (Eds.), *Proceedings of the 29th annual conference for the Psychology of Mathematics Education* (Vol. 2, pp. 265-272). Melbourne: University of Melbourne.

Davies, N., & Walker, K. (2005). Learning to notice: One aspect of teachers' content knowledge in the numeracy classroom. In P. Clarkson, A. Downton, D. Gronn, M. Horne, A McDonough, R, Pierce, & A. Roche (Eds.), *Building connections: Theory, research and practice* (Proceedings of the 28th annual conference of the Mathematics Education Research Group of Australasia, Melbourne, pp. 273-280). Sydney: MERGA.

Doerr, H. M., & English, L. D. (2006). Middle grade teachers' learning through students' engagement with modelling tasks. *Journal of Mathematics Teacher Education, 9*(1), 5-32.

English, L. D., & Doerr, H. M. (2004). Learning through interacting with students' ways of thinking. In I. Putt, R. Faragher, & M. McLean (Eds.), *Mathematics education for the third millennium: Towards 2010* (Proceedings of the 27th annual conference of the Mathematics Education Research Group of Australasia, Townsville, pp. 215-222). Sydney: MERGA.

Fitzallen, N. (2005). Integrating ICT into professional practice: A case study of four mathematics teachers. In P. Clarkson, A. Downton, D. Gronn, M. Horne, A. McDonough, R. Pierce, & A. Roche (Eds.), *Building connections: Theory, research and practice* (Proceedings of the 28th annual conference of the Mathematics Education Research Group of Australasia, Melbourne, pp. 353-360). Sydney: MERGA.

Goos, M. (2005). A sociocultural analysis of the development of pre-service and beginning teachers' pedagogical identities as users of technology. *Journal of Mathematics Teacher Education*, *8*, 35-59.

Goos, M. (2006). Creating learning spaces. In P. Grootenboer, R. Zevenbergen, & M. Chinnappan (Eds.), *Identities, cultures and learning spaces* (Proceedings of the 29th annual conference of the Mathematics Education Research Group of Australasia, Canberra, pp. 3-14). Adelaide: MERGA.

Goos, M., & Bennison, A. (2004). Emergence of a pre-service community of practice. In I. Putt, R. Faragher, & M. McLean (Eds.), *Mathematics education for the third millennium: Towards 2010* (Proceedings of the 27th annual conference of the Mathematics Education Group of Australasia, Townsville, Vol. 2, pp. 271-278). Sydney: MERGA.

Goos, M., Dole, S., & Makar, K. (2007). Supporting an investigative approach to teaching secondary school mathematics: A professional development model. In J. Watson & K. Beswick (Eds.), *Mathematics: Essential research, essential practice* (Proceedings of the 30th annual conference of the Mathematics Education Research Group of Australasia, Hobart, pp. 325-334). Adelaide: MERGA.

Grootenboer, P., & Zevenbergen, R. (2007). Identity and mathematics: Towards a theory of agency in coming to learn mathematics. In J. Watson & K. Beswick (Eds.), *Mathematics: Essential research, essential practice* (Proceedings of the 30th annual conference of the Mathematics Education Research Group of Australasia, Hobart, pp. 335-344). Adelaide: MERGA.

Guskey, T. (2002). Does it make a difference? Evaluating professional development. *Educational Leadership*, *59*(6), 45-51.

Hagan, F. (2005). Reflections on teaching mathematics in an exam-driven school: An autoethnography. In P. Clarkson, A. Downton, D. Gronn, M. Horne, A. McDonough, R. Pierce, & A. Roche (Eds.), *Building connections: Theory, research and practice* (Proceedings of the 28th annual conference of the Mathematics Education Research Group of Australasia, Melbourne, pp. 402-408). Sydney: MERGA.

Hawley, W. D., & Valli, L. (1999). The essentials of effective professional development: A new consensus. In L. Darling-Hammond & G. Sykes (Eds.), *Teaching as the learning profession: Handbook of policy and practice* (pp. 127-150). San Francisco: Jossey-Bass.

Higgins, J. (2005). Pedagogy of facilitation: How do we best help teachers of mathematics with new practices? In H. L. Chick & J. L. Vincent (Eds.), *Proceedings of the 29th annual conference for the Psychology of Mathematics Education* (Vol. 3, pp. 137-144). Melbourne: University of Melbourne.

Ho, K. F., & Hedberg, J. G. (2005). Teachers' pedagogies and their impact on students' mathematical problem solving. *Journal of Mathematical Behaviour*, *24*, 238-252.

Hunter, R. (2005). Reforming communication in the classroom: One teacher's journey of change. In P. Clarkson, A. Downton, D. Gronn, M. Horne, A. McDonough, R. Pierce, & A. Roche (Eds.), *Building connections: Theory, research and practice* (Proceedings of the 28th annual conference of the Mathematics Education Research Group of Australasia, Melbourne, pp. 451-458). Sydney: MERGA.

Irwin, K. C., & Britt, M. S. (2005). The algebraic nature of students' numerical manipulation in the New Zealand numeracy project. *Educational Studies in Mathematics, 58*(2), 169-188.

Irwin, K.C., & Irwin, R. J. (2005). Assessing development in numeracy of students from different socio-economic areas: A Rasch analysis of three fundamental tasks. *Educational Studies in Mathematics, 58*(3), 283-298.

Kensington-Miller, B. (2004). Professional development of mathematics teachers in low socio-economic secondary schools in New Zealand. In I. Putt, R. Faragher, & M. McLean (Eds.), *Mathematics education for the third millennium: Towards 2010* (Proceedings of the 27th annual conference of the Mathematics Education Group of Australasia, Townsville, Vol. 2, pp. 320-327). Sydney: MERGA.

Kensington-Miller, B. (2005). Mentoring mathematics teachers in low socio-economic secondary schools in New Zealand. In P. Clarkson, A. Downton, D. Gronn, M. Horne, A. McDonough, R. Pierce, & A. Roche (Eds.), *Building connections: Theory, research and practice* (Proceedings of the 28th annual conference of the Mathematics Education Research Group of Australasia, Melbourne, pp. 459-466). Sydney: MERGA.

Knight, N. (2005). The contested notion of sustainability: Possibility or pipe dream for numeracy reforms in New Zealand. In P. Clarkson, A. Dowton, D. Gronn, M. Horne, A. McDonough, R. Pierce, & A. Roche (Eds.), *Building connections: Research, theory and practice* (Proceedings of the 28th annual conference of the Mathematics Education Research Group of Australasia, Melbourne, pp. 467-474). Sydney: MERGA.

Lamb, J., & Booker, G. (2004). The impact of developing teacher conceptual knowledge on students' knowledge of division. In M. Johnsen-Høines & A. B. Fuglestad (Eds.), *Proceedings of the 28th conference of the International Group for the Psychology of Mathematics Education* (Vol. 3, pp. 177-184). Bergen, Norway: Bergen University College.

Llinares, S., & Krainer, K. (2006). Mathematics (student) teachers and teacher educators as learners. In A. Gutierrez & P. Boero (Eds.), *Handbook of research on the psychology of mathematics education: Past, present and future* (pp. 429-459). Rotterdam: Sense Publishers.

Makar, K. (2007). Connection levers: Developing teachers' expertise with mathematical inquiry. In J. Watson & K. Beswick (Eds.), *Mathematics: Essential research, essential practice* (Proceedings of the 30th annual conference of the Mathematics Education Research Group of Australasia, Hobart, pp. 483-492). Adelaide: MERGA.

McDonough, A., & Clarke, B. (2005). Professional development as a catalyst for changes in beliefs and practices: Perspectives from the Early Numeracy Research Project. In P. Clarkson, A. Downton, D. Gronn, M. Horne, A. McDonough, R. Pierce, & A. Roche (Eds.), *Building connections: Theory, research and practice* (Proceedings of the 28th annual conference of the Mathematics Education Research Group of Australasia, Melbourne, pp. 521-528). Sydney: MERGA.

McPhail, D. (2004). Professional learning in the teaching of area. In I. Putt, R. Faragher, & M. McLean. (Eds.), *Mathematics education for the third millennium: Towards 2010* (Proceedings of the 27th conference of the Mathematics Education Group of Australasia, Townsville, Vol. 2, pp. 359-366). Sydney: MERGA.

Muir, T. (2006). What does effective teaching for numeracy look like? The design of an observation schedule. In P. Grootenboer, R. Zevenbergen, & M. Chinnappan (Eds.), *Identities, cultures and learning spaces* (Proceedings of the 29th annual conference of the Mathematics Education Research Group of Australasia, Canberra, pp. 368-375). Adelaide: MERGA.

Nisbet, S. (2005). Regional differences in the professional development needs and preferences of teachers of primary mathematics. In P. Clarkson, A. Downton, D. Gronn, M. Horne, A. McDonough, R. Pierce, & A. Roche (Eds.), *Building connections: Theory, research and practice* (Proceedings of the 28th annual conference of the Mathematics Education Research Group of Australasia, Melbourne, pp. 577-584). Sydney: MERGA.

Nisbet, S., & Grimbeek, P. (2004). Primary teachers' beliefs and practices with respect to compulsory numeracy testing. In I. Putt, R. Faragher, & M. McLean. (Eds.), *Mathematics education for the third millennium: Towards 2010* (Proceedings of the 27th conference of the Mathematics Education Group of Australasia, Townsville, Vol. 2, pp. 406-413). Sydney: MERGA.

New South Wales Department of Education and Training (NSWDET). (2004). *Professional learning policy for schools*. Sydney: NSWDET.

New South Wales Institute of Teachers. (2005). *Professional teaching standards*. Sydney: NSW Institute for Teachers.

Panizzon, D., & Pegg, J. (2007). Enhancing student achievement in mathematics: Identifying the needs of rural and regional teachers in Australia. In J. Watson & K. Beswick (Eds.), *Mathematics:*

Essential research, essential practice (Proceedings of the 30th annual conference of the Mathematics Education Research Group of Australasia, Hobart, pp. 585-590). Adelaide: MERGA.

Pegg, J., & Panizzon, D. (2004). Addressing changing assessment agendas: Experiences of secondary mathematics teachers in rural NSW. In I. Putt, R. Faragher, & M. McLean (Eds.), *Mathematics education for the third millennium: Towards 2010* (Proceedings of the 27th conference of the Mathematics Education Group of Australasia, Townsville, Vol. 2, pp. 438-445). Sydney: MERGA.

Pritchard, R., & Bonne, L. (2007). Teachers research their practice: Developing methodologies that reflect teachers' perspectives. In J. Watson & K. Beswick (Eds.), *Mathematics: Essential research, essential practice* (Proceedings of the 30th annual conference of the Mathematics Education Research Group of Australasia, Hobart, pp. 621-630). Adelaide: MERGA.

Pritchard, R., & McDiarmid, F. (2006). Promoting change in teacher practices: Investigating factors which contribute to sustainability. In P. Grootenboer, R. Zevenbergen, & M. Chinnappan, *Identities, cultures and learning spaces* (Proceedings of the 29th annual conference of the Mathematics Education Research Group of Australasia, Canberra, pp. 432-439). Adelaide: MERGA.

Rogers, P. (2006). Mathematical teacher professional development incorporating an external critical friend. In P. Grootenboer, R. Zevenbergen, & M. Chinnappan (Eds.), *Identities, cultures and learning spaces* (Proceedings of the 29th annual conference of the Mathematics Education Research Group of Australasia, Canberra, pp. 447-454). Adelaide: MERGA.

Rogers, P. (2007). Teacher professional learning in mathematics: An example of a change process. In J. Watson & K. Beswick (Eds.), *Mathematics: Essential research, essential practice* (Proceedings of the 30th annual conference of the Mathematics Education Research Group of Australasia, Hobart, pp. 631-640). Aedelaide: MERGA.

Seah, W. T. (2005). Negotiating about perceived value differences in mathematics teaching: The case of immigrant teachers in Australia. In H. L. Chick & J. L. Vincent (Eds.), *Proceedings of the 29th annual conference for the Psychology of Mathematics Education* (Vol. 4, pp. 145-152). Melbourne: PME.

Siemon, D., & Virgona, J. (2003). *Identifying and describing teachers' scaffolding practices in mathematics.* Paper presented at the joint conference of the Australian Association for Research in Education/New Zealand Association for Research in Education, Auckland. Retrieved February 23, 2008 from: http://www.aare.edu.au/03pap/sie03241.pdf

Siemon, D., Virgona, J., Lasso, M., Parsons, V., & Cathcart, J. (2004). Elaborating the teacher's role: Towards a professional language. In M. Johnsen-Høines & A. B. Fuglestad (Eds.), *Proceedings of the 28th annual conference for the Psychology of Mathematics Education* (Vol. 4, pp. 193-200). Bergen, Norway: Bergen University College.

Smith, T. (2006a). Self-study through narrative inquiry: Fostering identity in mathematics teacher education. In P. Grootenboer, R. Zevenbergen, & M. Chinnappan (Eds.), *Identities, cultures and learning spaces* (Proceedings of the 29th annual conference of the Mathematics Education Research Group of Australasia, Canberra, pp. 471-478). Adelaide: MERGA.

Smith, T. (2006b). Exploring connections between teaching, learning and assessing mathematics for understanding. In J. Masingilia (Ed.), *Teachers engaged in research: Inquiry into mathematics classrooms, Grades 6-8*. Greenwich: Information Age Publishing Inc.

Sullivan, P., Mousley, J., & Zevenbergen, R. (2004). Describing elements of mathematics lessons that accommodate diversity in student background. In M. Johnsen-Høines & A. B. Fuglestad (Eds.), *Proceedings of the 28th annual conference for the Psychology of Mathematics Education* (Vol. 4, pp. 257-264). Bergen, Norway: Bergen University College.

Sullivan, P., Mousley, J., & Zevenbergen, R. (2006). Developing guidelines for teachers helping students experiencing difficulty in learning mathematics. In P. Grootenboer, R. Zevenbergen, & M. Chinnappan, *Identities, cultures and learning spaces* (Proceedings of the 29th annual conference of the Mathematics Education Research Group of Australasia, Canberra, pp. 496-503). Adelaide: MERGA.

Tait-McCutcheon, S., & Sherley, B. (2006). In the hands of the learner: The impact of self-assessment on teacher education. In P. Grootenboer, R. Zevenbergen, & M. Chinnappan (Eds.), *Identities, cultures and learning spaces* (Proceedings of the 29th annual conference of the Mathematics Education Research Group of Australasia, Canberra, pp. 353-359). Adelaide: MERGA.

Thomas, M. O. J., & Hong, Y. Y. (2005). Teacher factors in integration of graphic calculators into mathematics learning. In H. L. Chick & J. L. Vincent (Eds.), *Proceedings of the 29th annual conference for the Psychology of Mathematics Education* (Vol. 4, pp. 257-264). Melbourne: University of Melbourne.

Valsiner, J. (1997). *Culture and the development of children's action: A theory of human development* (2nd ed.). New York: John Wiley & Sons.

Warren, E. (2006). Supporting learning in early algebra: A model of professional learning. In P. Grootenboer, R. Zevenbergen, & M. Chinnappan (Eds.), *Identities, cultures and learning spaces* (Proceedings of the 29th annual conference of the Mathematics Education Research Group of Australasia, Canberra, pp. 535-542). Adelaide: MERGA.

Warren, E., Cooper, T., & Lamb, J. (2006). Teacher professional development in patterns and algebra: Being sensitive to a teacher's zone of proximal development. In P. Grootenboer, R. Zevenbergen, & M. Chinnappan (Eds.), *Identities, cultures and learning spaces* (Proceedings of the 29th annual conference of the Mathematics Education Research Group of Australasia, Canberra, pp. 542-550). Adelaide: MERGA.

Watson, J. M., Beswick, K., & Brown, N. (2006). Teachers' knowledge of their students as learners and how to intervene. In P. Grootenboer, R. Zevenbergen, & M. Chinnappan (Eds.), *Identities, cultures and learning spaces* (Proceedings of the 29th annual conference of the Mathematics Education Research Group of Australasia, Canberra, pp. 542-550). Adelaide: MERGA.

Watson, J., Beswick, K., Brown, N., & Callingham, R. (2007). Student change associated with teachers' professional learning. In J. Watson & K. Beswick (Eds.), *Mathematics: Essential research, essential practice* (Proceedings of the 30th annual conference of the Mathematics Education Research Group of Australasia, Hobart, Hobart, pp. 785-794). Adelaide: MERGA.

Watson, J., Beswick, K., Caney, A., & Skalicky, J. (2005/6). Profiling teacher change resulting from a professional learning program in middle school numeracy. *Mathematics Teacher Education and Development*, *7*, 3-17.

White, A. L. (2006). An examination of the international manifestation of Lesson Study that began as a Japanese model of teacher professional learning. In H. S. Dhindsa, I. J. Kyeleve, O. Chukwu, & J. S. H. Q. Perera (Eds.), *Shaping the future of science, mathematics and technical education* (Proceedings of the 11th international conference, pp. 129-138). Brunei: University of Brunei, Darussalam.

White, A. L., & Clements, M. A. (2005). Energising upper-primary mathematics classrooms in Brunei Darassalam: The active mathematics in classrooms (AMIC) project. In H. S. Dhindsa, I. J. Kyeleve, O. Chukwu, & J. S. H. Q. Perara (Eds.), *Future directions in science, mathematics and technical education* (Proceedings of the 10th international conference, pp. 151-160). Brunei: University of Brunei Darassalam.

White, P., Mitchelmore, M., Branca, N., & Maxon, M. (2004). Professional development: Mathematical content versus pedagogy. *Mathematics Teacher Education and Development*, *6*, 49-60.

Williams, A. (2006). Spreadsheets in year 5 chance and data: A professional development experience. In P. Grootenboer, R. Zevenbergen, & M. Chinnappan (Eds.), *Identities, cultures and learning spaces* (Proceedings of the 29th annual conference of the Mathematics Education Research Group of Australasia, Canberra, pp. 559-566). Adelaide: MERGA.

Yates, S. M. (2004). *The role of professional development in teacher renewal in mathematics.* Refereed paper presented at the annual conference of the Australian Association of Research in Education, Melbourne. Retrieved February 23, 2008 from: http://www.aare.edu.au/04pap/yat04983.pdf

AFFILIATIONS

Judy Anderson
University of Sydney, Australia

Janette Bobis
University of Sydney, Australia

Jenni Way
University of Sydney, Australia

M. A. (KEN) CLEMENTS

AUSTRALASIAN MATHEMATICS EDUCATION RESEARCH 2004-2007: AN OVERVIEW

The preceding chapters provide impressive evidence that between 2004 and 2007 Australasian mathematics education research was alive and well. The chapters draw attention to a wide range of research studies concerned with issues surrounding mathematics curricula, the teaching and learning of mathematics, at all levels, in Australia, New Zealand and Oceania. They also draw attention to research conducted by some expatriate Australians and New Zealanders who are working in other nations.

This four-yearly report is similar to, but different from, earlier four-yearly reports published by MERGA. One obvious difference is that new themes are addressed. For example, Anderson, Bobis and Way, in their chapter on teachers as learners, point out that this is the first time in a MERGA research review that "an entire chapter has been dedicated solely to the learning and development of practising teachers". In a similar vein, Galligan and Taylor, in their chapter on adults returning to study mathematics, indicate there has been no previous MERGA review of adults studying mathematics. Likewise, the chapter on university learners of mathematics points out that considering the number of students in university mathematics classes, it is surprising that so little research is being done on learners of mathematics at the university level, and that in particular there have been no Australian Research Council grants on issues associated directly with the teaching and learning of mathematics in universities.

It appears to be the case, then, that in the new millennium new areas of research interest have emerged within the MERGA community. MERGA's four-yearly reviews provide a summary of important recent research carried out by Australasian scholars, paying due attention to areas of importance – both emerging and continuing – relating to intended, implemented and attained mathematics curricula, at all levels, within Australasia.

In this closing chapter I will attempt to summarise, briefly, at least one of the themes addressed in each chapter in this book, and will comment on associated educational and research issues – particularly on wider scholarly issues in relation to the research that has been reported. Towards the end of this chapter I will comment on issues that arise as a result of some of the research findings that have been reported.

H. Forgasz et al. (eds.), Research in Mathematics Education in Australasia 2004-2007, 337-356.

ATTEMPTS AT CLARIFYING THE CONCEPT OF NUMERACY

Perry, Young-Loveridge, Dockett, and Doig's chapter revealed that there is still some doubt about whether "numeracy" can sensibly be regarded as being encompassed within a single, unidimensional trait. For example, some of the Numeracy Development Project (NDP) research on *Te Poutama Tau* drew attention to the distinctiveness of Mãori pedagogy and patterns of discourse, which would appear to suggest that numeracy should not be expected to be an all-embracing, essentially culture-free, concept. This doubt was encapsulated by White, in his chapter on research into the teaching and learning of mathematics in the middle years. After commenting that "definitions of numeracy abound", they went on to review a range of definitions, the intersections of which were, at best, fuzzy – not because of White and Southwell's summary, but because different researchers adopted quite different definitions of numeracy.

During the late 1980s, 1990s and early 2000s there continued to be considerable government funding for early childhood literacy and numeracy research. However, there has been considerable political pressure, via tagged funding, for the "numeracy" side of this research to be placed under an umbrella that situates numeracy beneath a larger "early literacy" banner. The Wickert and McGuirk (2005) study, for example, saw literacy as a general term embracing English language, numeracy, and IT literacy. It is hardly surprising, therefore, that I was interested to consider carefully how Perry et al. viewed numeracy in their chapter on early childhood mathematics education research. As it turned out, they ably summarised the findings of a number of major studies into numeracy, but they themselves never carefully defined the term itself. They did note, however, that a group of Australasian academics had not been able to agree on the components of number sense (Howell & Kemp, 2005). Galligan and Taylor, in their chapter on "Adults Returning to Study Mathematics", refer to Zevenbergen's conjecture that "in these changing times" there are different forms of numeracy, and that therefore "we may need to re-theorise knowledge, skills and dispositions to reconceptualise definitions of numeracy relevant to workplace needs". All of this would seem to imply that numeracy is not a unidimensional, culture-free, or context-free concept. The question arises, therefore, whether the term "numeracy" should continue to be employed as a "catch-all" term, despite apparent political advantages arising from using it in that way.

THE TEACHING AND LEARNING OF EARLY CHILDHOOD MATHEMATICS

One of the most exciting periods of my own life as a mathematics education researcher came during the period 1985-1987 when I served as Director of a Mathematics Learning Project, which was funded by the Federal Government as part of its Basic Learning in Primary Schools (BLIPS) initiative. It seemed to me, at that time, that State and Federal governments all over Australia had begun to recognise that what went on in the name of mathematics in kindergartens, in the lowest grades of primary schools, and in families, was crucially important in relation to the future mathematical health of the nation. Before BLIPS, although it had been a truism that the most crucial years, so far as mathematics development was concerned, were those for children up to eight years of age, governments and major research funding organisations had not been inclined to call for, or support, well designed early childhood investigations.

Notwithstanding the prevalence of vague and largely inadequate definitions of numeracy found in most official documents, Perry et al.'s chapter on early childhood mathematics education made it abundantly clear that Australian and New Zealand governments, and Australasian mathematics education researchers, have continued to recognise the importance, for the future of our nations, of developing the strongest possible early childhood mathematics programs. Furthermore, many Australasian mathematics education researchers have become involved in outstanding and comprehensive early childhood projects. Among these projects are the *Numeracy Development Project*, *Count Me In Too*, *Count Me Into Measurement*, the *Early Numeracy Research Project*, the *Australian Pattern and Structure Mathematics Awareness Program*, *Catch the Future*, and *Project Good Start* – to name just a few. Most of the projects reviewed in Perry et al.'s chapter are extraordinarily impressive.

Thus, for example, Thomas and Tagg (2004) reported on data collected from more than 31,000 students, aged from 5 to 8 years, who participated in New Zealand's *Numeracy Development Project*. Their report offered much fundamental baseline data – like, for example, 20% of students at the end of Years 0-1 were able to use counting on or back to solve addition and subtraction problems, and the percentage had increased to 58% by the end of Year 2. In addition to such quantitative data, there were also analyses of qualitative aspects associated with whether or not students used higher level strategies. In Australia, Heirdsfield and her colleagues used small-scale case studies to explore the extent to which teaching can positively influence young children's development of high-level mental addition and subtraction strategies. Yet, it was recognised that no single "fix-it" strategy is likely to be successful. As analyses of data from the *Early Numeracy Research Project* (ENRP), in Victoria, revealed, children often experience difficulties with the same piece of mathematics for a range of reasons. Therefore, teachers need to make decisions about what is best for individual learners (Gervasoni, 2006).

Clarke, Clarke, and Horne's (2006) important ongoing longitudinal investigation of early childhood and primary students' mental computational strategies has carried the sobering message that unless effective intervention programs are planned, and continually supported, it is easy for gains to be quickly lost. In New South Wales, the *Count Me In Too* (CMIT) project has developed and extended its impressive emphasis on researching real children in real schools through carefully planned programs.

TEACHING AND LEARNING MATHEMATICS IN THE MIDDLE YEARS OF SCHOOLING

The chapter summarising research on this theme was written by Allan White (in recognition of the late Beth Southwell). I feel privileged, here, to be able to make commentary on this final piece of work for MERGA that Dr Beth Southwell had agreed to co-author before her death. She was one of the greatest mathematics educators MERGA and the Mathematics Education Lecturers' Association (MELA) ever had. Her scholarship, modesty, wisdom, courtesy, diligence, friendship and longstanding service will never be forgotten by all who knew her well.

White drew attention to a result from Nisbet and Grimbeek's (2004) study suggesting that almost half of the teachers who responded to their survey instrument felt that compulsory state-wide or national numeracy tests for 3rd, 5th, and 7th year students were not educationally justifiable. Another study indicated that about one-third of responses to questions on such tests were such that an incorrect response was given when the student understood fully how to do the question, or a correct response was given when the student did not understand fully how to do the question. These kinds of results are not new, of course, but pragmatic officials and school administrators have never seemed to be overly concerned by them.

White reported data which indicated that, unlike many of their teachers, middle-years students are relatively "fearless users of information and communication technology". There is no doubt that that phenomenon is true, but there was not much research reported by White indicating that a sustained effort is being made to upgrade practising teachers' confidence and competence with respect to ICT. That said, Anderson, Bobis, and Way, in their chapter on "Teachers as Learners" reported several projects (for example, one by Yates, 2005) showing how teachers have worked together to incorporate a computer algebra system into a school mathematics program for the first time. White also summarised research papers relating to the teaching and learning of difficult middle-school topics (like rational numbers, number lines, ratio and proportion, scales, algebra and chance and data). They noted, though, that there was "a distinct paucity of research in the area of space and geometry. Although it would not be appropriate here to attempt to summarise this curriculum-related research, it should be noted that some of the research has generated interesting and rich findings that ought to be known by teachers. That raises the age-old question of how mathematics education research findings are best communicated, succinctly, accurately, and persuasively, to practising teachers.

PROFESSIONAL DEVELOPMENT OF TEACHERS OF MATHEMATICS

One of the striking features of much of the early childhood mathematics research carried out two decades ago in the BLIPS programs was that partnerships were often developed between large numbers of practising teachers, relevant Education

Ministry officials, and mathematics education researchers. The 1980s was a time when "action research" came to be widely used in association with professional development of teachers of mathematics. One of the most saddening aspects of the history of the subsequent use of the term "action research" is that it came to mean any project that a teacher decided to do to investigate his or her own practice. No tight definition of the term ever won wide acceptance among teachers or, for that matter, among university- and Ministry-based researchers.

For me, the Deakin University definition that called for collaborative groups of researchers, where membership of a research team was voluntary, and every member of a research team had as much authority within a group as any other member so far as the definitions of research questions, methodology, instruments, and evaluation strategies were concerned (see Carr & Kemmis, 1986), was always the tightest and most fruitful definition. The writings and work of Ian Robinson in the second half of the 1980s provided examples of how a strong, collaborative, and democratic view of action research could serve to empower, simultaneously, all participants in action research projects (Robinson, 1989). Robinson conjectured that the equal and active involvement of teacher researchers in collaborative action research projects was most likely to result in actual, long-term changes in curricula and teaching and learning practices. The partnerships between school teachers and university- or Ministry-based researchers – described in this present book resemble the collaborative mode of action research called for by Robinson (1989). The latent issue associated with large-scale collaborative research, however, is whether student learning improves as a result of changed teacher behaviours.

The chapter by Anderson, Bobis and Way in this volume deals with the theme "Teachers as Learners". Anderson et al. distinguish between "professional development" – "planned, focused activities and formal programs that teachers undertake to extend their professional learning" – and "professional learning" – "the individual growth of a teacher's expertise, which may be achieved through participation in professional development experiences, but may also be experienced through experience in classrooms, professional reading, postgraduate study and the like". For Anderson et al., professional learning is more inclusive than professional development because it fits more easily with the many activities in which teachers engage during the course of their careers.

Although Anderson et al. seem to be in no doubt regarding the distinction between professional development and professional learning, I find it difficult to make a sharp distinction. Thus, for example, a teacher who takes a master's unit on "Analysing Discourse Patterns in Mathematics Classrooms" would seem to be involved equally in both professional development and professional learning. Anderson et al. regard large-scale systemic or state-wide projects such as *Count Me In Too* (in New South Wales), the *Early Numeracy Research Project* and *Success in Numeracy Education* (in Victoria) and the *Numeracy Development Projects* (in New Zealand) as professional development projects, but it is difficult to imagine why these might not equally be regarded as professional learning projects. It would be an interesting research exercise to set out to determine the relative effects on *in Numeracy Education* (in Victoria) and the *Numeracy Development Projects* (in New Zealand) as professional development projects, but it is difficult to imagine why these might not equally be regarded as professional learning projects. It would be an interesting research exercise to set out to determine the relative effects on teacher learning, teacher behaviour, student learning, and student behaviour of the various forms of professional development and professional learning.

In the United States there is now a very strong impetus to design and implement professional development programs that expand the mathematics knowledge base of practising teachers (Ball, Hill, & Bass, 2005). Indeed, mathematics content knowledge is now being seen as an integral part of pedagogical content knowledge (PCK). The importance of improving practising teachers' mathematical content knowledge is an issue that has been taken up in Australia and New Zealand by many researchers, including White, Mitchelmore, Branca, and Maxon (2004). Although it would be inappropriate to quarrel with the desirability of teachers having a thorough and extended grasp of the mathematics they teach, the question arises what precisely are the components of the desired mathematics knowledge bases. In the United States it is common for Departments of Mathematics to believe that practising teachers will benefit from studying *any* kind of higher mathematics. In practice, Departments of Mathematics want to attract teachers so that they can make their graduate classes viable in numbers. Often, the mathematics courses are not intimately related to what the teachers are currently teaching, or even to what they are reasonably likely to teach, in schools. I would have liked to have seen a more careful discussion of this issue in at least one of the chapters in this volume.

Some Australasian researchers during the period under review attempted to measure the effects of teachers' participation in professional development projects on subsequent student performance. Bobis et al. (2005) claimed that there was evidence that students in trial schools significantly outperformed students in reference schools. That kind of result suggests that the large-scale numeracy professional development projects would stand up to scrutiny from a cost-benefit perspective. However, it is still not clear whether there were large Hawthorne effects associated with such evaluations, although the chapter by Anderson et al. did report that the effectiveness of projects was sustained even after benefits arising from funding for the projects were no longer available.

Anderson et al. admitted that an overarching characteristic of the studies they reviewed relied solely on self-report data as evidence of teacher change, and even those which did not, usually gathered observation from only a small number of teachers in a small number of classes. Some sceptics have little faith in self-report data, and there would appear to be a need for a much tougher test of the effectiveness of such projects. Anderson et al. also drew attention to some published reports in which the difficulty of designing and carrying out research which explored links between professional development and subsequent student performance (including one study that I conducted with Allan White in Brunei Darussalam) proved to be in the "too difficult" or "too expensive" categories.

TEACHING AND LEARNING MATHEMATICS IN NON-SCHOOL CONTEXTS

Researching the teaching and learning of mathematics within universities

In the chapter on "University Learners of Mathematics" the author, Leigh Wood, points out that it is surprising that so little research has been done on learners of mathematics at universities. In that chapter this state of affairs is explained away by the statement that mathematicians in universities "need to keep up a research profile in their mathematical discipline area and so tend to work on mathematics education research as a sideline". The author of the chapter also commented that most of the mathematicians "do not have formal qualifications in mathematics education and find it difficult to keep up with the educational literature and methodologies". That said, it must be acknowledged, that some Australasian mathematicians, for example, Professor Gary Davis and Professor Derek Holton – now living in the United States of America and New Zealand, respectively – have involved themselves in mathematics education research over many years.

When mathematics educators talk about the teaching and learning of mathematics at universities there is always the danger of precipitating unprofitable controversy, and so the language I have used in framing the following three observations was very carefully considered. First, mathematics educators recognise that mathematicians do not feel that their focus should be on mathematics education research. Second, few would deny there are many students at universities who struggle to learn mathematics well in the classes provided, and therefore research into how that situation can be improved ought to be welcomed. Third, cooperative research involving mathematicians and mathematics educators might help to develop needed bridges between the understandings and practices of mathematicians and mathematics education researchers. I currently work in the Department of Mathematics at Illinois State University in which there are 15 tenured or tenure-track mathematicians and 15 tenured or tenure-track mathematics educators, and there is a strong positive relationship between the two groups. There is much involvement by mathematicians in mathematics education research, and some involvement by mathematics educators in mathematics research.

There are many indications that in recent years traditional approaches to teaching and learning mathematics at the university level have begun to be examined, and their effectiveness problematised. In the preparation of the chapter on "University Learners of Mathematics" the author consulted about 80 papers that appeared between 2004 and 2007. All of these papers were concerned with intended, implemented and attained curriculum issues. There were quite a few studies on the use of technologies, and the main finding seemed to be that many students had difficulty finding the time to learn how to use the software. Other papers were concerned to investigate the effectiveness of less traditional modes of teaching tertiary-level mathematics, including distance methods. However, there have been no major research grants to support such research, and the chapter closes with the thought that perhaps the Carrick Institute might use its grants "to supply sufficient resources to help mathematicians develop a more cohesive picture of learning in mathematics at university".

In the penultimate paragraph of the chapter on university mathematics the author raised the possibility of developing a single coherent theory of how best to address the needs of all mathematics students. That raises the issue of the nature and role of theory in mathematics education research – which is a longstanding issue. Although history suggests that it is unlikely that it will be possible to develop a theory that will meet the needs of all, nevertheless, the effort to do so may be worthwhile. However, in any attempt to develop such a theory the main contribution should come from the relevant group of practitioners – in this case, those charged with the responsibility of teaching mathematics in universities. Unless the mathematicians are allowed a sense of ownership any attempt at developing a theory (or theories) could be divisive. Be that as it may, there can be little doubt that these days university mathematicians are pleased to provide instruction for service courses for students in many faculties. Any research in relation to these courses, including the development of appropriate theory, should become a joint enterprise between the mathematicians and members of the departments being serviced.

Research on the pre-service education of prospective mathematics teachers

Although there are many differences across Australia and New Zealand in the way pre-service mathematics teacher education is organised, most programs are of four years duration and combine professional studies in education, mathematics discipline studies, and field experience. It is generally acknowledged that both mathematics subject knowledge and pedagogical content knowledge (PCK) should be an integral part of the program. As Goos, Smith and Thornton's chapter on "Research on the Pre-Service Education of Teachers of Mathematics" reveals, there has been much recent research in Australia and New Zealand showing that many prospective early childhood and primary teachers have seriously defective knowledge in relation to basic areas of mathematics like mental arithmetic, fractions, decimals, place value, ratio and proportion, the metric system, graphs, probability, elementary algebra, and elementary geometry. Furthermore, research makes it clear that elementary word problems are particularly threatening for many would-be primary teachers. This kind of research has always been common, and it is heartening to note that there has been research – see, for example, Zevenbergen (2004) – on what might be needed to *improve* the state of affairs that is well known to exist.

Discussions of the concept of PCK, and details of attempts to assist prospective mathematics teachers to develop PCK, are provided in the chapter by Goos et al. I would have liked more discussion of the recent factor analytic approaches to PCK put forward by Ball et al. (2005) in the United States. I would also have liked more on the issue of who should be responsible for attempting to develop mathematics PCK in prospective teachers (at all levels). In these days, when mathematics departments often are experiencing declining enrolments for mathematics courses, there has been some controversy over whether mathematicians, as opposed to mathematics educators, should be responsible for teaching mathematics to prospective early childhood and primary teachers. If mathematics educators believe that they should be responsible, then they need to conduct very carefully designed research related to the issue. This research should generate both quantitative and qualitative data, gathered from multiple vantage points so that a healthy triangulation is achieved. That is because results of small-scale, single vantage point studies are unlikely to hold much sway in the political hurly burly often associated with the issues under consideration. Two points ought to be emphasised here: first, PCK is one of the key concepts that should, inevitably, enter into debates about who should teach mathematics to prospective teachers; second, research involving PCK ought to be based on a well-defined concept of PCK.

Most of the analyses of data in the research summarised in Goos et al.'s chapter on pre-service education were largely based on responses to questionnaires. Unless the designs of such studies incorporate data gathered from multiple perspectives, so that the questionnaire response analyses form only part of the overall analyses, one can only say that the analyses reveal, at best, the perceptions of those responding to the questionnaires. At worst, they may only reveal what the respondents feel they want to tell those carrying out the research. There can be no guarantee that the perceptions match the realities.

Research concerning the practicum component of the professional preparation of early childhood and primary teachers is also summarised by Goos et al. Several of the studies reported that many pre-service teachers felt that what they experienced during practicums was not consistent with what they were taught in the methods courses at university. This raises several possibilities, including the possibility that practicums can be vehicles for unduly reinforcing questionable aspects of the status quo in school mathematics classrooms. Perhaps a greater integration of on-campus coursework and school experience for pre-service teachers, called for by Beswick (2005), would help in this respect. My personal experience has been that this integration is notoriously difficult to achieve and, once achieved, can be very time-consuming for all concerned. There is also the issue of the extent to which it delivers what it promises.

Goos et al.'s chapter on pre-service education has a much greater emphasis on primary than on secondary pre-service teachers. Comments were made on relatively few research studies focusing on pre-service teacher education programs for prospective secondary teachers of mathematics. Assuming that this imbalance reflected the amount of research done with respect to the two fields, I want to know: What does Australasian research say about who should teach mathematics courses to prospective secondary teachers? Is the tension that primary pre-service students experience between on-campus and practicum components of their courses also a characteristic of pre-service secondary programs?

Research on adults returning to study mathematics

Not all adults who choose to study mathematics formally do so within university mathematics departments. In their chapter on "Adults Returning to Study Mathematics", Galligan and Taylor state that there has been a resurgence of research interest, especially in relation to the vocational education sector, in what was previously an under-theorised and research-scant field. There have been attempts to develop appropriate theories for adult mathematics learning in non-school contexts. Thus, for example, Taylor and Galligan (2006) developed and used a "conversation theory framework." Much of the research has been government-led, especially in relation to the National Council of Vocational Educational Research (NCVER), in Australia, but this research has often been carried out under the literacy banner, and hence led by literacy-oriented researchers rather than by mathematics education researchers.

There has been a healthy increase in research associated with "everyday situations" in which people can apply mathematical principles – if they know them – to situations arising in their own lives. The Commonwealth Bank Foundation (2005) in Australia, for example, funded research into whether adults know how best to reduce credit card debts, or take control of their mobile phone bills. Other research has been reported on the development of "a one stop shop to manage the safe administration of drugs". Which is more effective: Supply appropriate formula, or allow adults to employ a problem-solving strategy of their choice from a suite of possible strategies? Although the importance of such research is recognised, the issue of who should plan and conduct the research needs to be considered, and the role of mathematics education researchers within research teams, would appear to be an issue that MERGA might profitably address.

Galligan and Taylor close their chapter with the statement that appropriate theories relating to "adults returning to study mathematics" are hard to find, and need to be developed if the quality of research is to progress to the "next level". By this statement it is presumed that they believe that theory-based research is preferred to the more grounded "theory-developing" research. As I will discuss towards the end of this chapter, the issue of whether, at this stage in most areas of mathematics education, we should be more concerned with theory-developing than theory-based research, is problematic.

SOCIOCULTURAL, LANGUAGE AND AFFECTIVE FACTORS

In this section, five chapters will be considered together, with links being made on the basis of related themes identified in the chapters. The five chapters are: "Sociocultural Perspectives in Mathematics Teaching and Learning" by Seah, Atweh, Clarkson, and Ellerton; "Mathematics Education and Indigenous Students" by Meaney, McMurchy-Pilkington, and Trinick; "Gender and Mathematics" by Vale and Bartholomew; "The Affective Domain and Mathematics Education" by Grootenboer, Lomas, and Ingram; and "Characteristics of Effective Pedagogy for Mathematics Education" by Anthony and Walshaw.

Socio-economic status/class and the teaching and learning of mathematics

This section of Seah et al.'s chapter strongly asserts that politicians' and others' judgments of the effectiveness of mathematics programs in classrooms, schools, and systems should not be based solely on analyses of data arising from large-scale pencil-and-paper testing regimes. As is well known, students from poorer families attending schools in low SES areas do not perform as well on standardised mathematics tests as students from high SES families who attend schools within affluent communities. That raises the thorny issue of whether the tests are culture- and class-free instruments.

The chapter asks mathematics education researchers to provide leadership in moves to make school mathematics programs more equitable. Various research papers reporting inequalities and lack of expected inequalities are summarised, and the relevant section in Seah et al.'s chapter concludes by deploring the "dearth of studies that address structural variables directly in mathematics education research in recent times."

Research on gender and mathematics teaching and learning, and on mathematics education experienced by indigenous students

Vale and Bartholomew begin their chapter on "Gender and Mathematics" by providing details of Bishop and Forgasz's (2007) careful distinction between two main competing concepts of equity in education. The first approach regards equity as being associated with equal outcomes, equal opportunities and equal treatment, and with the second, equity is considered to be fundamentally a quality involving fairness and justice. Anthony and Walshaw (2007), on the other hand, argue that equity is mainly concerned with interactions between contexts and people, and is not so much concerned with equal outcomes and approaches, or about equal access to people and curriculum materials. In Meaney et al.'s chapter on "Mathematics Education for Indigenous Students" the issue of equity is tackled in a different way. Meaney et al. emphasise the "under-achievement of indigenous students" and draw attention to four types of interaction which affect the positioning of indigenous students as knowers, doers, and learners of mathematics. These four types of interactions are societal interactions, community interactions, student interactions, and teacher interactions. This approach seems to be in line with the second view of equity put forward by Anthony and Walshaw in which equity is mainly concerned with interactions between contexts and people.

What I liked about Meaney et al.'s chapter was how the authors carefully related the whole paper to their thesis on educational disadvantage, constantly reminding the reader that the word "disadvantage" is only appropriate because, from the point of view of school mathematics, indigenous people have been more or less compelled to study mathematics in a way that is inconsistent with their world views. The often-repeated aim is "equal outcomes", but those outcomes represent outcomes valued in someone else's world view. Meaney et al. emphasise that the ethnomathematics literature makes it clear that mathematical content and problem solving used by children are quite different in and out of school, and that success on everyday mathematical activities does not necessarily translate to mathematical competence in school.

Meaney et al. refer to research by Owens and Kaleva (2007) in Papua New Guinea in which measurement activities were linked to purpose, and to cultural activities like bride price, canoe making, garden building or hole making. But what teachers offer in the name of school classrooms, especially where high-stakes national examinations are in place, usually does not take full advantage of obvious curriculum opportunities. It is not possible to provide details, here, of research into societal, community interactions, teacher interactions, and student interactions that are reported in Meaney et al. Suffice to say, that in Australasia the problems associated with equity with respect to indigenous students have not gone away, and are unlikely to do so while the main criterion for equity and equality is assumed to be equal performance on, what could be argued to be, inappropriate educational measurement instruments.

Vale and Bartholomew's chapter on gender and mathematics revealed that in large-scale comparative studies (like TIMSS and PISA) sometimes girls perform at a higher level than boys, and sometimes the reverse is the case. Often, there are no statistically significant differences. There is a tendency for primary school girls in New Zealand to outperform boys, but in Australia, secondary school boys tend to outperform girls.

I especially looked for summaries of studies comparing performances of boys and girls on the Australian Mathematics Competitions (AMC), for in the 1980s and 1990s analyses of very large AMC data sets revealed substantial differences in favour of males. Leder (2006) did analyse recent AMC data, and found that, once again, males were more highly represented than females among the highest achievers. In the 1990s Fred Bishop and I reported analyses of data from large annually-reported Mathematics Competition data for primary school students in and around Newcastle, in New South Wales. Our analyses showed that males clearly outperformed females at the upper primary school level. The persistent mathematics competition results in favour of male students seemed to be in line with the statement by Vale and Bartholomew that "males were consistently more highly represented among the highest achievers in mathematics for studies involving primary, junior secondary, and senior secondary students".

Vale and Bartholomew's chapter reports research confirming what has been known for years: boys seem to be much more confident in school mathematics environments than girls, despite the fact that their greater confidence and mathematics self-concept were not usually matched by higher performance on standard mathematics tests. The question arises: What needs be done, if anything, to change this state of affairs? Vale and Bartholomew report research on that question, but possible "solutions", such as single-sex classrooms, were not strongly supported by findings from analyses of the research data. Vale and Bartholomew's conclusion begins with the statement that although there had been "a very wide range of studies reviewed… it is hard to make generalisations about them". As with most chapters in this volume, the authors lamented there being "no explicit theoretical framework" in the majority of studies reviewed.

Research on affect: The influences of emotions, attitudes, beliefs and values

Grootenboer, Lomas and Ingram's chapter carried the title "The Affective Domain and Mathematics Education", and one might have expected discussion of research dealing with attitudes, beliefs and values to be largely confined to that chapter. However, Seah et al.'s chapter has a major section on the same affective variables. Here I will briefly summarise the directions taken by Grootenboer et al. and also those taken in the pertinent section of Seah et al.'s chapter. One of the important achievements during the period under review was the development of a theory relating emotions (or feelings), attitudes, beliefs and values (Leder & Grootenboer, 2005). These qualities are seen as essentially being placed along a linear continuum, with emotions – characterised as intense but unstable, and without a strong cognitive base, being at one end of the continuum, and attitudes, values, and finally beliefs – which are less intense, more stable, and have a stronger cognitive base – being towards the other end of the continuum.

According to both Grootenboer et al. and Seah et al., internationally there has been a large number of recent studies on the influence of beliefs and values on mathematics teaching and learning, but relatively few on emotions or attitudes. The authors of the chapters regard that as a positive trend, but mathematics educators should never dismiss emotions and attitudes as unimportant, for as McLeod noted in the 1990s, they represent the hot end of affect, the components which have immediate influence – either positive or negative – in actual classroom environments. According to Thomson, Cresswell, and de Bortoli (2004), multilevel analysis of PISA data showed that, of all factors incorporated into a model to investigate Australian students' mathematical literacy, attitudes and beliefs explained the greatest amount of variance.

Seah, Bishop and others have invested much effort into sharpening the definition of the term "values" and the influence of values on those who define mathematics curricula and those who teach and learn mathematics. The Australasian research on values in mathematics education seems to have a different emphasis from much of the American and Asian research, where a major pre-occupation has been on clarifying family values in Confucian-heritage cultures that seem to be associated with superior performance in international comparative studies (such as TIMSS and PISA).

Research on language factors

The section on language factors in Seah et al.'s chapter focused on language associated with mathematics curricula, communication in mathematics classrooms, mathematics problem solving, and sociolinguistics. Many researchers were associated with the articles reviewed in this section, and criteria for analysing data varied considerably – from criteria arising out of poststructuralist and feminist literatures, to criteria associated with the emerging "cognitive emptying" literature, to the more straightforward statistical comparisons of the numbers of words spoken in a class by teachers and students, or the numbers of high-, medium- or low-level questions asked in mathematics lessons of different kinds of students (for example, of male or female students).

In their chapter on "Characteristics of Effective Pedagogy for Mathematics Classrooms" Anthony and Walshaw refer to a classroom scenario – described in a paper by Walls (2005) – in which teacher instructions for problem-solving activities, as well as task-related protocols, demanded procedural compliance by students. That scenario furnishes an example of "cognitive emptying" – to use terminology introduced in the language section in Seah et al.'s chapter. A potentially challenging problem-solving task can be emptied of its cognitive challenge if too many instructions or too much advice on procedures are offered, or too many leading questions are asked, by the teacher. Similarly, as Anthony and Walshaw (2007) report, "a shift in focus from understanding to correctness or completeness, inappropriate allocation of time, relaxing of accountability requirements, and a lack of alignment between the task and students' prior knowledge, interest and motivation, can all contribute to the lowering of task demands".

Anthony and Walshaw also comment that teaching practice founded on an ethic of care takes pains to avoid permanent dependency from students towards teachers. That is because "it pays attention to developing interrelationships that are oriented towards enhancing students' capacity to think, reason, communicate, reflect upon, and teaching strategies and associated classroom discourse patterns by which such qualities are most likely to be achieved". Nevertheless, it is recognised that hard-nosed cynics are unlikely to be impressed with the findings of small, qualitative research studies. Anthony and Walshaw's chapter summarised a wide variety of research in which many strategies for better teaching of mathematics were presented and evaluated. The issue of which strategies are best for which teachers, and for which classes, is one which any researcher must face every time a classroom study is being planned.

The section on language factors in Seah et al.'s chapter also drew attention to the growth in the use of new (or, rather, newish) modes of communication in mathematics classes (like, for example, computer-based information communicated through spreadsheets). There were also reports on research aimed at improving students' mathematics vocabularies and ability to cope with any difficult semantic structures within the language with which mathematical problems are presented to learners.

In Meaney et al.'s chapter on "Mathematics and Indigenous Students" studies into the effects of receiving mathematics instruction in a language which is not a learner's first language are reviewed. Young-Loveridge (2005), for example, showed that Mãori boys' performance improved dramatically when they were assessed through diagnostic interviews rather than through pencil-and-paper tests. Young-Loveridge commented that "given many indigenous students are not learning in their first language, the impact of different assessment types may well affect indigenous students' performance". In most nations the choice of the language of instruction is usually a political decision rather than one made at the school or classroom level. In Samoa, for example, the teaching of mathematics switches from the vernacular language to English at the upper-primary level, but many upper-primary teachers continue to code-switch between the two languages. Samoan teachers believe that students often perform poorly in Year 8 examinations because they experience difficulty decoding and comprehending questions written in English (Afamasaga-Fuata'i, 2005). This important issue has been taken up in the past by MERGA researchers, and it is pleasing to see that it is still being investigated.

TEACHING AND LEARNING WITH TECHNOLOGY

The chapter by Thomas and Chinnappan, on research into the impact of technology on mathematics teaching and learning, provides comprehensive evidence that mathematics education researchers across Australia and New Zealand have embraced the need to investigate how students, at all levels, can be assisted to work productively and creatively in technologically rich environments. The chapter features a dazzling array of acronyms – some now well known, others less so – such as CAS-CAT, GC, IWB, META, PTK, RIPA, RITEMATHS, TRTLE, ZFM, and ZPA. Perhaps we are in danger of creating a special class of learners, and mathematics education researchers who understand the language and symbols being used, while at the same time excluding others who do not. That said, as Galbraith (2006) has pointed out, we should not be transfixed by technological blindness in our haste to harness the power of technology in mathematics learning environments.

There are now many mathematics education research projects in Australia and New Zealand in which new kinds of questions are being investigated as a consequence of the possibilities inherent in our new technological worlds. The findings of these projects should have strong implications for the intended, implemented, and attained mathematics curricula of Australasian education institutions, at all levels. One challenge is to avoid being carried away by unreal expectations. Another, is to resist adopting head-in-the-sand, luddite attitudes regarding what might be achieved. Thomas and Chinnappan close their chapter with the timely reminder that, in technology-rich mathematics learning environments, "it is not the technology per se that illuminates the mathematics, but rather the illumination is in its role in epistemic mediation, in the mathematical thinking that a trained teacher can assist students to access through technology".

TEACHING, LEARNING AND APPLYING MATHEMATICAL MODELLING

It is interesting to note that of all the chapters in this volume, the chapter by Stillman, Brown, and Galbraith comes closest to focusing on issues directly concerned with the mathematics content taught and learned in classrooms. There are no chapters in the volume dedicated to research on the teaching and learning of number sense, or algebra, or geometry, or chance and data, or calculus, despite Australasian researchers having carried out and reported much research, published in high-level outlets, on such matters.

I recall that in the early 1990s the Victorian Ministry of Education jointly funded the University of Melbourne and Deakin University to develop and offer courses on mathematical modelling for practising secondary teachers. Since that time there has been increased emphasis on the need for school curricula to offer opportunities in mathematical modelling. Furthermore, possibly as a result of the increasing popularity of the Freudenthal Institute's Realistic Mathematics Education (RME), with its modelling orientation, as well as the prolonged impact of findings from investigations carried out by leading international scholars – including the three authors of the pertinent chapter in this volume – interest in the role and potential of modelling approaches to school mathematics has increased across Australasia.

It was encouraging to observe that "the most concentrated areas of Australasian research activity has been in modelling competencies, teaching/learning practice of modelling in primary and lower secondary classrooms, and implementation of applications and modelling in practice." The Stillman et al. chapter lists a wide range of questions currently being researched by Australasian scholars. Questions such as "What pedagogies are most productive for the successful implementation of modelling approaches?" and "How have efforts at incorporating modelling at a curriculum level unfolded, and what impediments have impacted on the outcomes achieved?" are being researched, and the results are keenly awaited. It has been pleasing to note how modelling at the primary and middle school levels, as well as at secondary and university levels, has been researched. That said, it is disappointing to read in the Stillman et al. chapter that "no research was identified as reported in the period in the contexts of modelling in vocational/technical education and teacher education".

SOME CLOSING COMMENTS: FROM A PERSONAL PERSPECTIVE

Every four years MERGA prepares and publishes a review of recent mathematics education research in Australasia. The 2004-2007 review, which is presented to you in this book, is the latest effort. The word "effort" is appropriate, for MERGA is to be highly commended for committing itself to preparing and publishing such a review without receiving any additional funds to assist in the preparation and publication. MERGA members are invited to write reviews of areas in which they have known expertise and research experience. Those who accept the invitation then have the difficult task of trying to locate, and summarise all the educationally significant research articles published by Australasian researchers over the past four years. Sometimes, those who accept the challenge are unable to complete their review. That can happen for a number of reasons: illness, inability to find the time to do what needs to be done, or, after seeking after relevant publications, a decision is made that there are simply not enough significant publications in the agreed area to warrant going ahead with preparing the chapter.

It therefore comes as no surprise, then, that not all areas of research are represented, in this book. Unfortunately, in a few instances, research conducted on key areas of concern has not been summarised in this book. In the remaining part of this chapter I shall introduce, briefly, three issues which, from my own perspective, did not receive the amount of attention I would have liked in this volume. The issues are all controversial.

Issue 1: Australia's and New Zealand's Performances on Recent International Comparative Performance Studies

Recently, Jan Thomas (2007), of the Australian Mathematical Sciences Institute, commented that although the late 1980s was an optimistic and enthusiastic time in mathematics education in Australia, "by any measure the optimism of the 1980s has not been realised" (p. 25). Thomas maintained that Australian TIMSS results have a long tail of underachievement, and that, when compared with the average for OECD nations, a lower percentage of students study advanced or intermediate courses at Year 12. Thomas also pointed out that only 0.4% of Australian graduates major in mathematics or statistics compared with an OECD average of 1%.

Until recently I would not have had much sympathy with Thomas's argument, because I interpreted the 1995 TIMSS results as showing that of all the English-speaking nations participating in TIMSS, except Singapore, Australian students had the best results, and New Zealand students were not far behind them. But while Australia's Year 8 mean performances on two subsequent TIMSS studies in mathematics in Year 8 have remained about "the same" as their TIMSS 1994/1995 mean results, the performance of other countries has improved – so that half of the countries outscored by Australia in TIMSS 1994/1995 performed at a similar level to Australia in TIMSS 2002/2003.

I wish there had been more discussion in this volume about this apparent decline in relative performance. I would like to know why other nations have, apparently, improved but we have not. Were there extenuating circumstances about which we need to be informed? Are there changes which have been introduced into school mathematics programs in other nations which have not been introduced in Australia or New Zealand?

Issue 2: Should Mathematics Education Research Be Expected to Generate Important and Measurable Improvements in Mathematics Curricula and Mathematics Teaching and Learning?

From a slightly different but related perspective, I wonder whether all the mathematics education research conducted and reported in Australasia over the past four years has resulted in better mathematics curricula, and better teaching and learning of mathematics in classrooms in Australasian schools, colleges and universities. Although such a question is impossible to answer definitively, it is important, nevertheless, that mathematics education researchers think about it and discuss it.

I have always found it somewhat paradoxical that the nation that has conducted the most mathematics education research studies – the United States of America – does not seem to perform well on international comparative studies. Yes, for members of MERGA high quality mathematics education research is an end in itself: but, also, we need to be concerned with making sure that our results are known and, where possible, applied for the common good.

Issue 3: What Is the Role of Theory in Mathematics Education Research?

Mathematics education researchers tend to have wide differences of opinion on what they believe to be the role of theory in mathematics education research. In case anyone believes that their point of view on this is *the* correct one, they should consult the two-volume collection of essays edited by Anna Sierpinska and Jeremy Kilpatrick and published by Kluwer Education Publishers in 1998 titled, *Mathematics Education as a Research Domain: A Search for Identity*. In Nerida Ellerton's and my contribution to that collection we argued that at this time in the development of the field of mathematics education we need far more theory-generating, grounded research, than we need research tightly based on set theories. In other words, we should be thinking of developing theory from our data rather than fitting our data to existing theory (Ellerton & Clements, 1998).

Of course, there would be many readers who disagree with that perspective on research, but that is only to be expected. Here I want to point out that in many of the chapters in this volume (more than half of them) the authors lament there not being sufficiently established theories on which they could hang the discussions of the research reports they were reviewing. My reaction to that is: Well, of course that is the case. It is our task to work hard at developing theory that can be tested against further data generated in other studies. If anyone involved with mathematics education research is certain that he or she knows *the correct* theory for a proposed research study, then he or she is probably wrong. Or am *I* wrong on that?

REFERENCES

Afamasaga-Fuata'i, K. (2005). Mathematics education in Sāmoa: From past and current situations to future directions. *The Journal of Sāmoan Studies, 1*, 147-165.

Anthony, G., & Walshaw, M. (2007). *Effective pedagogy in mathematics/Pāmgarau: Best Evidence Synthesis Iteration (BES)*. Wellington: Ministry of Education, New Zealand.

Ball, D. L., Hill, H. H., & Bass, H. (2005). Knowing mathematics for teaching: Who knows mathematics well enough to teach third grade, and how can we decide? *American Mathematical Educator, Fall*, 14-46.

Beswick, K. (2005). Pre-service teachers' understandings of relational and instrumental understanding. In H. Chick & J. L. Vincent (Eds.), *Proceedings of the 29th conference of the International Group for the Psychology of Mathematics Education* (Vol. 1, pp. 161-168). Melbourne: University of Melbourne.

Bishop, A. J., & Forgasz, H. (2007). Issues in access and equity in mathematics education. In F. Lester (Ed.), *Second handbook of research in mathematics teaching and learning* (Vol. 2, pp. 1145-1167). Reston, VA: National Council of Teachers of Mathematics.

Bobis, J., Clarke, B., Clarke, D., Thomas, G., Wright, R., Young-Loveridge, J., & Gould, P. (2005). Supporting teachers in the development of young children's mathematical thinking: Three large-scale cases. *Mathematics Education Research Journal, 16*(3), 27-57.

Carr, W., & Kemmis, S. (1986). *Education, knowledge, and action research.* Geelong, Australia: Deakin University.

Clarke, B., Clarke, D. M., & Horne, M. (2006). A longitudinal study of children's mental computation strategies. In J. Novotná, H. Moraová, M. Krátká & N. Stehliková (Eds.), *Proceedings of the 30th conference of the International Group for the Psychology of Mathematics Education* (Vol. 2, pp. 329-336). Prague: Charles University.

Commonwealth Bank Foundation (2005). *Improving financial literacy in Australia: Benefits for the individual and the nation.* Clayton, Victoria: Monash University.

Ellerton, N. F., & Clements, M. A. (1998). Transforming the international mathematics education research agenda. In A. Sierpinska & J. Kilpatrick (Eds.), *Mathematics education as a research domain: A search for identity* (pp. 153-175). Dordrecht, The Netherlands: Kluwer Academic Publishers.

Galbraith, P. (2006). Students, mathematics, and technology: Assessing the present – challenging the future. *International Journal of Mathematical Education in Science and Technology, 37*(3), 277–290.

Gervasoni, A. (2006). The challenge of meeting the instructional needs of grade 1 and grade 2 children who are vulnerable in number learning. In J. Novotná, H. Moraová, M. Krátká & N. Stehliková (Eds.), *Proceedings of the 30th conference of the International Group for the Psychology of Mathematics Education* (Vol. 3, pp. 177-184). Prague: Charles University.

Howell, S., & Kemp, C. (2005). Defining number sense: A participatory Australian study. *Educational Psychology, 25,* 555-571.

Leder, G. C. (2006). Catering for individual differences: Lessons learnt from the Australian Mathematics Competition. In P. Grootenboer, R. Zevenbergen & M. Chinappan (Eds.), *Identities, cultures and learning spaces* (Proceedings of the 29th annual conference of the Mathematics Education Research Group of Australasia, Canberra, pp. 336-343). Sydney: MERGA.

Leder, G., & Grootenboer, P. (2005). Affect and mathematics education. *Mathematics Education Research Journal, 17*(2), 1-8.

Nisbet, S., & Grimbeek, P. (2004). Primary teachers' beliefs and practices with respect to compulsory numeracy testing. In I. Putt, R. Faragher, & M. McLean (Eds.), *Mathematics education for the third millennium: Towards 2010* (Proceedings of the 27th annual conference of the Mathematics Education Research Group of Australasia, Townsville, pp. 406-413). Sydney: MERGA.

Owens, K., & Kaleva, W. (2007). Changing our perspective on measurement: A cultural case study. In J. Watson & K. Beswick (Eds.), *Mathematics: Essential research, essential practice* (Proceedings of the 30th annual conference of the Mathematics Education Research Group of Australasia, Hobart, Vol. 2, pp. 571-580). Sydney: MERGA.

Robinson, I. (1989). The empowerment paradigm for the professional development of teachers of mathematics. In N. F. Ellerton & M. A. Clements (Eds.), *School mathematics: The challenge to change* (pp. 269-283). Geelong, Australia: Deakin University.

Taylor, J. A., & Galligan, L. (2006). Mathematics for maths anxious tertiary students: Integrating the cognitive and affective domains using interactive multimedia. *Literacy and Numeracy Studies, 15*(1), 23-42.

Thomas, G., & Tagg, A. (2004). *An evaluation of the Early Numeracy Project 2003: Exploring issues in mathematics education.* Wellington: Ministry of Education, New Zealand.

Thomas, J. (2007). Reshaping school mathematics in Australia 1788-1988 (review). In G. C. Leder & H. J. Forgasz (Eds.), *Stepping stones for the 21st century: Australasian mathematics education research* (p. 25). Rotterdam, The Netherlands: Sense Publishers.

Thomson, S., Cresswell, J., & de Bortoli, L. (2004). *Facing the future: A focus on mathematical literacy among Australian 15-year-old students in PISA 2003.* Camberwell, Victoria: Australian Council for Educational Research.

Walls F. (2005). Challenging task-driven pedagogies of mathematics. In P. Clarkson, A. Downton, D. Gronn, M. Horne, A. McDonough, R. Pierce & A. Roche (Eds.), *Building connections: Research, theory and practice* (Proceedings of the 28th annual conference of the Mathematics Education Research Group of Australasia, Melbourne, Vol. 2, pp. 751-758). Sydney: MERGA.

White, P., Mitchelmore, M., Branca, N., & Maxon, M. (2004). Professional development: Mathematical content versus pedagogy. *Mathematics Teacher Education and Development, 6,* 41-51.

Wickert, R., & McGuirk, J. (2005). *Integrating literacies: Using partnerships to build literacy capabilities in communities.* Adelaide: National Council of Vocational Educational Research.

Yates, S. M. (2005). The role of professional development in teacher renewal in mathematics. In *Proceedings of the 34th Annual Conference of the Australian Association of Research in Education.* Melbourne: Australian Association of Research in Education.

Young-Loveridge, J. (2005, December). Seeing the mathematics of Mãori boys in a new light: The benefits of diagnostic interviews. Paper presented at *New Zealand Association for Research in Education (NZARE) National Conference.* Dunedin, New Zealand.

Zevenbergen, R. (2004). Study groups as a tool for enhancing pre-service students' content knowledge. *Mathematics Teacher Education and Development, 6,* 4-22.

AFFILIATION

M. A. (Ken) Clements
Illinois State University

www.ingramcontent.com/pod-product-compliance
Lightning Source LLC
Chambersburg PA
CBHW071415230926
51846CB00001B/98
9789087904999